AGRICULTURAL EXPERIMENTATION

Design and Analysis

AGRICULTURAL EXPERIMENTATION

Design and Analysis

Thomas M. Little
Extension Biometrician, Emeritus
University of California, Riverside

F. Jackson Hills
Extension Agronomist
University of California, Davis

John Wiley and Sons
New York • Chichester • Brisbane • Toronto • Singapore

Library of Congress Cataloging in Publication Data:

Little, Thomas Morton, 1910–
Agricultural experimentation.

Includes index.
1. Agriculture—Experimentation. I. Hills, Frederick Jackson, 1919– joint author. II. Title.

S540.A2L57 630'.7'2 77-26745
ISBN 0-471-02352-3

Printed in the United States of America

20 19 18 17 16 15 14

PREFACE

Few agricultural research workers have the time to master the details of abstract and sophisticated mathematics, yet they would like to gain a general understanding of the logic and reasoning involved in the designing and conducting of experiments. In short, they would like to learn enough of the basic principles of statistics to be able to design experiments properly and to draw valid conclusions from the results. This book is written to satisfy these needs. Mathematics beyond simple arithmetic has been kept to a minimum. Many of the mathematical relations are presented simply as facts without formal proof. Yet, every effort has been made to make the discussion mathematically correct and to avoid the dangers of oversimplification.

Separate chapters are devoted to each of seven experimental designs that probably constitute over 90 percent of the designs used in agricultural research. Four chapters are devoted to correlation and regression (linear, curvilinear, and multiple). Regression is presented in the context of the analysis of variance as well as a technique used in survey-type research. A special feature is a chapter on the use of shortcut methods for handling regression when the experimental treatments or observations are equally spaced.

For each technique emphasis is on a detailed, step-by-step procedure for computing the essential statistics. The spiral binding has been chosen so the book will lay open to enable users to follow a procedure in relation to their own work. Yet, this is much more than a "cookbook." In each case, the logic and reasoning behind the analysis is explained. An entire chapter is devoted to the assumptions underlying the analysis of variance and the ways of handling data that do not satisfy these assumptions.

Special emphasis is given to the subject of mean separation (determining which of several means are significantly different). It is evident from the current agricultural research literature that there is a great deal of misunderstanding about this subject. As a result, important conclusions justified by the data are often overlooked. This is especially true with regard to the method known as the functional analysis of variance or the method of orthogonal coefficients. This extremely simple, yet powerful technique is unfamiliar to a large proportion of agricultural research workers. This and other methods of mean separation are fully discussed.

The forerunner of this book, *Statistical Methods in Agricultural Research*, was used for several years to teach a methods course to extension agents and other professional agriculturists. The revisions and additions in the development of the present book make it more useful for this purpose. Special features are an improved presentation of mean separation, instructions for the use of preprogrammed calculators to simplify calculations in the analysis of variance, a discussion and example of a response surface, and a chapter, "Improving Precision," which discusses covariance, and the determination of the number of replications required in experiments.

We are grateful to the Literary Executor of the late Sir Ronald A. Fisher, F.R.S., to Dr. Frank Yates, F.R.S. and to Longman Group Ltd., London, for permission to reprint Tables A.2, A.6, A.7, and A.8 from their book *Statistical Tables for Biological, Agricultural and Medical Research*. (6th edition, 1974.)

T. M. Little
F. J. Hills

September 1977

CONTENTS

1. LOGIC, RESEARCH, AND EXPERIMENT

DEDUCTIVE AND INDUCTIVE REASONING 1
The Researcher's Problem 2
The Element of Chance 3
The Need for Statistical Evaluation 5
RESEARCH, SCIENTIFIC METHOD, AND THE EXPERIMENT 6
STEPS IN EXPERIMENTATION 7
SUMMARY 9

2. SOME BASIC CONCEPTS

THE NORMAL DISTRIBUTION 13
STATISTICAL NOTATION, MEANS, AND STANDARD DEVIATIONS 14
Variates in a Two-Way Table 18
SAMPLING FROM A NORMAL DISTRIBUTION 20
The Distribution of a Sample Means 21
The t Distribution and Confidence Limits 22
STATISTICAL HYPOTHESES AND TESTS OF SIGNIFICANCE 24
The F Distribution 25
SUMMARY 27

3. THE ANALYSIS OF VARIANCE AND t TESTS

ANOVA WITH TWO SAMPLES 31
A Cookbook Procedure 33
A POPULATION OF MEAN DIFFERENCES 37
t Tests for Significance 38
ROUNDING AND REPORTING NUMBERS 41
FACTORIAL EXPERIMENTS 42
THE ANALYSIS OF VARIANCE AND EXPERIMENTAL DESIGN 44
SUMMARY 44

4. THE COMPLETELY RANDOMIZED DESIGN

RANDOMIZATION 47
ANALYSIS OF VARIANCE 48

Sources of Variation and Degrees of Freedom **48**
Correction Term **49**
Sums of Squares and Mean Squares **49**
F Value **50**
THE WHAT AND WHY OF THE ANALYSIS **50**
SUMMARY **52**

5. THE RANDOMIZED COMPLETE BLOCK DESIGN

RANDOMIZATION **54**
ANALYSIS OF VARIANCE **54**
Sources of Variation and Degrees of Freedom **55**
Correction Term **56**
Sums of Squares and Mean Squares **56**
THE WHAT AND WHY OF THE ANALYSIS **57**
Mean Square for Blocks **57**
Mean Square for Treatments **58**
Mean Square for Error **58**
F Values **60**
SUMMARY **60**

6. MEAN SEPARATION

LEAST SIGNIFICANT DIFFERENCE **61**
MULTIPLE-RANGE TESTS **63**
Duncan's Multiple-Range Test **63**
PLANNED F TESTS **65**
Orthogonal Coefficients **65**
Class Comparisons **68**
Trend Comparisons **70**
SUMMARY **75**

7. THE LATIN SQUARE DESIGN

RANDOMIZATION **78**
ANALYSIS OF VARIANCE **80**
Sources of Variation and Degrees of Freedom **80**
Correction Term **80**
Sums of Squares and Mean Squares **80**
F Values **82**

MEAN SEPARATION **82**
SUMMARY **85**

8. THE SPLIT-PLOT DESIGN

RANDOMIZATION **90**
ANALYSIS OF VARIANCE **90**
Sources of Variation and Degrees of Freedom **91**
Correction Term **92**
Sums of Squares and Mean Squares **92**
F Values **93**
MEAN SEPARATION **94**
Pertinent F Tests **94**
Standard Errors and LSDs **98**
SUMMARY **100**

9. THE SPLIT-SPLIT PLOT

ORGANIZATION OF DATA **101**
ANALYSIS OF VARIANCE **101**
Sources of Variation and Degrees of Freedom **105**
Correction Term **106**
Sums of Squares and Mean Squares **106**
The Standard Deviation Key **107**
F Values **107**
MEAN SEPARATION **107**
Partitioning Interaction **108**
Standard Errors and LSDs **110**
SUMMARY **113**

10. THE SPLIT-BLOCK

ANALYSIS OF VARIANCE **118**
Degrees of Freedom **118**
Correction Term **118**
Sums of Squares **119**
Mean Squares **120**
The Standard Deviation Key **121**
F Values and Mean Separation **121**
STANDARD ERRORS **123**
SUMMARY **124**

11. SUBPLOTS AS REPEATED OBSERVATIONS

ANALYSIS FOR EACH SET OF OBSERVATIONS 125
Annual Analysis 128
F Values and Mean Separation 130
COMBINING TWO OR MORE YEARS 132
The Analysis for Each Year 132
Putting the Years Together 133
SUMMARY 137

12. TRANSFORMATIONS

ASSUMPTIONS OF THE ANALYSIS OF VARIANCE 139
Normality 139
Homogeneity of Variances 140
Independence of Means and Variances 142
Additivity 143
TESTS FOR VIOLATIONS OF THE ASSUMPTIONS 144
THE LOG TRANSFORMATION 150
THE SQUARE ROOT TRANSFORMATION 154
THE ARCSINE OR ANGULAR TRANSFORMATION 159
PRETRANSFORMED SCALES 162
SUMMARY 164

13. LINEAR CORRELATION AND REGRESSION

THE IDEA 167
MEASURING CORRELATION 168
REGRESSION 169
CORRELATION VERSUS REGRESSION 170
CALCULATING LINEAR CORRELATION 171
Quick Shortcut Method 173
Standard Method 174
STATISTICAL SIGNIFICANCE 176
THE REGRESSION LINE 177
CONFIDENCE LIMITS 182
REGRESSION IN REPLICATED EXPERIMENTS 185
PITFALLS 187
SUMMARY 192

14. CURVILINEAR RELATIONS

DECIDING WHAT CURVE TO FIT **195**
The Power Curve **196**
The Exponential Curve **202**
Asymptotic Curves **206**
The Polynomial Type **207**
Combining Curve Types **218**
The Periodic Type **220**
SUMMARY **225**

15. SHORTCUT REGRESSION METHODS FOR EQUALLY SPACED OBSERVATIONS OR TREATMENTS

POLYNOMIAL CURVE FITTING **229**
Partitioning the Sums of Squares **235**
Comparison of Shortcut and Regular Methods **237**
Unequally Spaced Treatments **238**
PERIODIC CURVE FITTING **238**
Partitioning the Sum of Squares **241**
SUMMARY **243**

16. CORRELATION AND REGRESSION FOR MORE THAN TWO VARIABLES

CORRELATION COEFFICIENTS **247**
REGRESSION COEFFICIENTS **249**
AN EXAMPLE WITH THREE VARIABLES **250**
MORE THAN THREE VARIABLES **256**
RESPONSE SURFACES **258**
SUMMARY **264**

17. ANALYSIS OF COUNTS

CHI-SQUARE **268**
Yates Correction for Continuity **270**
GUIDES FOR USING CHI-SQUARE **272**
INTERPRETING RESULTS **272**
TESTING FOR INDEPENDENCE **274**

HETEROGENEITY 279
SUMMARY 282

18. IMPROVING PRECISION

INCREASED REPLICATION 283
SELECTION OF TREATMENTS 284
REFINEMENT OF TECHNIQUE 285
SELECTION OF EXPERIMENTAL MATERIAL 285
SELECTION OF THE EXPERIMENTAL UNIT 285
TAKING ADDITIONAL MEASUREMENTS--COVARIANCE 285
Adjusting More than One Source of Variation 290
Adjusting the Treatment Means 290
Comparing Two Adjusted Treatment Means 292
Interpretation of the Covariance Analysis 293
PLANNED GROUPING OF EXPERIMENTAL UNITS--DESIGN 293
SUMMARY 293

SELECTED REFERENCES

APPENDIX TABLES

A1. RANDOM NUMBERS 296

A2. DISTRIBUTION OF t 297

A3. 10%, 5%, and 1% POINTS FOR THE F DISTRIBUTION 299

A4. SIGNIFICANT STUDENTIZED FACTORS (R) TO MULTIPLY BY LSD FOR TESTING MEANS AT VARIOUS RANGES, 5% LEVEL 307

A5. SIGNIFICANT STUDENTIZED FACTORS (R) TO MULTIPLY BY LSD FOR TESTING MEANS AT VARIOUS RANGES, 1% LEVEL 308

A6. DISTRIBUTION OF χ^2 (CHI-SQUARE) 309

A7. VALUES OF THE CORRELATION COEFFICIENT, r, FOR CERTAIN LEVELS OF SIGNIFICANCE 310

A8. THE ANGULAR TRANSFORMATION OF PERCENTAGES TO DEGREES 311

A9. LOGARITHMS 312

A10. SQUARES AND SQUARE ROOTS 316

A11. COEFFICIENTS, DIVISORS, AND K VALUES FOR FITTING UP TO QUARTIC CURVES TO EQUALLY SPACED DATA, AND PARTITIONING THE SUM OF SQUARES **331**

A11a. COEFFICIENTS AND DIVISORS FOR SOME SELECTED SETS OF UNEQUALLY SPACED TREATMENTS **341**

A12. COEFFICIENTS FOR FITTING PERIODIC CURVES AND PARTITIONING SUMS OF SQUARES FOR DATA TAKEN AT EQUAL TIME INTERVALS THROUGHOUT A COMPLETE CYCLE **342**

Index **345**

1

LOGIC, RESEARCH, AND EXPERIMENT

"The purpose of statistical science is to provide an objective basis for the analyses of problems in which the data depart from the laws of exact causality. A general logical system of inductive reasoning has been devised, is applicable to data of this kind, and is now widely used in scientific research. Some understanding of its principles is, therefore, important both for research workers and for those whose interests lie in the employment of technological advances resulting from research. Especially is this true of the agricultural and biological sciences."

D. J. Finney,
An Introduction to Statistical Science in Agriculture

The above quotation is a concise statement of the importance of statistical science in agriculture. To grasp fully what is meant by a "logical system of inductive reasoning," we must review some elementary concepts of logic. When we classify problems according to the system of reasoning employed in their solution, we find that there are just two kinds of problems.

DEDUCTIVE AND INDUCTIVE REASONING

First, there is the kind of problem in which we are given some general principle or set of principles and asked to determine what would happen under a specific set of conditions. The type of reasoning employed, from the general to the particular, is called *deductive reasoning*. A few examples will serve to make this concept clear.

Given the general formula for the area of a circle $A = \pi r^2$, what is the area of a circle whose radius is 6 inches?

Given a key and descriptions of the weeds of California, to what species does a certain weed belong?

Given Boyle's and Charles' laws, how do we expect a certain volume of gas to change when subjected to certain changes in pressure and temperature?

Given some general principles of disease control, what yield response do we expect from the application of a given dose of a fungicide to an acre of a particular crop?

Given an unbiased coin whose probability of coming up heads when tossed is one-half, what will happen when this coin is tossed 10 times?

Nearly all the problems encountered during our formal education were of this type, where the solution required deductive reasoning. It is frequently said that agriculturalists should be "well grounded in basic fundamentals." This implies that they should have at their command a large store of general principles and the skills of deductive reasoning to apply these to specific cases.

The second type of problem is the opposite of the first. We are given some specific cases and asked to arrive at some general principles that will apply to all members of the class represented by these cases. The reasoning employed, from the specific to the general, is called *inductive reasoning*. The following examples of problems requiring inductive reasoning are analogous to those given above to illustrate the deductive type of problems.

Given the areas and radii of several circles, what general formula can we give expressing the relation between the areas and radii of *all* circles?

Given several specimens of an undescribed weed species, how would we describe the species as a whole and express its relation to other species in a key?

Given a series of observations on the volume of a gas under different conditions of pressure and temperature, what general laws will account for these observations?

Given the results of a series of disease control trials, what general recommendations can we make regarding the use of control methods?

Given the results of tossing a coin 10 times, what conclusions can we draw regarding the bias or lack of bias of the coin?

Notice that all problems of this type have one thing in common—they start with a *group of observations*. In some cases, as in the description of a new species, the observations are simply made of phenomena as they occur in nature. Usually, however, the observations are made under controlled conditions. The factors being studied are made to vary in some systematic fashion by the application of *treatments*. Other factors that might influence the observations are minimized as much as is practical. We then have an *experiment*.

The Researcher's Problem

We have said that nearly all problems encountered in our formal schooling are of the type requiring deductive reasoning. We can also say that nearly all problems encountered by an agriculturalist are those requiring inductive reasoning.

What is the typical problem that confronts the agricultural researcher? It could be stated in these general terms: Will the use of a new or different practice affect the outcome of some particular segment of agricultural enterprise, and if so, to what extent? Since this problem can never be answered with 100% certainty, we must also consider the risk and cost of making an incorrect decision. This will become clearer as we go along.

To answer such a problem, an experiment is generally required. In the simplest experiment there may be only two treatments—the new practice and the old. A more complicated experiment might include several rates or methods of applying the new practice. Still more complex are those experiments in which the effects of several practices are studied simultaneously.

Whatever the design of the experiment, its purpose is to provide a means of making observations (probability sampling) that can be used for making plausible generalizations about the practice under study. Arriving at such generalizations is a typical problem in inductive reasoning.

The reader should not gain the impression that inductive reasoning involves an independent line of thought distinct from deductive reasoning. Inductive conclusions must always be checked by precise deductive methods.

The Element of Chance

Another phrase that appears in the quotation at the beginning of this chapter requires some clarification. What is meant by "problems in which the data depart from the laws of exact causality"?

Looking at the examples of problems given before, we note that there are some important differences among them. In the problem of finding the area of a circle there is no uncertainty regarding the answer. For any given radius, there can be only a single answer.

The coin-tossing problem is quite different. The general assumption is that the coin is not biased; but even with a single toss we are uncertain as to the result. One of two results may be obtained, both being equally probable. The question of what will happen when the coin is tossed 10 times has an even more uncertain answer, for there are 11 possible results as to the number of heads that will turn up, and these results differ in their probability of occurrence. Obviously, sampling vagaries will occur in this case, for there is not a simple one-to-one relation between cause and effect.

Such a situation is almost universal in the field of agriculture. No matter how much scientists know about nutrition and physiology, they cannot predict precisely what will be the gain in weight of a steer or the yield of a plot of potatoes under given sets of conditions. Chance variations resulting from a multitude of causes always make the results vary, no matter how much effort was put into controlling all known factors.

The term *chance* is hard to define, but even without a clear definition, its meaning is understood well enough to appreciate its importance in affecting biological results. When the element of chance enters into a problem, real difficulties are introduced. These are much more serious in the field of inductive reasoning than in deductive reasoning.

Consider the deductive problem of tossing an unbiased coin 10 times. By deductive methods we can enumerate all 11 possible results and calculate the probability of each fairly easily. For example, suppose we ask, "What is the probability of getting the result of five heads and five tails?" This answer can be found by calculating the value of

$$\frac{10!}{5!(5!)(2^{10})}$$

which turns out to be 0.246, or 24.6%. As the number of tosses is increased, or as the initial assumptions are modified to include certain degrees of bias in the coin, the calculations become more laborious, but they are still straightforward, and the results are simple and definite. Fortunately, the theory of probability has been developed by mathematicians, so that short-cut methods and tables are available to reduce greatly the necessary calculations in complicated cases.

Now consider the inductive problem. If a coin is tossed 10 times and comes up five heads and five tails, what can we say about the bias or lack of bias of the coin? All we can say with certainty is that the coin was neither two-headed nor two-tailed. If it were not biased, we would expect this result about 25% of the times the trial was repeated. We can say with a high degree of probability of being correct that the coin is not strongly biased in favor of either heads or tails. We must remember that we can never make such a statement with complete certainty. Even with a strongly biased coin (one that comes up heads 90% of the time), the observed result of five heads and five tails would have been *possible* but not very *probable*.

The only other statement we can make about the coin is that we feel fairly confident that its degree of bias was somewhere between a slight bias in favor of tails and a slight bias in favor of heads. Notice there is an infinity of possibilities in this interval, and that zero bias is one of these. It is very important to realize that with no other knowledge about the coin than the results of these 10 tosses, we are not justified in concluding that the coin was unbiased. With more tosses, we can narrow the interval of biases that could reasonably be expected to produce our observed result, but we will never be able to state with certainty that the coin was unbiased.

We have purposely avoided defining the terms *strong* and *slight* bias for the sake of simplicity. However, it is possible by statistical methods to determine what ranges of bias we will accept or reject depending on the degree of confidence we wish to have in our conclusions.

We can now see that the answer to our question, "What can we say about the bias of the coin?" was rather vague. The reader who is accustomed only to the precise answers of deductive mathematics may be disappointed at the vagueness of the answer. Yet, unsatisfactory as this may seem, the very nature of inductive reasoning is such that the answer is the best we can give. As Alfred North Whitehead, the great mathematical philosopher, has said, "The Theory of Induction is the despair of philosophy—and yet all our activities are based upon it."

The researcher should not despair in attempts to answer questions through observations and experiments. However, it should be realized that answers can never be absolute, and generalizations must be made with caution and only after making careful observations and exercising the best systems of reasoning at one's command.

The Need for Statistical Evaluation

Most agriculturists readily see the need for statistical analysis to provide an objective basis for evaluation, but some examples may be useful. If one harvests two equal areas of wheat from a field, the grain yield from the two areas, whether they be rod rows in length or halves of the entire field, will seldom be equal; the weight of fruit from adjacent trees in an orchard is seldom the same; rates of weight gain of any two animals of the same species and breed nearly always differ. Differences of this sort among crop or animal units result from genetic and environmental differences beyond the control of an experimenter. Although they are not errors in the sense of being wrong, they represent the variability among experimental units we call *experimental error*.

Once we recognize the existence of this variability, we realize the difficulty in evaluating a new practice by applying it to a single experimental unit and then comparing this unit to one that is similar but *nontreated*. The effect of the new practice is confounded with unaccounted variability. Thus, an experiment with a single replication provides a very poor measure of treatment effect; further, since there are no two experimental units treated alike, it provides no measure of experimental error. Statistical science overcomes these difficulties by requiring the collection of experimental data that will allow an unbiased estimate of treatment effects and the evaluation of treatment differences by tests of significance based on measuring *experimental error*.

Treatment effects are estimated by applying treatments to at least two experimental units (usually more) and averaging the results for each treatment. Tests of significance assess the probability that treatment differences could have occurred by chance alone.

There are three important principles inherent in all experimental designs that are essential to the objectives of statistical science:

1. *Replication*. Replication means that a treatment is repeated two or more times. Its function is to provide an estimate of experimental error and to provide a more precise measure of treatment effects. The number of replications that will be required in a particular experiment depends on the magnitude of the differences you wish to detect and the variability of the data with which you are working. Considering these two things at the beginning of an experiment will save much frustration.

2. *Randomization*. Randomization is the assignment of treatments to experimental units so that all units considered have an equal chance of receiving

a treatment. It functions to assure unbiased estimates of treatment means and experimental error.

3. *Local control.* This principle of experimental design allows for certain restrictions on randomization to reduce experimental error. For example, in the randomized complete block design, treatments are grouped into blocks that are expected to perform differently, allowing a *block effect* to be removed from the total variation in the trial.

RESEARCH, SCIENTIFIC METHOD, AND THE EXPERIMENT

Research can be broadly defined as systematic inquiry into a subject to discover new facts or principles. The procedure for research is generally known as the scientific method which, although difficult to define precisely, usually involves the following steps:

1. *Formulation of an hypothesis*—a tentative explanation or solution.

2. *Planning an experiment to objectively test the hypothesis.*

3. *Careful observation and collection of data from the experiment.*

4. *Interpretation of the experimental results.* A consideration of the results in the context of other known facts concerning the problem leads to confirmation, rejection, or alteration of the hypothesis.

The experiment is an important tool of research. Some important characteristics of a well-planned experiment are given below.

1. *Simplicity.* The selection of treatments and the experimental arrangement should be as simple as possible, consistent with the objectives of the experiment.

2. *Degree of precision.* The probability should be high that the experiment will be able to measure differences with the degree of precision the experimenter desires. This implies an appropriate design and sufficient replication.

3. *Absence of systematic error.* The experiment must be planned to ensure that experimental units receiving one treatment in no systematic way differ from those receiving another treatment so that an unbiased estimate of each treatment effect can be obtained.

4. *Range of validity of conclusions.* Conclusions should have as wide a range

of validity as possible. An experiment replicated in time and space would increase the range of validity of the conclusions that could be drawn from it. A factorial set of treatments is another way for increasing the range of validity of an experiment. In a factorial experiment the effects of one factor are evaluated under varying levels of a second factor.

5. *Calculation of degree of uncertainty*. In any experiment there is always some degree of uncertainty as to the validity of the conclusions. The experiment should be designed so that it is possible to calculate the probability of obtaining the observed results by chance alone.

STEPS IN EXPERIMENTATION

The selection of a procedure for research depends, to a large extent, on the subject matter in which the research is being conducted and on the objectives of the research. The research might be descriptive and involve a sampling survey, or it might involve a controlled experiment or series of experiments. When an experiment is involved there are a number of considerations that should be carefully thought through if it is to be a success. The following are some of the more important steps to be taken:

1. *Definition of the problem*. The first step in problem solving is to state the problem clearly and concisely. If the problem cannot be defined, there is little chance of it ever being solved. Once the problem is understood, you should be able to formulate questions which, when answered, will lead to solutions.

2. *Statement of objectives*. This may be in the form of questions to be answered, the hypothesis to be tested, or the effects to be estimated. Objectives should be written out in precise terms. This allows the experimenter to plan the experimental procedures more effectively. When there is more than one objective, they should be listed in order of importance, as this might have a bearing on the experimental design. In stating objectives, do not be vague or too ambitious.

3. *Selection of treatments*. The success of the experiment rests on the careful selection of treatments, whose evaluation will answer the questions posed.

4. *Selection of experimental material*. In selecting experimental material, the objectives of the experiment and the population about which inferences are to be made must be considered. The material used should be representative of the population on which the treatments will be tested.

5. *Selection of experimental design.* Here again a consideration of objectives is important, but a general rule would be to choose the simplest design that is likely to provide the precision you require.

6. *Selection of the unit for observation and the number of replications.* For example, in field experiments with plants, this means deciding on the size and shape of field plots. In experiments with animals, this means deciding on the number of animals to consider as an experimental unit. Experience from other similar experiments is invaluable in making these decisions. Both plot size and the number of replications should be chosen to produce the required precision of treatment estimate.

7. *Control of the effects of the adjacent units on each other.* This is usually accomplished through the use of border rows and by randomization of treatments.

8. *Consideration of data to be collected.* The data collected should properly evaluate treatment effects in line with the objectives of the experiment. In addition, consideration should be given to collection of data that will explain why the treatments perform as they do.

9. *Outlining statistical analysis and summarization of results.* Write out the sources of variation and associated degrees of freedom in the analysis of variance. Include the various F tests you may have planned. Consider how the results might be used, and prepare possible summary tables or graphs that will show the effects you expect. Compare these expected results to the objectives of your experiment to see if the experiment will give the answers you are looking for.

   At this point it is well to provide for a review of your plans by a statistician and by one or more of your colleagues. A review by others may bring out points you have overlooked. Certain alterations or adjustments may greatly enrich your experiment and make it possible to learn considerably more from the work you are about to undertake.

10. *Conducting the experiment.* In conducting the experiment, use procedures that are free from personal biases. Make use of the experimental design in collecting data so that differences among individuals or differences associated with order of collection can be removed from experimental error. Avoid fatigue in collecting data. Immediately recheck observations that seem out of line. Organize the collection of your data to facilitate analysis and to avoid errors in recopying. If it is necessary to copy data, check the copied figures against the originals immediately.

11. *Analyzing data and interpreting results*. All data should be analyzed as planned and the results interpreted in the light of the experimental conditions, hypothesis tested, and the relation of the results to facts previously established. Remember that statistics do not prove anything and that there is always a probability that your conclusions may be wrong. Therefore, consider the consequences of making an incorrect decision. Do not jump to a conclusion, even though it is *statistically significant* if the conclusion appears out of line with previously established facts. In this case, investigate the matter further.

12. *Preparation of a complete, readable, and correct report of the research.* There is no such thing as a *negative result.* If the null hypothesis is not rejected, it is *positive* evidence that there may be no real differences among the treatments tested. Again, check with your colleagues and provide for review of your conclusions.

Although most of the above steps are nonstatistical, statistical analysis is an important part of experimentation. Statistical science helps the researcher design the experiment and objectively evaluate the resulting numerical data. As experimenters, few of us will have the time or the inclination to become competent biometricians, but we can all learn and practice the three "R's" of experimentation.

1. *Replicate*. This is the only way you will be able to measure the validity of your conclusions from an experiment.

2. *Randomize*. Statistical analysis depends upon the assignment of treatments to plots in a purely objective, random manner.

3. *Request help*. Ask for help when in doubt about how to design, execute, or analyze an experiment. You are not expected to be an expert statistician, but you should know enough to understand the important principles of scientific experimentation, to be on guard against the common pitfalls, and to ask for help when you need it.

SUMMARY

Reasoning that proceeds from a general principle to a specific conclusion is a *deductive process*. *Inductive* reasoning arrives at a general principle from a specific conclusion. *Experiments* are conducted to provide specific facts from which general conclusions or principles are established and thus involve *inductive* reasoning.

Variability is a characteristic of biological material and creates the problem of deciding whether differences between experimental units result from unaccounted variability or real treatment effects. Statistical science helps overcome this difficulty by requiring the collection of data to provide unbiased estimates of treatment effects and the evaluation of treatment differences by tests of significance based on measuring unaccounted variability.

Three important principles of experimental design are *replication*, *randomization*, and *local control*.

The *scientific method* involves a flow process from known facts to hypothesis to experimentation which furnishes more facts that will cancel, strengthen, or alter the hypothesis.

A well-conceived and properly designed experiment should be as simple as possible, have a high probability of achieving its objective, and avoid systematic and biased errors. Its conclusions should have a wide range of validity, and data collected from it must be analyzable by valid statistical procedures.

The procedure for experimentation involves defining a problem, stating objectives, selecting treatments, selecting experimental material, selecting an experimental design, selecting the experimental units and number of replications, controlling the effects of adjacent units on each other, collection of data, and analyzing, interpreting, and reporting results.

2

SOME BASIC CONCEPTS

An *experimental unit* refers to the unit of experimental material to which a treatment is applied. It can be a single leaf, a whole plant, an area of ground containing many plants, a pot or a flat in the greenhouse, a single animal, several animals, or an entire herd. The term *plot* is synonymous with experimental unit and is frequently used in referring to plant experimental units. "Plot" is sometimes incorrectly used in referring to an entire experiment that really consists of several plots. A measurable characteristic of an experimental unit is called a *variable*. A variable can be *discrete* (discontinuous), assuming only specific values, the number of diseased plants per plot for example, or it can be *continuous* and assume any value between certain limits, for example, the yield of grain from a plot of barley. Individual measurements of a variable are called *variates*.

A *treatment* is a dosage of material or a method that is to be tested in the experiment. A crop variety is a kind of a treatment. When a treatment is applied to more than one experimental unit we have *replication* of that treatment. Two experimental units treated alike constitute two replications (or replicates). Experimental units receiving different treatments that have been replicated and arranged in a suitable design constitute an *experiment* (or trial or test).

In a statistical sense, a *population* is a set of measurements or counts of a single variable taken on all the units specified to be in the population. The population may be relatively small, such as the grain production per acre of all the barley fields in a specified area in a specified year, or it may be large, for example, the heights of all men over 20 years of age in the United States or the yields that would result from all possible plots of a given shape that could be arranged on an experimental area. Even a *small* population usually involves a measurement on a very large number of individuals or experimental units. We may have a population of a variable from individual experimental units, a population of means of samples of the variable, or a population of differences between pairs of sample means.

A *sample* is a set of measurements that constitutes part of a population. We obtain information and make inferences about a population from a sample. For this reason it is important that the sample be representative of the population. To obtain a representative sample we use the principle of randomness. A *random sample* is one in which any individual measurement is as likely to be included as any other.

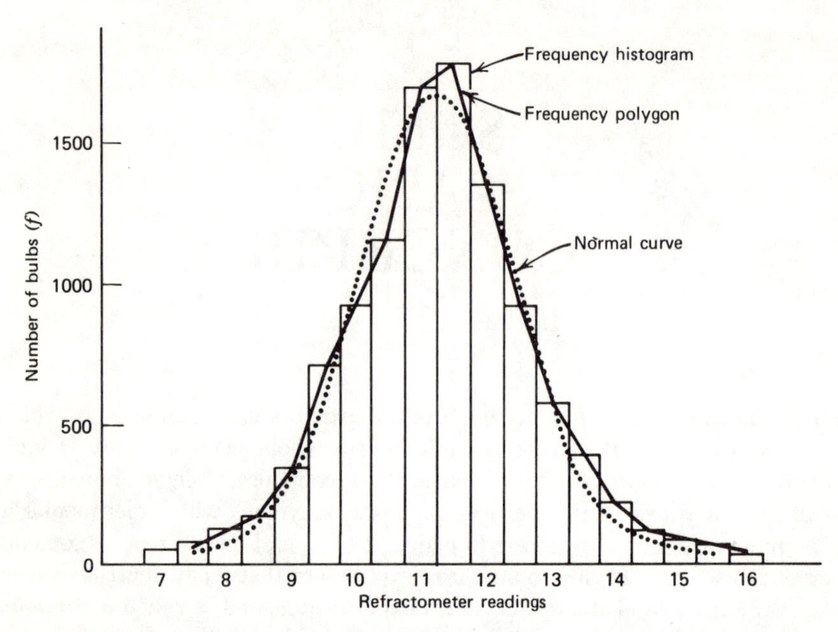

Figure 2.1. Frequency distribution of refractometer readings of 10,000 onion bulbs with the theoretical curve of normal distribution.

Populations are described by characteristics called *parameters*. Parameters are fixed values. For example, the arithmetic mean of all the variates in a population is a parameter. It has only one value, although we seldom know what it is. Samples are described by the same characteristics, but when applied to samples they are called *statistics*. The mean of a sample is a statistic. We calculate statistics from samples to estimate population parameters. Statistics vary from sample to sample.

Different values of a variable have different *frequencies* of occurrence in the population. To conveniently describe (characterize) a population, data from a large sample are commonly organized by the construction of a *frequency table*, a *frequency histogram*, and a *frequency polygon*. In a frequency table (Table 2.1), variates are tallied as to the several class intervals in which they fall. The totals can then be plotted as frequencies of occurrence for each class interval and a frequency histogram constructed (see Fig. 2.1). Connecting the midpoints of the class intervals gives a frequency polygon.

If we were to plot the frequency of yields of grain from many plots of barley, the percentage of butterfat in milk from many cows, the gains in weight of many groups of lambs, the number of scab lesions per potato in a thousand potatoes, or the refractometer readings of many onion bulbs, the resulting graphs would show several important features in common. The curves would all be approximately bell-shaped, with the high point near the middle, representing the most common class. They would slope off rather symmetrically on either side to rare, exceptional classes at the two ends.

Most biological data (and, in fact, data in many other fields of application), when plotted in a frequency curve, closely fit a mathematically defined curve

TABLE 2.1.
A frequency table. Refractometer readings of 10,000 onion bulbs

Class Interval	Midpoint	Tabulation	Frequency
6.8– 7.2	7.0	~~1111~~ ~~1111~~	10
7.3– 7.7	7.5	~~1111~~ ~~1111~~ ~~1111~~ 1111	19
7.8– 8.2	8.0		60
⋮			
10.8–11.2	11.0		1600
11.3–11.7	11.5		1700
⋮			
14.3–14.7	14.5		65
14.8–15.2	15.0		50
15.3–15.7	15.5	~~1111~~ ~~1111~~ ~~1111~~ ~~1111~~ ~~1111~~	25
15.8–16.2	16.0	~~1111~~ ~~1111~~ ~~1111~~ ~~1111~~	20
16.3–16.7	16.5	~~1111~~ ~~1111~~ 11	12

called the *normal frequency curve*. In Figure 2.1 a normal frequency curve has been superimposed over the frequency histogram and polygon of onion bulb refractometer readings. Note how well the curve fits the distribution of the sample.

THE NORMAL DISTRIBUTION

The imposing formula for describing a normal frequency curve is

$$f = \frac{N}{(\sigma\sqrt{2\pi})} e^{-(y-\mu)^2/2\sigma^2}$$

where f is the frequency of occurrence of any given variate, y is any given variate, N is the number of variates in the population, μ is the population mean, and σ is the population standard deviation. Note that the normal curve describing the frequency of occurrence of variates of different sizes can be plotted by the calculation of just two parameters, μ and σ.

Normal distributions only vary from one another with respect to their mean and/or standard deviation. The mean determines the position of a curve on the horizontal axis. The standard deviation determines the amount of spread or dispersion among the variates. Figure 2.2*a* shows two normal distributions with identical standard deviations but different means. The two normal distributions in Figure 2.2*b* have identical means but different standard deviations.

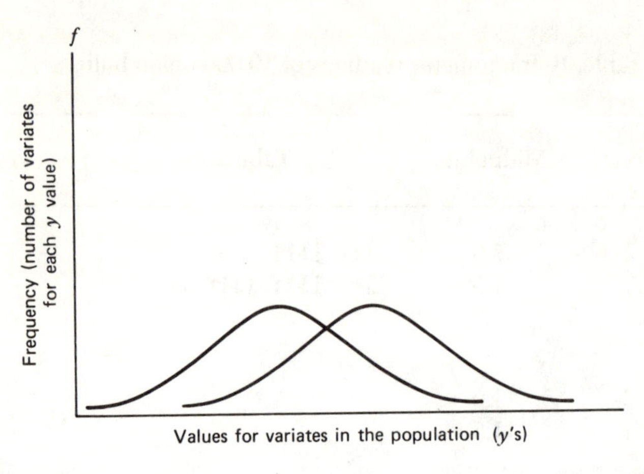

Figure 2.2*a*. Normal distributions—standard deviations equal, means different.

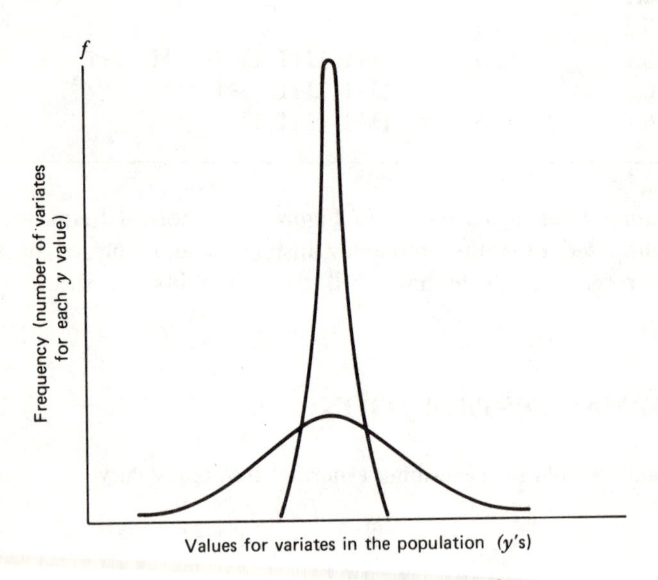

Figure 2.2*b*. Normal distributions—means equal, standard deviations different.

STATISTICAL NOTATION, MEANS, AND STANDARD DEVIATIONS

To deal mathematically with variates, means, and totals, it is necessary to have a system of notation to express procedures and relationships. In this book, complicated notation is avoided as much as possible, since it is confusing to most students. Nevertheless, if you continue to study statistics in other books, a brief introduction to the more or less standard system of notation may be helpful. We say "more or less standard system of notation" because there is considerable variability from book to book—to the frustration of all students. First, we discuss the mean and standard deviation and in so doing learn some simple notation.

The most common and usually the best measure of central tendency is the *arithmetic mean*. The symbols used to represent the arithmetic mean (hereafter shortened to *mean*) are the Greek μ for the mean of a population and $\bar{Y}$ or $\bar{X}$ for the mean of a sample. Mu (μ) is a parameter, a fixed value, that we seldom know, and $\bar{Y}$ is a statistic, a value that varies from sample to sample drawn from the same population.

The population mean is defined as

$$\mu = \frac{Y_1 + Y_2 + Y_3 + \cdots + Y_N}{N}$$

where Y_1, Y_2, and so forth are the variates of the population, and N is the number of variates in the population. Thus, Y_N is the Nth variate of the population.

Many books use X rather than Y to stand for a variable. However, this leads to some confusion when you first study regression. In regression you consider the values of the variable you are studying as Y values as they are plotted on the Y (ordinate) axis of the graph; the X values, plotted on the X (abscissa) axis, are the treatments of your experiment, for example, levels of fertilization. Thus it avoids some confusion to call variates Y values at the start.

The mean (μ) can be defined by a shorthand notation called a *summation notation*.

$$\mu = \frac{\sum_{i=1}^{N} Y_i}{N}$$

In this shorthand, the Greek capital Σ (sigma) tells you to sum all the values of Y_i. The summation index, $i = 1 \ldots N$, says that the values of Y_i go from the value of Y_1 to that of Y_N.

Since we seldom, if ever, know the value of μ, we estimate it from a sample mean, $\bar{Y}$, which is defined as

$$\bar{Y} = \frac{\sum_{i=1}^{r} Y_i}{r}$$

where r represents the number of variates in the sample. When it is clear what Y's are to be summed, the notation is frequently shortened to ΣY_i or even ΣY.

For the sample of Table 2.2, $\bar{Y} = \Sigma Y_i / r = (3+4+5+2+1)/5 = 15/5 = 3$ g/plant. Often, we wish to denote the difference between a variate (Y) and a mean ($\bar{Y}$). Such deviates are often represented by an italicized lowercase y or x. Thus $y = Y - \bar{Y}$, or $x = X - \bar{X}$.

TABLE 2.2.
Dry weight of five plants, $\bar{Y}=3$

	Grams per plant Y	$Y-\bar{Y}$	$(Y-\bar{Y})^2$
	3	0	0
	4	1	1
	5	2	4
	2	−1	1
	1	−2	4
Σ	15	0	10

There are two important properties of the mean: the sum of its deviates is zero (column 2, Table 2.2), and the sum of squares of the deviates (column 3, Table 2.2) is minimal, that is, the sum of squares of deviates from any other value will be larger.

Other measures of central tendency, which we will not use in this book, are: the median—the value situated at the center of the variates when these are arranged in order of magnitude; if the number of variates is even, the median is the average of the two central values; and the mode—the value of most frequent occurrence. In a normal distribution the mean, median, and mode are equal.

The most common measure of dispersion, and the best for most purposes, is the *standard deviation* and its square, the *variance*. The standard deviation of a population, σ, and the variance, σ^2, when estimated from a sample are symbolized as s and s^2, respectively.

The population variance is defined as

$$\sigma^2=\frac{\Sigma(Y_i-\mu)^2}{N}$$

where N is the number of variates in the population. The best estimate of σ^2 from a small sample (where r is less than 60), is defined as

$$s^2=\frac{\Sigma\left(Y_i-\bar{Y}\right)^2}{r-1}$$

where r is the number of variates in the sample. Why use $r-1$ rather than r as the divisor? If we know the value of μ, the best estimate of σ^2 from a sample is

$$s^2=\frac{\Sigma(Y_i-\mu)^2}{r}$$

r being the number of variates in the sample. However, we seldom, if ever, know the value of μ, so in the numerator, we replace it with its estimator, $\bar{Y}$. Now, while $\bar{Y}$ is on the average equal to μ, it varies from sample to sample and seldom is exactly equal to μ. We saw previously that $\Sigma(Y_i - \bar{Y})^2$ is less than the sum of squares of deviates from any value other than $\bar{Y}$. Therefore, if $\bar{Y}$ is not exactly equal to μ, $\Sigma(Y_i - \bar{Y})^2$ is less than $\Sigma(Y_i - \mu)^2$. This means that $\Sigma(Y_i - \bar{Y})^2/r$ will give too small an estimate of σ^2. It turns out that the proper correction can be made by using $r-1$ in the denominator instead of r. In other words, on the average,

$$\frac{\Sigma(Y_i - \bar{Y})^2}{r-1} \cong \frac{\Sigma(Y_i - \mu)^2}{r} \cong \sigma^2$$

The numerator, $\Sigma(Y_i - \bar{Y})^2$, is a *sum of squares* in this case the sum of the squares of deviations of individual variates from their mean. The denominator, $r-1$, is referred to as the *degrees of freedom* for the sample, usually one less than the number of observations.

We will use the small sample in Table 2.2 to illustrate the calculation of s^2 and s.

$$s^2 = \frac{\Sigma(Y_i - \bar{Y})^2}{r-1} = \frac{(3-3)^2 + (4-3)^2 + (5-3)^2 + (2-3)^2 + (1-3)^2}{5-1}$$

$$= \frac{(0)^2 + (1)^2 + (2)^2 + (-1)^2 + (-2)^2}{4} = \frac{0+1+4+1+4}{4} = \frac{10}{4} = 2.5$$

$$s = \sqrt{2.5} = 1.58 \text{ g/plant}$$

For small samples without decimals where the mean happens to be a whole number, s^2 and s can easily be calculated by the definition formula, but for larger samples there is a shortcut method that is much easier to perform, especially when a desk calculator is used. It can be proven that

$$\Sigma(Y_i - \bar{Y})^2 = \Sigma Y_i^2 - \frac{(\Sigma Y_i)^2}{r}$$

Therefore, a convenient working formula for s^2 is

$$s^2 = \frac{\Sigma Y_i^2 - \frac{(\Sigma Y_i)^2}{r}}{r-1}$$

The right-hand term of the numerator is called the *correction term* or *correction factor* and will be denoted in the book as C. $C = (\Sigma Y_i)^2/r$. The denominator $(r-1)$ is called the degrees of freedom (denoted by df) on which the variance is based—in this case, one less than the number of variates in the sample.

Applying this formula to the data of Table 2.2 gives

$$s^2 = \frac{3^2+4^2+5^2+2^2+1^2-\dfrac{(3+4+5+2+1)^2}{5}}{5-1} = \frac{55-\dfrac{15^2}{5}}{4}$$

$$= \frac{55-45}{4} = \frac{10}{4} = 2.5, \text{ as before}$$

Many desk and pocket calculators are programmed to compute a standard deviation by depressing a key after entering a sample of variates. A calculator with this capability greatly facilitates the computations in the analysis of variance (acronym ANOVA). One caution—know whether your calculator computes s^2 using r or $r-1$ as a divisor. The divisor r is only used when the sample is large, that is, when it contains at least 60 variates.

Other measures of dispersion are the range and the mean deviation. However, they will not be discussed here because of the far greater utility of s^2 and s.

The variability among experimental units of experiments involving different units of measurements and/or plot sizes can be compared by *coefficients of variation*, which express the standard deviation per experimental unit as a percent of the general mean of the experiment; thus $CV=(s/\bar{Y})100$. For example, consider two experiments—one involving sugar beet root yield, where $s=1.18$ tons per acre and the mean of all the plots is 30.5 tons per acre, and the other involving lima beans, where the variable was seedlings per plot and $s=5.8$ and $\bar{Y}=82.7$. The coefficients of variation are $(1.18/30.5)100=3.9\%$ and $(5.8/82.7)100=7.0\%$. A comparison of the two indicates that there was 1.8 times (7.0/3.9) more variability among the plots within a treatment of the lima bean experiment.

Variates in a Two-Way Table

Because of the design of the experiment or to facilitate computations, variates often are arranged in a two-way table and symbolized as in Table 2.3. The symbol for any variate in such a two-way table is Y_{ij} or, in some books, X_{ij}. The i subscript refers to the rows of the table that go from 1 to n, and j refers to the columns that go from 1 to r. A particular variate is indicated by the intersection of a row and column; for example, Y_{23} is the variate of row 2 and column 3.

Note the use of the dot subscript to indicate an operation over all the variates in a row or column. $Y_{1.}$ means the sum of all the variates of row 1. To indicate an operation involving all the row totals, we use the symbol $Y_{i.}$; for example, $\Sigma Y_{i.}^2$ indicates that you should square each row total and sum the squares. The mean of row 1 is $\bar{Y}_{1.}$ which equals $Y_{1.}/r$ which also equals $\Sigma_{j=1}^{r} Y_{1j}/r$. This last formula merely says "sum all the j's of row 1 and divide by r, the number of j's."

The use of such a system of notation (when you finally get used to it) saves much space in indicating operations and relationships. We will use it sparingly and almost always with a numerical example for illustration. To practice it a little, we will use the real numbers of Table 2.4 along with the symbols of Table 2.3.

TABLE 2.3
Symbolic presentation of variates in a two-way table

Rows (i, Treatments)	Columns (j, Replications) 1	2	3	··· r	Totals, $Y_{i.}$	Means, $\bar{Y}_{i.}$
1	Y_{11}	Y_{12}		······ Y_{1r}	$Y_{1.}$	$\bar{Y}_{1.}$
2	Y_{21}	Y_{22}		Y_{2r}	$Y_{2.}$	$\bar{Y}_{2.}$
⋮	⋮	⋮		⋮	⋮	⋮
n	Y_{n1}	Y_{n2}		······ Y_{nr}	$Y_{n.}$	$\bar{Y}_{n.}$
Totals, $Y_{.j}$	$Y_{.1}$	$Y_{.2}$		······ $Y_{.r}$	$Y_{..}$	
Means, $\bar{Y}_{.j}$	$\bar{Y}_{.1}$	$\bar{Y}_{.2}$		$\bar{Y}_{.r}$		$\bar{Y}_{..}$

TABLE 2.4.
Sugar beet root yields (tons per acre) from an experiment with five treatments in four replications

Treatments (Row)	Replications (Columns) 1	2	3	4=r	Totals $Y_{i.}$	Means $\bar{Y}_{i.}$
1	15	18	17	18	68	17.0
2	16	15	13	16	60	15.0
3	23	25	22	24	94	23.5
4	20	16	14	16	66	16.5
5=n	20	17	15	16	68	17.0
Totals, $Y_{.j}$	94	91	81	90	$356=Y_{..}$	
Means, $\bar{Y}_{.j}$	18.8	18.2	16.2	18.0		$17.8=\bar{Y}_{..}$

To indicate the computation of the sum of squares of all the variates in a two-way table we would write,

$$SS = \sum_{i=1}^{n} \sum_{j=1}^{r} \left(Y_{ij} - \bar{Y}_{..}\right)^2$$

The summation indices are often omitted, and sometimes one of the summation signs is also omitted to shorten the expression to

$$SS = \Sigma\left(Y_{ij} - \bar{Y}_{..}\right)^2$$

The first formula is more complete, as it identifies the summation limits for both rows and columns, but when this is understood the second formula is sufficient.

To compute the SS by this formula for the data of Table 2.4,

$$SS = (15-17.8)^2 + (18-17.8)^2 + \cdots + (15-17.8)^2 + (16-17.8)^2 = 223.2$$

The series of dots, ..., means to continue the indicated operation throughout the table, ending with the last two variates, 15 and 16.

To compute the sum of squares among column means, we write, $SSC = n\Sigma(\bar{Y}_{.j} - \bar{Y}_{..})^2$. This indicates that we take each column mean $(\bar{Y}_{.j})$, subtract the general mean $(\bar{Y}_{..})$, square each difference $[(\ )^2]$, sum the squares (Σ), and multiply by the number of variates (n) in each column. The significance of multiplying by n will be pointed out shortly. For now, we are only interested in trying to follow this confusing shorthand.

For Table 2.4,

$$SSC = 5\left[(18.8-17.8)^2 + (18.2-17.8)^2 + (16.2-17.8)^2 + (18.0-17.8)^2\right]$$

$$= 5(3.76)$$

$$= 18.8$$

SSC can also be computed from column totals as $SSC = (\Sigma Y_{.j}^2/n) - (Y_{..}^2/nr)$. For Table 2.4, this says

$$SSC = \frac{94^2 + 91^2 + 81^2 + 90^2}{5} - \frac{356^2}{5(4)}$$

$$= 6355.0 \quad 6336.8$$

$$= 18.8$$

Now we return to the normal distribution and other relationships important to statistical procedure.

SAMPLING FROM A NORMAL DISTRIBUTION

We commonly expose a number of plots or animals to a certain treatment. The treatment effect is estimated by calculating the mean of the sample. We know that repetitions of the experiment (in effect, drawing other samples) will produce a series of different means. One problem then is how well does a single mean represent the true treatment effect? One approach to this problem is to calculate *confidence limits*, a range of values within which the true mean of the treatment effect will fall unless we have drawn a very unusual sample. Before we calculate confidence limits for a treatment mean we should look at the relationship between certain parameters of a population of *individuals* and a population of *means* generated by repeated sampling from the parent population.

The Distribution of Sample Means

If all possible samples of a given size are drawn from a normally distributed population of individual variates, the means of these samples will form a much larger population than the parent population; the mean of the new population will be the same as the parent population, but the standard deviation will be smaller. In this kind of sampling, each variate of the parent population is identified, and after a sample is drawn and the mean determined the sample is returned and another one drawn. The process is repeated until all possible combinations of variates appear together in a sample.

The standard deviation of the population of means is called the *standard error of a mean*, or just *standard error*, and is symbolized by $\sigma_{\bar{y}}$. When $\sigma_{\bar{y}}$ is estimated from a sample its symbol is $s_{\bar{y}}$.

There is an important and very useful mathematical relationship between the variance of the parent population and the variance of a population of means drawn from it: $\sigma_{\bar{y}}^2 = \sigma^2/r$, where r is the sample size on which the population of means is based. Figure 2.3 illustrates this relationship. With increasing sample size

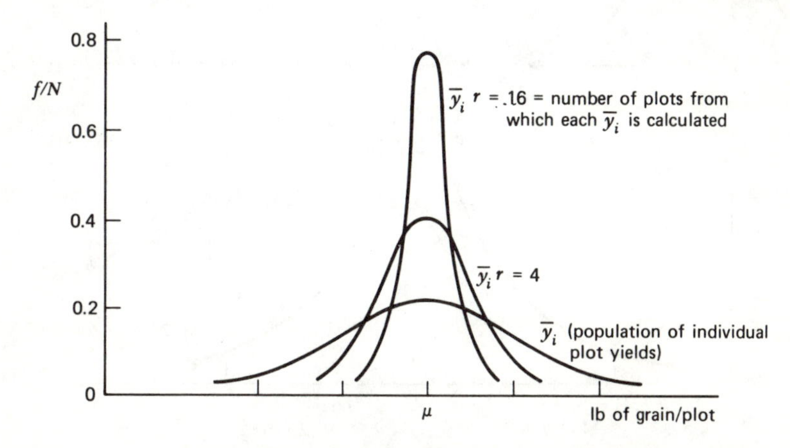

Figure 2.3. Frequency distributions of populations of means, varying in sample size, generated by repeated sampling from the same normally distributed population of plot grain yields. The distributions (all normal) become narrower and taller as sample size increases according to the relationship $\sigma_{\bar{y}}^2 = \sigma^2/r$.

(r) the distribution of means becomes narrower and taller, that is, the standard deviation becomes smaller, but the mean remains the same. Because of this relationship, $\sigma_{\bar{y}}^2 = \sigma^2/r$, we can estimate $\sigma_{\bar{y}}^2$ from only a single sample by $s_{\bar{y}}^2 = s^2/r$. We use this relationship when we calculate a confidence interval about a sample mean. The relationship is also used repeatedly in the ANOVA when we wish to estimate the variance per plot, s^2, form a series of means when we assume each mean is from a sample drawn from the same population. In this case we compute $s_{\bar{y}}^2$ from the sample means as $s_{\bar{y}}^2 = \Sigma(\bar{Y}_{i.} - \bar{Y}_{..})^2/(n-1)$ and then estimate s^2 by solving $s_{\bar{y}}^2 = s^2/r$ for $s^2 = rs_{\bar{y}}^2$. We will discuss this in more detail later.

The t Distribution and Confidence Limits

Consider another repeated drawing of samples of a given size, say $r=5$ as in Figure 2.4. For each sample compute $\bar{Y}$, s, $s_{\bar{y}}$, and another statistic, t, where $t = (\bar{Y} - \mu)/s_{\bar{y}}$. Now imagine organizing the large population of t values in a frequency distribution. The frequency curve will look like the curve in Figure 2.4.

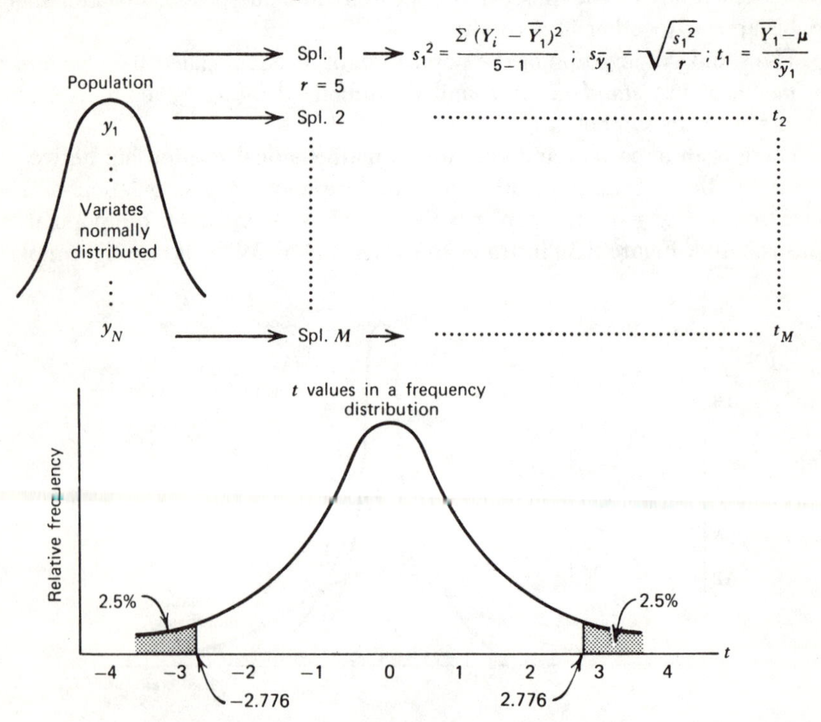

Figure 2.4. Generation of the t distribution for sample size of 5. A t value is computed for each of all the possible samples of five variates. Plotting the frequencies of the t values gives a distribution that has fewer values near the center and more toward the tails than is the case with the normal distribution.

There is a unique t distribution for each sample size. For a sample size of 5, 2.5% of the t values will be equal to or greater than 2.776, and 2.5% will be equal to or less than −2.776. Table A.2 in the Appendix is a two-tailed t table where probabilities are shown for obtaining ±t values for the degrees of freedom for different sample sizes. For example, for df=10, find that the ±t value to be expected with a probability of 0.01 (1%) is 3.169.

Figure 2.5 shows the t distribution for a sample size of 5 compared to the normal distribution. Note that the t distribution is more variable than the normal distribution. The larger the sample size, the closer t approaches a normal distribution. When t values are based on samples containing 60 or more variates, they are approximately normally distributed, as they closely estimate a normally distributed statistic, Z, which is calculated as $Z=\bar{Y}-\mu/\sigma_{\bar{y}}$; t and Z only differ in the denominator. With small samples, $s_{\bar{y}}$ is quite variable from sample to sample, and therefore t is more variable than Z, whose denominator, $\sigma_{\bar{y}}$, is a constant. With larger samples, however, $s_{\bar{y}}$ is less variable, and therefore t values more closely estimate Z values. For the last line of most t tables, where degrees of freedom are infinite, t=Z (Table A.2). A table of areas under the normal curve corresponding to Z values is not included in this book, as we seldom deal with samples large enough to justify its use.

CONFIDENCE LIMITS. From any random sample, confidence limits (CL) can be calculated within which μ will fall with a specified confidence. This is done by solving $\pm t=(\bar{Y}-\mu)/s_{\bar{y}}$ for μ and calling the resulting two values confidence limits: $CL=\bar{Y}\pm ts_{\bar{y}}$. If we wish to be 95% confident that CL will contain μ, we

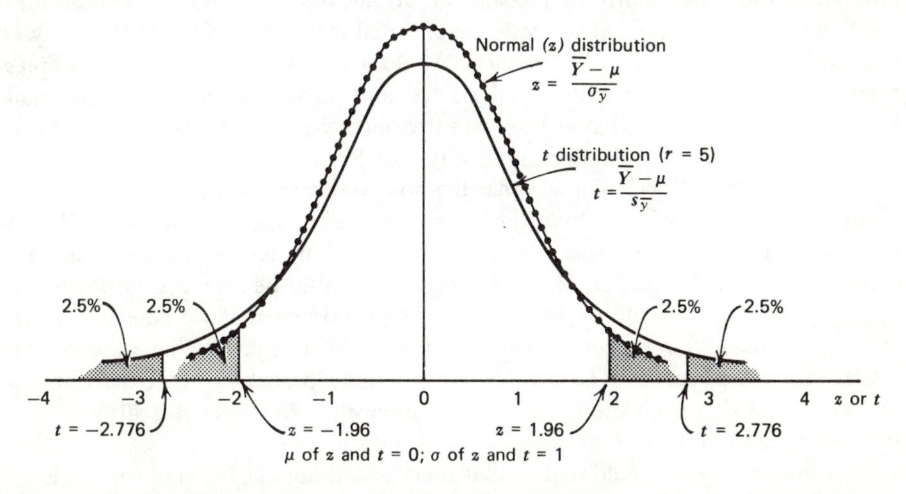

Figure 2.5. Distribution of z compared to the t distribution based on a sample size of 5. As sample size increases the t distribution approaches the normal z distribution. (Values of t and z that exclude 5% of the area under each curve are indicated.)

multiply $s_{\bar{Y}}$ by a tabular t value depending on $n-1$ degrees of freedom and the 5% level of probability (Table A2). For a sample where $r=5$, $s_{\bar{Y}}$ is multiplied by 2.776.

To illustrate, consider the sample of Table 2.2 where $r=5, \bar{Y}=3$, and $s^2=2.5$. Then

$$s_{\bar{Y}}=\sqrt{\frac{2.5}{5}}=0.707 \qquad \text{and} \qquad CL_{.95}=3\pm 2.776(0.707)=4.96 \text{ to } 1.04 \text{ gm/plant}$$

Thus, with a confidence of 95% we can say that μ lies in this range. It is incorrect to say that the *probability* is 95% that μ lies within these confidence limits because, based on the statistics of the particular sample, μ *will* or *will not* lie in the calculated interval. We may have drawn a sample whose $\bar{Y}$ and/or s^2 deviates sufficiently from μ and/or σ^2 so that $CL_{.95}$ will not contain μ. However the chance of drawing such a sample is only 5%.

STATISTICAL HYPOTHESES AND TESTS OF SIGNIFICANCE

The statistical procedure for comparing two or more treatment means employs the use of an assumption called the *null hypothesis*, which assumes that the treatments have no effect. We then proceed to test the probability that means as divergent as those of our samples would occur by chance alone if the samples were indeed random samples from normally distributed populations with equal means and variances. If our analysis leads to the conclusion that we could expect such mean differences quite frequently by chance, we do not reject the null hypothesis and conclude that we have no good evidence of a real treatment effect. If the analysis indicates that the observed differences would rarely occur in random samples drawn from populations with equal means and variances, we reject the null hypothesis and conclude that at least one treatment had a real effect. At least one of the means is said to be *significantly* different from the others.

If the probability is 5% or less that the observed variation among means could occur by chance, we say that the means are *significantly different*. If the probability is 1% or less that the observed variation among means could be expected to occur by chance, the differences are said to be *highly significant*.

The fact that the null hypothesis is not rejected and that we conclude there are no significant differences among the means does not prove that some of the treatments had no effect. There is always a definite probability that there was a real effect but that the experiment was too insensitive to detect the difference at the desired level of probability.

At this point you should realize that there is nothing magic about the 5% level of significance. The conlcusions you make concerning an experiment are your own, not the statistician's, and should be based on more than statistical evidence. The logic of the conclusions should be considered in the light of what is already known about the subject. Do not be too ready to accept a *significant* result if it does not make sense in the light of other known facts. There is always a chance that your significant result occurred by chance and that you have made an error in rejecting the null hypothesis.

Consider the consequences of being wrong. If the consequences are serious, such as being wrong in recommending a change that would require a considerable expense for a relatively small increased profit, you may hesitate to reject the null hypothesis on the basis of a single test even though the results are significant at the 5% level. In such a situation additional testing is clearly in order.

On the other hand, if the consequences of being wrong are not serious, you might reject the null hypothesis even though statistical analysis says you could expect such a result by chance as often as 1 out of 15 or even 1 out of 10 times. Consider, for example, the testing of a new inexpensive seed treatment when the combined analysis of several field experiments falls just short of being significant at the 5% level. Further, suppose that the results of several greenhouse experiments have indicated that the new treatment gave significantly better protection against the major pathogens that attack seedlings of the crop in question. In such a situation you might be justified in rejecting the null hypothesis, even to the point of recommending the practice to farmers, while you proceed to further test your conclusions in additional field experiments.

The F Distribution

An F test is a ratio between two variances and is used to determine whether two independent estimates of variance can be assumed to be estimates of the same variance. This ratio was called F by George W. Snedecor in honor of Ronald A. Fisher, a pioneer in the use of mathematical statistics in agriculture. In the analysis of variance, the F test is used to test equality of means; that is, to answer the question, Can it reasonably be assumed that the treatment means resulted from sampling populations with equal means? This can be illustrated by a description of how a portion of the F table could be determined.

Consider the following: From a normally distributed population (Fig. 2.6), draw five samples ($n=5$), each containing a specified number of variates, nine for example ($r=9$). Calculate the means of these five samples. Estimate σ^2 by calculating s^2 *for each sample* to give $s_1^2 \ldots s_5^2$. Sum these estimates of σ^2 to obtain an average (pooled) estimate: $s^2=(s_1^2+\ldots+s_5^2)/5$.

Now estimate the variance of means ($\sigma_{\bar{y}}^2$) from the *means* of the five samples: $s_{\bar{y}}^2=\Sigma(\bar{Y}_{i.}-\bar{Y}_{..})^2/(5-1)$. From $s_{\bar{y}}^2$, again estimate σ^2, using the relationship $s^2=rs_{\bar{y}}^2$, where r is the number of variates in each sample. Compute the variance ratio F, where

$$F=\frac{s^2\text{, calculated from samples means}}{s^2\text{, calculated by pooling sample variances}}$$

The degrees of freedom for the numerator are $n-1=4$ (where n is the number of samples) and for the denominator $n(r-1)=5(8)=40$ (where r is the number of

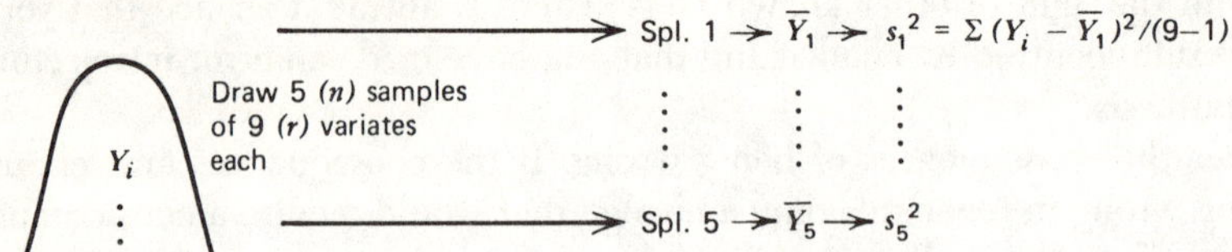

To compute F for a single drawing of 5 samples of 9 variates:

(1) Estimate $\sigma_{\bar{y}}^2$ as $s_{\bar{y}}^2 = \Sigma(\bar{Y}_{i.} - \bar{Y}_{..})^2/5-1$ where $\bar{Y}_{..} = \Sigma\bar{Y}_{i.}/5$

(2) Estimate σ^2 from the variability among the sample means as: $s_b^2 = rs_{\bar{y}}^2 = 9s_{\bar{y}}^2$

(3) Estimate σ^2 from the variability within the samples as: $s_w^2 = (s_1^2 + \dots + s_5^2)/5$

then $F = \dfrac{s_b^2}{s_w^2}$ with df $= \dfrac{5-1}{5(9-1)} = \dfrac{4}{40}$

Repeating this sampling procedure many times generates a population of F values which when plotted looks like the curve below.

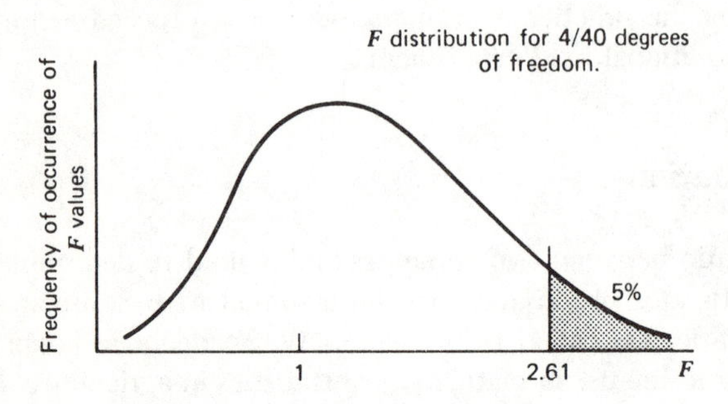

Figure 2.6. Repeated drawing of 5(n) samples of 9(r) variates each from a population of normally distributed variates ($Y_1 \dots Y_N$) to generate an F distribution. Five percent of the F values will be 2.61 or larger (see text).

variates in each sample). Now imagine that this sampling procedure is repeated until all possible sets of samples have been drawn and recorded, the frequencies of obtaining F values of various sizes have been recorded, and the frequency curve has been plotted. The F value 2.61 is the value beyond which 5% of the calculated values fall. This is the value for the 5% level found in an F table for 4 and 40 degrees of freedom (Table A.3). Similarly, F values can be determined for other sample sizes, numbers of samples, and for other levels of probability (2.5%, 1%, etc.).

Since both variances in the F ratio are estimates of the same variance (σ^2), the ratio will be close to 1 unless an unusual set of samples has been drawn. The F distribution for the sample size we are considering ($n=5, r=9$) will look like the graph in Figure 2.6. The area under the curve represents the frequency of obtaining any given F value. For any given draw of a set of samples of $n=5$ and $r=9$ the chances of the calculated F value being equal to or greater than 2.61 are 5%. Or, the chances are 95% that any given draw of such a set of samples will produce an F value of less than 2.61. Note that the F test is a one-tailed test. That is, we are not interested in the probability that F is equal to some value less than 1.

The above hypothetical sampling experiments are intended to show how t and F distributions can be obtained by sampling from a population of normally distributed variates. Tables for t and F are not determined by these laborious sampling procedures but are calculated from precise and rather complicated mathematical relationships. The use of F ratios in the ANOVA will be discussed in the next and subsequent chapters.

SUMMARY

Experimental unit (or plot, for an area of ground in the field). The unit of experimental material to which a treatment is applied.
Variable. A measurable characteristic of an experimental unit.
Variate. A specific measurement of a variable.
Population. A set of measurements (or counts) of a variable taken on all the individuals specified to be in the population.
Sample. A set of measurements (variates) that constitute a part of a population.
Parameter. A characteristic of a population (e.g., the mean). A parameter is a fixed value we seldom know. Parameters are estimated from samples. Parameters are usually symbolized by Greek letters (μ, σ, etc.).
Statistic. A characteristic of a sample—often used to estimate a parameter; generally symbolized by Roman letter ($\bar{Y}$, s, etc.).
Normal distribution. A mathematically defined, bell-shaped curve resulting from plotting the frequencies of occurrence of values of a variate against the range of the variate values. A normal distribution is uniquely described by its mean and standard deviation.
The mean of a population of individual variates, μ.

$$\mu = \frac{\sum Y_i}{N}, \text{ where N is the number of individuals in the population.}$$

The estimate of μ from a sample, $\bar{Y}$.

$$\bar{Y} = \frac{\sum Y_i}{r}, \text{ where r is the number of individuals in the sample.}$$

The variance of a population of individual variates, σ^2.

$$\sigma^2 = \frac{\sum (Y_i - \mu)^2}{N}$$

The standard deviation of a population of individual variates, σ.

$$\sigma = \sqrt{\sigma^2}$$

The estimate of σ^2 *from a sample,* s^2.

$$s^2 = \frac{\sum (Y_i - \bar{Y})^2}{r-1} \text{(definition formula).} \qquad s^2 = \frac{\sum Y_i^2 - \frac{\left(\sum Y_i\right)^2}{r}}{r-1} \text{(working formula)}$$

Correction term, used in the working formula, C.

$$C = \frac{\left(\sum Y_i\right)^2}{r}$$

Estimate of σ *from a sample,* s.

$$s = \sqrt{s^2}$$

Coefficient of variation, CV.

$$CV = \frac{s}{\bar{Y}}(100)$$

A population of means. The population of all possible means ($\bar{Y}$'s) of a specified sample size (r) drawn from a population of individuals.

The mean of a population of means, $\mu_{\bar{y}}$.

$$\mu_{\bar{y}} = \frac{\sum \bar{Y}_i}{M} =, \text{ where M is the number of sample means.}$$

The variance of a population of means, $\sigma_{\bar{y}}^2$.

$$\sigma_{\bar{y}}^2 = \frac{\sum (\bar{Y}_i - \mu)^2}{M}$$

The standard deviation of a population of means, or standard error, $\sigma_{\bar{y}}$.

$$\sigma_{\bar{y}} = \sqrt{\sigma_{\bar{y}}^2}$$

The relation between σ^2 *and* $\sigma_{\bar{y}}^2$.

$$\sigma_{\bar{y}}^2 = \frac{\sigma^2}{r},$$

where r is the number of variates in each sample mean (sample size).

The estimate of $\sigma_{\bar{y}}^2$ *from* n *samples,* $s_{\bar{y}}^2$.

$$s_{\bar{y}}^2 = \frac{\sum(\bar{Y}_{i.} - \bar{Y}_{..})^2}{n-1}$$

The estimate of $\sigma_{\bar{y}}^2$ *from a single sample of size* r.

$$s_{\bar{y}}^2 = \frac{s^2}{r} = \frac{\sum(Y_i - \bar{Y})^2}{r-1}\left(\frac{1}{r}\right)$$

Estimate of σ^2 *when* $s_{\bar{y}}^2$ *is known.*

$\sigma^2 \cong r s_{\bar{y}}^2$, where r is the number of variates in each sample.

t, a statistic computed from a sample that expresses the difference between the sample mean and the population mean in standard error units.

$$t = (\bar{Y} - \mu)/s_{\bar{y}}$$

Confidence limits of μ, *small sample.*

$$CL = \bar{Y} \pm t s_{\bar{y}}$$

F, the ratio between two estimates of σ^2.

$$F = \frac{s^2\text{, calculated from sample means}}{s^2\text{, calculated by pooling sample variances}}$$

3

THE ANALYSIS OF VARIANCE AND t TESTS

We now have the statistical concepts needed to understand the analysis of variance. But before discussing complicated experiments, it will be informative to see how we can use these concepts to analyze the simplest case of two treatments when each has been randomly assigned to 5 of 10 experimental units. First we explain what is done in the analysis of variance procedure, and then we show a routine procedure for carrying out the computations.

ANOVA WITH TWO SAMPLES

We will use the data of Table 3.1 to illustrate the ANOVA procedure.

To determine the variability called *experimental error*, we compute the variance of each sample (s_1^2 and s_2^2), assume they both estimate a common

TABLE 3.1.
Yields (100 lb/acre) of wheat varieties 1 and 2 from plots to which the varieties were randomly assigned

Varieties	Replications					$Y_{i.}$	$\bar{Y}_{i.}$	
1	19	14	15	17	20	85	17	$\bar{Y}_{1.}$
2	23	19	19	21	18	100	20	$\bar{Y}_{2.}$
						185	18.5 =	$\bar{Y}_{..}$

variance (σ^2), and then estimate this common variance by pooling the sample variances.

$$s_1^2 = \frac{\Sigma\left(Y_{1j}-\bar{Y}_{1.}\right)^2}{r-1}$$

$$= \frac{(19-17)^2 + \ldots + (20-17)^2}{5-1}$$

$$= \frac{26}{4} = 6.5$$

$$s_2^2 = \frac{\Sigma\left(Y_{2j}-\bar{Y}_{2.}\right)^2}{r-1}$$

$$= \frac{(23-20)^2 + \ldots + (18-20)^2}{5-1}$$

$$= \frac{16}{4} = 4.0$$

Pooling s_1^2 and s_2^2 gives an estimate of σ^2 based on variability *within* the samples, which we will designate as s_w^2:

$$s_w^2 = \frac{s_1^2 + s_2^2}{2} = \frac{6.5+4.0}{2} = 5.25$$

Assuming the null hypothesis that these two samples are random samples drawn from the same population and that, therefore, $\bar{Y}_{1.}$ and $\bar{Y}_{2.}$ both estimate the same population mean (μ), we estimate the variance of means ($\sigma_{\bar{y}}^2$) from the means of samples 1 and 2.

$$s_{\bar{y}}^2 = \frac{\Sigma\left(\bar{Y}_{i.}-\bar{Y}_{..}\right)^2}{n-1} = \frac{(17-18.5)^2+(20-18.5)^2}{2-1} = \frac{(-1.5)^2+(1.5)^2}{1} = 4.5$$

We again estimate σ^2 using the relationship $s_{\bar{y}}^2 = s^2/r$ and solving for s^2. Remember, r is the number of variates on which each sample mean is based. We will designate this estimate of σ^2 as s_b^2.

$$s_b^2 = rs_{\bar{y}}^2 = 5(4.5) = 22.5$$

We now have two estimates of σ^2: s_w^2 based on the variability *within* each sample and s_b^2 based on the variability *between* the samples. Assuming the null hypothesis to be true, we would expect s_w^2 and s_b^2 to be nearly alike since they both estimate the same variance (σ^2). We can determine the probability of obtaining divergent estimates of σ^2 by calculating an F ratio and referring to a table of F values. For this F ratio we always put the variance estimated from the sample (treatment) means (s_b^2) as numerator and the variance estimated from the individual variates as denominator. Thus, $F = s_b^2/s_w^2$.

If the two treatments (samples) come from populations having different means, s_b^2 will contain a component reflecting this difference and will be larger than s_w^2. For our experiment, $F = 22.5/5.25 = 4.29$.

The numerator, s_b^2, is based on 1 degree of freedom, since there are two sample means. The denominator, s_w^2, is based on pooling the degrees of freedom *within* each sample. Each sample has 5 variates and therefore 4 df so the degrees of freedom for s_w^2 are $4+4=8$.

From an F table (Table A.3), we look up the F values we would expect with a specified probability if the null hypothesis is true and our sample means differ only by chance. For degrees of freedom 1 (numerator) and 8 (denominator) we would expect an F value of 4.29 or larger with a probability of about 7%. To put it another way, if the true mean difference is zero ($\mu_1 - \mu_2 = \mu_{\bar{d}} = 0$), the chance of obtaining an estimate of $\mu_{\bar{d}} = 3$ cwt per acre is about 7%. Usually, we are not willing to gamble that this event (which has a 7% probability of occurrence) did not occur; therefore it would be unwise to reject the null hypothesis and conclude that the mean of variety 1 is really different from the mean of variety 2. On the other hand, a mean variety difference of 3 cwt per acre, if real, represents a considerable economic gain. Therefore, we might decide to evaluate the two varieties in additional experiments.

A Cookbook Procedure

The following is a stepwise procedure for completing the ANOVA for the data of Table 3.1 using a desk or pocket calculator.

1. *Outline the ANOVA table* (*Table* 3.2) *by listing the sources of variation and degrees of freedom.* There are 10 experimental units in the experiment and, therefore, $10-1$ or 9 df in total. These total degrees of

TABLE 3.2.
ANOVA for the data of Table 3.1

Source of Variation	Degrees of Freedom, df	Sum of Squares, SS	Mean Square, MS	Observed F	Required F	
					10%	5%
Total	9	64.5				
Varieties	1	22.5	22.5	4.29	3.46	5.32
Error	8	42.0	5.25			

freedom are then partitioned according to the experimental design. There are two treatments; therefore, $2-1=1$ df. Degrees of freedom for error can always be obtained by subtraction, $9-1=8$, but also, in this case, by pooling degrees of freedom within each sample. There are 5 variates in each sample, and therefore $5-1=4$ df; $4+4=8$ df for error.

2. *Compute the sum of squares for varieties (SSV) and the mean square for varieties (MSV).*

$$SSV = \frac{\sum Y_{i.}^2}{r} - \frac{Y_{..}^2}{nr}$$

$$= \frac{85^2 + 100^2}{5} - \frac{185^2}{2(5)}$$

$$= 3445.0 - 3422.5 = 22.5$$

$$MSV = \frac{SSV}{(df)V} = \frac{22.5}{1} = 22.5$$

Note that we use totals, not means, in computing SSV. With a large computer it is easy to use means in computing sums of squares, but with desk calculators it is much easier and more accurate to use totals, since you avoid taking differences and the rounding off of decimals in computing means. The following bit of algebra illustrates why totals can be used in place of means to calculate sums of squares.

Based on the hypothesis that our two varieties are not different and both are samples from the same population, we learned that a second estimate of σ^2 is obtained by $s_b^2 = rs_{\bar{y}}^2$, where $s_{\bar{y}}^2$ is the variance of variety means and r is the number of replications in each variety mean. The mean square for treatments of Table 3.2 (varieties in this case) is s_b^2, that is $MSV = rs_{\bar{y}}^2$. Note that $s_{\bar{y}}^2 = \Sigma(\bar{Y}_{i.} - \bar{Y}_{..})^2/(n-1)$ and thus

$$MSV = r\left[\frac{\sum(\bar{Y}_{i.} - \bar{Y}_{..})^2}{n-1}\right]$$

Since $MSV = SSV/(n-1)$, $SSV = r[\Sigma(\bar{Y}_{i.} - \bar{Y}_{..})^2]$. In Chapter II we saw that $\Sigma(Y_{i.} - \bar{Y}_{..})^2 = \Sigma Y_{i.}^2 - (\Sigma Y_{i.})^2/n$, so we can now write

$$SSV = r\left[\sum \bar{Y}_{i.}^2 - \frac{(\sum \bar{Y}_{i.})^2}{n}\right]$$

Now we replace means with totals, noting that $\bar{Y}_{i.} = Y_{i.}/r$ and that $\Sigma\bar{Y}_{i.} = Y_{..}/r$ and thus

$$SSV = r\left[\frac{\sum Y_{i.}^2}{r^2} - \frac{Y_{..}^2}{r^2}\left(\frac{1}{n}\right)\right]$$

Carrying out the indicated multiplication, r[ ], gives $SSV = (\Sigma Y_{i.}^2/r) - (Y_{..}^2/rn)$, which is the formula previously given. This formula involves some basic rules you should learn in order to compute sums of square from totals.

(a) The first term, $\Sigma Y_{i.}^2/r$, tells you to sum the squares of the totals (variety totals in this case) and divide by r, *the number of variates making up each total in the numerator.* Students most often err in deciding on the divisor and divide by the number of totals being squared rather than by the number of variates in each total.

(b) The second term, $Y_{..}^2/nr = (\Sigma Y_{ij})^2/nr$, is known as the *correction term* or correction factor. *It is the square of the sum of all the variates in the totals of the first term divided by the number of variates in the sum* ($Y_{..}$) *being squared.*

(c) *If all treatments do not have the same number of replications, each total must be squared and divided by the number of variates it contains before summing.* Thus

$$SST = \sum \frac{Y_{i.}^2}{r_i} - \frac{Y_{..}^2}{\sum r_i} = \left(\frac{Y_{1.}^2}{r_1} + \ldots + \frac{Y_{n.}^2}{r_n}\right) - \left(\frac{Y_{..}^2}{r_1 + \ldots + r_n}\right)$$

For example, if $Y_{25} = 18$ of Table 3.1 is missing, the total for variety 2 is $100 - 18 = 82$ and $Y_{..}$ is $85 + 82 = 167$. Then

$$SSV = \left(\frac{85^2}{5} + \frac{82^2}{4}\right) - \frac{167^2}{5+4} = 3126 - 3099 = 27.$$

Now we continue with step 3 of the cookbook procedure.

3. *Compute the total sum of squares* (SS). This step is done just before computing the sum of squares for error. With SS in the calculator, the error sum of squares is then obtained by subtraction.

$$SS = \sum Y_{ij}^2 - \frac{Y_{..}^2}{nr}$$

$$= (19^2 + 14^2 + \ldots + 18^2) - \frac{185^2}{2(5)}$$

$$= 3487.0 - 3422.5 = 64.5$$

4. *Compute the sum of squares and mean square for error* (SSE and MSE).

$$\text{SSE} = \text{SS-SST} = 64.5 - 22.5 = 42.0$$

$$\text{MSE} = \frac{\text{SSE}}{\text{dfE}} = \frac{42.0}{8} = 5.25$$

5. *Calculate F ratio for varieties.*

$$F = \frac{\text{MST}}{\text{MSE}} = \frac{22.5}{5.25} = 4.29$$

6. *In Table A.3, look up the required F values for the levels of significance you wish to compare.* Degrees of freedom pertaining to the numerator of the F ratios are read across the top of Table A.3 and degrees of freedom for the denominator are read down the left side.

THE STANDARD DEVIATION KEY. Desk and pocket calculators that are preprogrammed to compute a standard deviation or variance simplify computations in the ANOVA and eliminate the use of a correction term when treatment replications are equal. First, be sure the calculator computes s or s^2 by dividing the sum of squares by one less than the number of totals or means you enter, that is $s = \sqrt{(Y_i - \bar{Y})^2/(r-1)}$. Use the following set of variates to check: 19, 14, 15, 17, 20. Enter each in turn with the appropriate key—often marked ΣX. After the last variate (20) is entered, depress the standard deviation key, usually marked σ. If the calculator divides by $n-1$, the answer is 2.5495. If the divisor is n, the answer is 2.2803.

Using a σ key, MST and SST are computed as follows from the totals of Table 3.1:

Enter 85, enter 100.

Depress σ(ans. = 10.6066+), square σ (ans. = 112.5)

Divide by the number of variates in each total you entered, that is 5, Answer = 22.5 = MST

Multiply MST by the degrees of freedom for treatment, that is 1. Answer = 22.5 = SST.

The total SS is calculated by entering each variate (19, 14, ..., 21, 18), depressing σ (ans. = 2.677+), squaring σ (ans. = 7.166+), and multiplying by degrees of freedom for total (9), answer = 64.5.

With a little practice you can learn these simple rules and easily and quickly do an ANOVA on a desk or pocket calculator. Remember, *enter treatment or other totals* (or individual variates for calculating the total sum of squares), *depress the standard deviation key, square σ to obtain σ^2, divide by the number of variates in each total you entered* (divide by 1 when calculating the total sum of squares)—*the answer is the mean square* for the source of variation you are computing.

A POPULATION OF MEAN DIFFERENCES

In addition to an F test we can also use a t test to evaluate the odds that two means are significantly different. First we need to see how a population of mean differences is generated from a population of normally distributed variates; in particular, we need to know how parameters of this new population are related to parameters of the parent populations and to the populations of means also generated in obtaining the population of mean differences.

If from two normally distributed populations, $X_1, X_2, \ldots, X_N$ and $Y_1, Y_2, \ldots, Y_N$, we draw all possible samples of a given size and calculate their means, we will have two additional populations, $\bar{X}_1, \bar{X}_2, \ldots, \bar{X}_M$ and $\bar{Y}_1, \bar{Y}_2, \ldots, \bar{Y}_M$. Now if we take all possible pairs of means and subtract, thus, $\bar{X}_1 - \bar{Y}_1, \bar{X}_1 - \bar{Y}_2, \ldots, \bar{X}_1 - \bar{Y}_M, \bar{X}_2 - \bar{Y}_1, \ldots, \bar{X}_2 - \bar{Y}_M, \ldots, \bar{X}_M - \bar{Y}_M$, we will have a fifth population, that of *mean differences* (see Fig 3.1). The number of mean differences (Q) of this population will be much larger than the populations of $\bar{X}_i$ and $\bar{Y}_i$. If the number of means in these two populations both equal M, then $Q = M^2$. The following relationships among the means and standard deviations of these populations can be proven mathematically but will merely be stated here. The mean of the mean differences equals the difference between the means of the sample means from populations X and Y, and this difference also equals the difference between the mean of population X and population Y:

$$\mu_{\bar{d}} = \mu_{\bar{x}} - \mu_{\bar{y}} = \mu_x - \mu_y. \text{ If the } \mu_x = \mu_y, \text{ then } \mu_{\bar{d}} = 0.$$

The variance of the population of mean differences is

$$\sigma_{\bar{d}}^2 = \frac{\sum (\bar{d}_i - \mu_{\bar{d}})^2}{Q}$$

and is equal to the sum of the variances of the respective means. Thus, $\sigma_{\bar{d}}^2 = \sigma_{\bar{x}}^2 + \sigma_{\bar{y}}^2$. From two samples, $\sigma_{\bar{d}}^2$ is estimated by $s_{\bar{d}}^2$ from the variances of sample means: $s_{\bar{d}}^2 = s_{\bar{x}}^2 + s_{\bar{y}}^2$. Since $s_{\bar{x}}^2 = s_x^2/r_x$ and $s_{\bar{y}}^2 = s_y^2/r_y$, $s_{\bar{d}}^2 = (s_x^2/r_x) + (s_y^2/r_y)$.

The square root of the variance of mean differences is often called the *standard error of a difference*. Often in statistical analyses, one variance is estimated from another.

Important relationships among variances that you will use frequently are

$$s_{\bar{y}}^2 = \frac{s^2}{r} \qquad s_{\bar{d}}^2 = s_{\bar{x}}^2 + s_{\bar{y}}^2 \qquad s_{\bar{d}}^2 = \frac{s_x^2}{r_x} + \frac{s_y^2}{r_y}$$

and when $r_x = r_y = r$ and $s_x^2 = s_y^2 = s^2$, then

$$s_{\bar{d}}^2 = \frac{2s^2}{r}$$

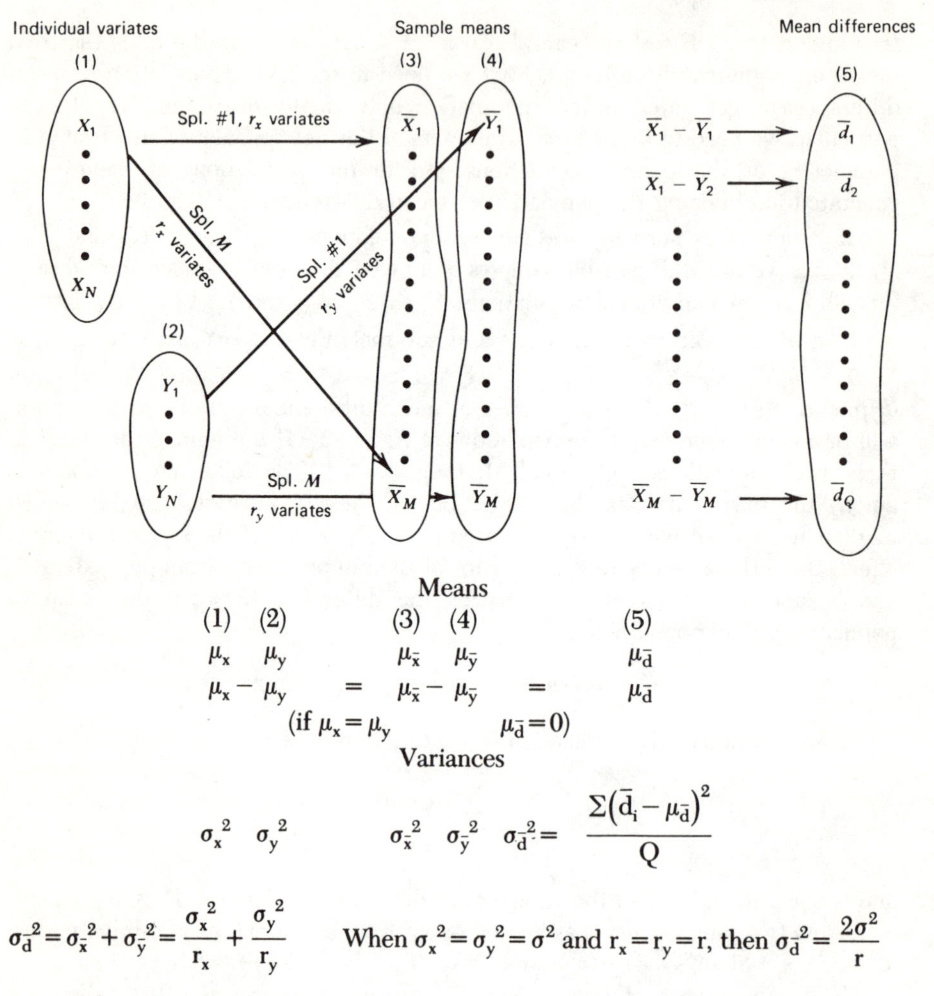

Figure 3.1 The generation of populations of means and mean differences from two populations of individual variates and relationships among parameters (see text).

t Tests for Significance

The formula for t as applied to a population of *mean differences* is $t = (\bar{d} - \mu_{\bar{d}})/s_{\bar{d}}$. For the experiment of Table 3.1 we want to know the probability that samples 1 and 2 could have come from populations having identical means ($\mu_1 = \mu_2$). This is analogous to the discussion above where we referred to populations X and Y, only

now we are calling them Y_1 and Y_2. The mean difference of our sample means is $\bar{d} = 17 - 20 = 3(10^2)$ lb/acre.
The standard error of the difference is

$$s_{\bar{d}} = \sqrt{s_{\bar{y}_1}^2 + s_{\bar{y}_2}^2}, = \sqrt{\frac{s_1^2}{r_1} + \frac{s_2^2}{r_2}} = \sqrt{\frac{6.5}{5} + \frac{4}{5}} = \sqrt{\frac{10.5}{5}} = \sqrt{2.10} = 1.449$$

Assuming the null hypothesis that $\mu_1 = \mu_2, (\mu_{\bar{d}} = 0)$, t is calculated as

$$t = \frac{\bar{d} - \mu_{\bar{d}}}{s_{\bar{d}}} = \frac{3 - 0}{1.449} = 2.07$$

From Table A.2 we can find the lowest value of t that has a 5% chance of occurring. If we assume that $\sigma_1^2 = \sigma_2^2$ we look up t based on the pooled degrees of freedom within the samples, in this case $4 + 4 = 8$. The expected t value for the 5% level of probability is 2.306, and thus our treatment difference is again judged not significant. Note that $t^2 = F$, that is, $2.07^2 = 4.285$. Allowing for rounding errors this equals our previously calculated F of 4.29.

A point to be emphasized is that the analysis of variance procedure and the calculation of an F value leads to the same conclusions as the t test. Researchers often express the idea that there is something unique and more powerful about the t test compared to the F test of the analysis of variance. The tests are equivalent, while the analysis of variance procedure is usually easier to carry out.

One additional point should be made with regard to the use of a t test: a t test is appropriate when $\sigma_1 \neq \sigma_2$. In this case the F test of the analysis of variance is not valid. When $\sigma_1 \neq \sigma_2$ and $r_1 = r_2 = r$, the t value required for significance is for $r - 1$ degrees of freedom. In our example $r = 5$ and the required t value at the 5% level would be the tabular value for 4 df, or 2.776. When $r_1 \neq r_2$, the required t value must be calculated as it is somewhere between the tabular t for $r_1 - 1$ and $r_2 - 1$ df. When $\sigma_1 \neq \sigma_2$ and $r_1 \neq r_2$ the required t is approximated by

$$t = \frac{t_1 s_{\bar{y}_1}^2 + t_2 s_{\bar{y}_2}^2}{s_{\bar{y}_1}^2 + s_{\bar{y}_2}^2}$$

where t_1 and t_2 are tabular t values for $r_1 - 1$ and $r_2 - 1$ df, respectively.

CONFIDENCE LIMITS FOR A MEAN DIFFERENCE. For our example, we have an estimate of the population mean difference, namely 3(100) lb per acre, and might wish to calculate a confidence interval within which the true population mean difference will fall unless the samples we have drawn are very unusual. With a confidence of 95%, we can say that $\mu_{\bar{d}}$ lies within $\bar{d} \pm t_{.05} s_{\bar{d}}$, where $t_{.05}$ is a tabular value from Table A.2 for the degrees of freedom for error (Table 3.2). The 95%

confidence limits are therefore

$$CL_{95}=3\pm2.306(1.449)=3\pm3.34=-0.34 \text{ to } 6.34(10^2) \text{ lb/acre}$$

Note that this confidence interval includes zero, which is another way of showing that the means of varieties 1 and 2 are not significantly different.

LEAST SIGNIFICANT DIFFERENCE. Least significant difference (LSD) is discussed at greater length in Chapter 6, "Mean Separation," but it is mentioned here, since it is a form of the the t test we have been considering. The formula for calculating the LSD between two means is: $LSD=t\ s_{\bar{d}}$, which is the second term of the CL equation above. For experiments involving two treatments only, there is no need to calculate LSD, as there is only one mean difference to consider and an F or a t test tells whether the difference is significant.

A t TEST FOR PAIRED PLOTS. If we assume that the replicates of Table 3.1 are paired, we can determine the difference between each pair and analyze the differences. Subtracting treatment 1 from treatment 2, we have the paired plot differences 4, 5, 4, 4, −2. The mean of the differences is 3, that is, $\bar{d}=3$; the variance of the differences is

$$s_d^2=\frac{(4-3)^2+(5-3)^2+\cdots+(-2-3)^2}{5-1}=\frac{32}{4}=8$$

The variance of the mean difference is estimated by $s_{\bar{d}}^2=s_d^2/r=8/5=1.6$, and the standard error of the mean difference is, $s_{\bar{d}}=1.265$. The appropriate t test for significance of the mean difference is

$$t=\frac{\bar{d}-\mu_{\bar{d}}}{s_{\bar{d}}}=\frac{3-0}{1.265}=2.37$$

After you have completed Chapter V and understand the randomized complete block design, assume that the replications of Table 3.1 are also blocks and do the ANOVA and show that MST=22.5, MSE=4, F for varieties=5.62, and that the standard error of a mean difference is

$$s_{\bar{d}}=\sqrt{\frac{2s^2}{r}}=\sqrt{\frac{2(4)}{5}}=1.265 \text{ as above}$$

Note that $t^2=F$, that is, $2.37^2=5.62$. The point is that a t test for paired plots leads to the identical statistical conclusion as the F test for the randomized complete block design with two treatments. The latter is usually easier to compute.

ROUNDING AND REPORTING NUMBERS

The terms *precision* and *accuracy* are often used synonymously, but in a statistical sense, they have different meanings. Precision refers to the magnitude of the difference between two treatments that an experiment is capable of detecting at a given level of significance, while accuracy refers to the closeness with which a particular measurement can be made. In a later chapter we will consider methods for increasing the precision of an experiment, but here we will briefly discuss accuracy in data collection and computations.

Whenever possible, original records should be collected in a manner to avoid recopying. If electronic processing equipment is to be used, the collection of data can be organized so that the original figures are used to punch data cards. This prevents errors in recopying. If figures must be transferred, they should be rechecked immediately.

At the time data are collected, they should be examined for out-of line figures, and all such entries rechecked to prevent possible errors. There is enough variation in biological data without allowing more to creep in through avoidable mistakes.

In taking weights or other measurements on experimental units it is seldom worthwhile to record figures to a number place less than one-fourth the standard deviation per unit. If s is 6.96 lb per experimental unit, $6.96/4=1.74$. As the first place is in the one's position, data can be recorded to the closest pound. If s were 2.5 lb/unit, $2.5/4=0.625$, the first place is the tenth position, and data could be recorded to the closest tenth of a pound.

The instrument used for weighing and measuring need be no more accurate than required by the precision of the experiment. For example, if a series of weighings are to be made and rounded off to the closest pound, the scale used can be in whole pound units rather than divisions of a pound.

It is not incorrect to carry more digits than the variability of the data justify, and with modern data-processing equipment this can be done easily, but in reporting final results, superfluous digits should be dropped. Apply the above rounding rule to treatment means and round them to the place indicated by taking one-fourth of the standard error of a mean. If the standard deviation per experimental unit is 6.96 lb and each treatment mean is based on five replications, $s_{\bar{y}}=6.96/\sqrt{5}=3.11$ and $3.11/4=0.68$, indicating that means should be rounded off to one decimal place.

In doing an analysis of variance, it is best to carry the full number of figures obtained from the uncorrected sum of squares; for example, if original data contain one decimal, the sum of squares will contain two decimal places. Do not round closer than this until reporting final results.

When rounding numbers the digit to be retained is rounded upward if the digit to be dropped is greater than 5 or is 5 followed by a digit greater than zero. If the amount following the 5 is zero, the digit to be rounded is rounded upward if odd or left as is if it is even. For example, rounding 21.550 to the closest tenth gives 21.6, but rounding 21.450 would give 21.4.

FACTORIAL EXPERIMENTS

In a factorial experiment the effects of two or more factors are investigated simultaneously. If the behavior of one factor is suspected of changing with changes in another factor, this behavior can be tested by a factorial set of treatments laid out in a suitable experimental design.

When two or more factors (each may be at two or more levels) are tested in all possible combinations, the resulting treatments are said to be factorial. Differential effects of one factor on another are called interactions. The discovery of interactions broadens the conclusions of an experiment. The range of validity of the experiment is increased—a desirable characteristic of a well-planned experiment. Even if interactions do not occur in factorial experiments, the results are more widely applicable because the main treatment effects have been shown to hold over a wider range of conditions.

Examples of combinations of factors in an experiment are: testing varieties at varying levels of soil fertility and evaluating the effect of a hormone on the gaining ability of male versus female lambs.

A factorial set of treatments is illustrated in Table 3.3. The nine *treatments* are all possible combinations of three dosage levels of an insecticide and three dosage levels of a fungicide used as seed treatments for lima beans.

This set of treatments makes it possible to evaluate the relative contribution of fungicide and insecticide to the emergence of lima bean seedlings. See Table 3.4 for treatment averages and Figure 3.2 for a graphic presentation of the results illustrating the meaning of interaction.

In Figure 3.2, note the decrease in emergence with increased dose of insecticide when the insecticide was used without the fungicide. That decrease did not occur when a fungicide was added to the seed treatment. The differential effect of insecticide, depending on whether or not a fungicide was used we call *interaction*. If an interaction does not occur, the factorial arrangement multiplies

TABLE 3.3.
Lima bean seed treatments. A factorial combination of three dosage levels of a fungicide with three dosage levels of an insecticide

	Insecticide dose		
Fungicide dose	I_0 (none)	I_1	I_2
F_0 (none)	F_0I_0	F_0I_1	F_0I_2
F_1	F_1I_0	F_1I_1	F_1I_2
F_2	F_2I_0	F_2I_1	F_2I_2

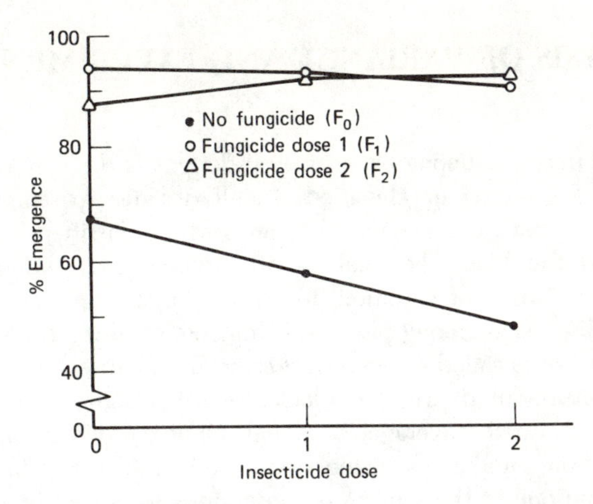

Figure 3.2. Graphic presentation of averages of treatments in Table 3.4.

the number of replications for testing overall average effects of treatment components. Note that there is no appreciable differential effect of insecticide on doses F_1 and F_2 of the fungicide. In other words, there is no interaction of I×F with respect to doses F_1 and F_2 of the fungicide. In this case, the best estimate of the effect of fungicide doses F_1 and F_2 are the averages for these doses over all levels of the insecticide. The resulting averages, $F_1=92\%$ and $F_2=91\%$ (Table 3.4) are based on 3×the number of replications of an individual treatment. No superiority of the higher dose of the fungicide is indicated.

Occasionally you may read about *factorial designs*. This terminology is not strictly correct; it is the treatment combination that is factorial—not the design.

TABLE 3.4.
The effect of levels of fungicide and insecticide seed treatment on emergence of lima bean seedlings (values given are seedlings per 100 seeds).

Fungicide (oz per 100 lb seed)	Insecticide (oz per 100 lb seed) 0 (I_0)	$\frac{1}{6}$(I_1)	$\frac{1}{3}$(I_2)	Average effect of fungicide
	Insecticide × Fungicide Means			
0 (F_0)	68	58	48	59
$1\frac{1}{3}$(F_1)	94	93	90	92
$2\frac{2}{3}$(F_2)	89	92	92	91

THE ANALYSIS OF VARIANCE AND EXPERIMENTAL DESIGN

The principal difference among experimental designs is the way in which experimental units are grouped or classified. In all designs, experimental units are classified by treatments, but in some they are further classified into blocks, rows, main plots, and the like. The analysis of variance uses the means of these groupings, called sources of variation, to estimate mean squares. A mean square estimating the dispersion among plot measurements resulting from random causes is also calculated—it is called *experimental error*. In the absence of real differences resulting from means of treatments, blocks, or other sources of variation, these mean squares will, on the average, be equal. Only rarely will one mean square deviate greatly from another by chance alone. When an F test indicates that the mean square from one of the sources of variation is significantly greater than the mean square resulting from random effects, we say that there are real differences among the means of that particular source of variation. But remember—there is always a definite chance that we will be wrong in such a conclusion. It is up to the experimenter to select the odds at which it is believed there are real effects.

It is customary to describe results that would be expected by chance 5% or less as *significant* and those expected 1% or less as *highly significant*. When an experimenter uses the phrase "the treatments are significantly different," what is really meant is that if the null hypothesis is true, the odds of obtaining such mean treatment differences are only 5%. The experimenter is gambling that there was no such chance occurrence in the experiment and that, therefore, the significant result was due to a real treatment effect.

In the following chapters the principal features of each of the experimental designs commonly used in field research are explained, an example is given for each, and the procedure to follow in analyzing data is shown. The same set of data is used for the first two designs, the completely randomized design and the randomized complete block design. It illustrates the possible advantage of one design over another, and it keeps the computations simple so that you can concentrate on what is being done and why.

SUMMARY

The ANOVA in its simplest form of two treatments randomly assigned to an equal number of experimental units involves the following procedure:

1. Calculating experimental error as the pooled variance of the two samples, for example, $MSE = (s_1^2 + s_2^2)/2$.

2. Computing a mean square for treatments (MST) based on the null hypothesis that both sample means estimate a common population mean, that is, $MST = r s_{\bar{y}}^2$, where r is the number of variates in each treatment mean.

3. Computing the F ratio: MST/MSE and comparing the calculated F value to a tabular F value to indicate the probability of obtaining the calculated F value by chance if the null hypothesis is true and both sample means represent a common mean.

The statistical significance of a *difference* between two sample means can be tested by the F ratio in an *analysis of variance* or by a *t test*. Both tests are statistically equivalent, $t^2 = F$. The analysis of variance and the F test are usually easier to compute.

The *means of differences* between all possible pairs of sample means from two populations, X and Y, is symbolized by $\mu_{\bar{d}}$ and is related to the means of parent populations of means and individual variates as follows:

$$\mu_{\bar{d}} = \mu_{\bar{x}} - \mu_{\bar{y}} = \mu_x - \mu_y$$

The *variance of mean differences,* $\sigma_{\bar{d}}^2$ is estimated from two samples by $s_{\bar{d}}^2$.

$$s_{\bar{d}}^2 = s_{\bar{x}}^2 + s_{\bar{y}}^2 = \frac{s_x^2}{r_x} + \frac{s_y^2}{r_y} \text{ and when } s_x^2 = s_y^2 = s^2 \text{ and } r_x = r_y = r,$$

$$s_{\bar{d}}^2 = \frac{2s^2}{r}.$$

Avoid superfluous digits in reporting results. Round treatment means to the number place indicated by one-fourth of the standard error of a mean.

A *factorial experiment* is one in which two or more factors, each at two or more levels, are compared in all possible combinations.

Experimental designs arise from the way in which experimental units are grouped or classified.

4

THE COMPLETELY RANDOMIZED DESIGN

This design, the simplest type possible, is set up by assigning treatments at random to a previously determined set of experimental units. The design is the most efficient in situations in which there is little variability among the units associated with position in the experimental area, age, vigor, or other identifiable sources. It is flexible with regard to the physical arrangement of the experimental units, maximizes the degrees of freedom for estimating experimental error, and minimizes the F value required for statistical significance. A disadvantage is that there are often identifiable sources of variation among the experimental units, so that other designs, when skillfully employed, usually are capable of reducing the variability we call experimental error, which makes it possible to detect smaller, significantly different treatment effects.

Any number of treatments may be tested in this design. It is desirable, but not essential, to assign the same number of experimental units to each treatment. The experiment of Table 3.1 is an example of this design with only two treatments.

RANDOMIZATION

A number can be arbitrarily assigned to each of the required number of field plots or animals to be used in the experiment. The number of experimental units will be the number of treatments × the number of replications. A table of random numbers is convenient to decide the experimental units to receive each treatment. If each treatment is to be replicated four times, the first four random numbers drawn will be assigned to treatment 1, the second four random numbers to treatment 2, and so on. For example, suppose we wish to test three different hormones, each at a single dose, to determine their effects on the weight-gaining ability of lambs. Thus, including the control, we have four treatments. Assuming the experimental unit to be a single lamb and that we will assign 4 lambs to each treatment, we will use 16 lambs. The 16 lambs selected for the trial are each given an ear tag with a number from 1 to 16. Using Table A.1, begin at a random two digit point, for example, columns 5 and 6. Proceed down this column of two digit

TABLE 4.1.
Weight gains of lambs grouped by treatment (pounds per animal per 100 days). Numbers in parentheses are ear tag numbers of the 16 lambs assigned to the trial and randomly selected to receive the indicated treatment

Treatment	Replications								Treatment Total ($Y_{i.}$)	Treatment Mean ($\bar{Y}_{i.}$)
1 (check)	47	(14)	52	(13)	62	(9)	51	(8)	212	53
2	50	(12)	54	(11)	67	(6)	57	(5)	228	57
3	57	(2)	53	(7)	69	(1)	57	(15)	236	59
4	54	(3)	65	(4)	75	(10)	59	(16)	252	63
									$928 = Y_{..}$	$58 = \bar{Y}_{..}$

numbers, up columns 6 and 7 and down 7 and 8, assigning the first four numbers (lambs) between 1 and 16 to treatment 1 (14, 13, 9, 8), the second four to treatment 2 (12, 11, 6, 5), the third four to treatment 3 (2, 7; 1, 15). The remaining four lambs (3, 4, 10, 16) are assigned to treatment 4. After a feeding period, the weight gains of the lambs are organized for analysis as in Table 4.1.

ANALYSIS OF VARIANCE

Sources of Variation and Degrees of Freedom

An analysis of variance table is started (Table 4.2), and the first two columns are completed. There are only two sources of variation in the completely randomized design, among experimental units *within* a treatment, which we call *experimental error*, and that *among* treatment means.

Degrees of freedom are one less than the number of observations for each source of variation: there are four treatments, therefore 3 df; there are four experimental units per treatment, therefore, 3 df for each treatment $\times$ 4 treatments gives 12 df for error. R/T means replications within treatments. The degrees of freedom associated with the total variability in the experiment is one less than the total number of experimental units: $16 - 1 = 15$ df. Note that the degrees of freedom associated with the sources of variation are additive. This makes it easy to determine the degrees of freedom for error by subtraction from degrees of freedom for total: $15 - 3 = 12$.

To facilitate calculation of degrees of freedom and sum of squares for *error*, we place *total* variation first in the analysis of variance table, but calculate its sum of squares after the treatment sum of squares has been determined.

Correction Term (C)

$$C=\frac{(Y_{..}^{2})}{rn}=\frac{928^2}{4(4)}=53824$$

Sums of Squares and Mean Squares

TREATMENT: SST AND MST. $SST=(\Sigma Y_{i.}^{2}/r)-C$, where $Y_{i.}$ = treatment totals and r = number of replications in each treatment.

$$SST=\frac{212^2+228^2+\ldots+252^2}{4}-C=54032-53824=208$$

SST is entered in Table 4.2. Mean square for treatment (MST) is obtained by dividing SST by df for treatment. MST = SST/df(T) = 208/3 = 69.3, which is entered in Table 4.2.

For calculators with standard deviation keys, $MST=s_T^2/r$, where s_T^2 is the variance of a series of totals (in this case 212...252) and r is the number of variates in each total. Then SST = df(MST). For this example, $s_T^2=277.33$, MST = 277.333/4 = 69.3 as before, and SST = 3(69.3) = 208 as before.

TABLE 4.2.
Analysis of variance

Source of variation	Degrees of freedom (df)	Sums of squares (ss)	Mean squares (ms)	Observed F	Required F 5%	Required F 1%
Total	15	854				
Treatments	3	208	69.3	1.29	3.49	5.95
Error (R/T)	12	646	53.8			

TOTAL: SS. We do not need a MS for total as this contains variances for all the sources of variation.

$$SS = \Sigma Y_{ij}^2 - C = 47^2 + 50^2 + \ldots + 59^2 - C = 54678 - 53824 = 854.$$

ERROR. $SSE = SS - SST = 854 - 208 = 646$. $MSE = SSE/df(E) = 646/12 = 53.8$.

With a standard deviation key, $SS = s_{ij}^2(nr-1)$ where s_{ij}^2 is the variance of all the variates in the experiment (47...59) and is 56.933; $nr - 1 = 4(4) - 1 = 15$, the degrees of freedom for "total." Thus, $SS = 56.933(15) = 854$ as before.

F Value

An F value for treatments is calculated by dividing MST by MSE: $F = MST/MSE = 69.3/53.8 = 1.29$. F values required for significance are found in Table A.3 for degrees of freedom associated with MST across the top and MSE down the left-hand side. Since the observed F value of 1.29 is considerably less than the required F for the 5% level of significance, we would be wise to accept the null hypothesis and conclude that there are no real differences among the treatments. But remember, this does not prove that there are no differences among the treatments. It may be that real treatment differences do exist but that the experiment was not sensitive enough to detect them at the desired level of probability.

THE WHAT AND WHY OF THE ANALYSIS

In testing the null hypothesis, we assume that there are no treatment effects, and therefore the treatment means only vary as would be expected of samples drawn from the same population. Thus the variance per experimental unit we call error (σ^2) can be estimated from the variability among the sample means using the relationship $s_{\bar{y}}^2 = s^2/r$ and solving for s^2, thus,

$$s^2 = rs_{\bar{y}}^2 = MST$$

The variance of means is

$$s_{\bar{y}}^2 = \frac{\Sigma(\bar{Y}_{i.} - \bar{Y}_{..})^2}{n-1} = \frac{(53-58)^2 + \ldots + (63-58)^2}{4-1} = 17.33$$

where n = number of treatment means. Then $MST = rs_{\bar{y}}^2 = 4(17.33) = 69.3$, an estimate of the variability per experimental unit (σ^2) based on variability among treatment means.

The variance within each treatment gives an independent estimate of σ^2 and a weighted average of these variances is our best estimate based on the variability within treatments. Thus,

$$s^2 = \frac{(r_1-1)s_1^2 + (r_2-1)s_2^2 + (r_3-1)s_3^2 + (r_4-1)s_4^2}{(r_1-1)+(r_2-1)+(r_3-1)+(r_4-1)}$$

Note that each estimate of $\sigma^2(s_1^2,\ s_2^2$ etc.) is weighted by its degrees of freedom. When all treatments have the same number of replications, that is, $r_1 = r_2 = \ldots = r_4 = r$, then

$$s^2 = \frac{s_1^2 + s_2^2 + s_3^2 + s_4^2}{n}$$

where n is the number of treatments. In our experiment the variances within treatments are:

$$s_1^2 = \frac{\Sigma\left(Y_{1j} - \bar{Y}_{1.}\right)^2}{r_1 - 1} = \frac{(47-53)^2 + \ldots + (51-53)^2}{4-1} = 40.67$$

$$s_2^2 = \frac{(50-57)^2 + \ldots + (57-57)^2}{4-1} = 52.67$$

$$s_3^2 = \frac{(57-59)^2 + \ldots + (57-59)^2}{4-1} = 48.0$$

$$s_4^2 = \frac{(54-63)^2 + \ldots + (59-63)^2}{4-1} = 74.0$$

The average of these variances gives the estimate of σ^2 we call *experimental error*,

$$s^2 = \frac{\left(s_1^2 + s_2^2 + s_3^2 + s_4^2\right)}{n} = \frac{40.67 + 52.67 + 48.0 + 74.0}{4} = \frac{215.34}{4} = 53.8.$$

Now we have estimated the variance per experimental unit (σ^2) in two ways: by pooling variances within treatments (MSE) and by the variability among treatment means, $s_{\bar{y}}^2$, to obtain $\text{MST} = rs_{\bar{y}}^2$. If the null hypothesis is true—that is, if all four samples are random samples from the same population—we would expect MST to be close to MSE and the ratio MST/MSE (the F value) to be close to 1 unless we have drawn a very unusual set of samples. In this case, the ratio is 1.29, a value that has a greater than 25% chance of occurring if there are no real treatment differences. Thus, we choose not to reject the null hypothesis and conclude that there are no *significant* differences. When we do find a significant difference among treatment means, the next step is to decide which means are different. This is called mean separation. A discussion of this problem is given in Chapter 6.

SUMMARY

The completely randomized design is most useful where there are no identifiable sources of variation among the experimental units other than treatment effects. It is the most flexible with regard to the physical arrangement of experimental units. It maximizes the degrees of freedom available for estimating the variance per experimental unit (experimental error); and minimizes the F value required for statistical significance.

5

THE RANDOMIZED COMPLETE BLOCK DESIGN

In this design the treatments are assigned at random to a group of experimental units called the block or replication. Block is the preferable term, as it avoids confusion with replications of the completely randomized design. The object is to keep the variability among experimental units within a block as small as possible and to maximize differences among blocks. If there are no block differences, this design will not contribute to precision in detecting treatment differences.

A block should consist of experimental units that are as uniform as possible. To achieve uniformity, experimental units may be classified on the basis of age, weight, general vigor, prior knowledge of gaining or yielding ability, or some other characteristic that will provide uniformity within the classification. With crops, adjacent field plots usually yield more alike than those separated by some distance. Blocks can be kept compact by placing the plots, usually long and narrow in shape, close together. The number of treatments should be as few as possible and still meet the objectives of the experiment. As the block size increases, so does the within-block variability. It is not necessary that each block be the same shape, but in field experiments with crops, this is usually desirable, as differences in block shapes usually increase within-block variability.

I	II	III	IV
D	A	C	C
A	D	D	B
B	C	B	D
C	B	A	A

Low fertility ⟶ High fertility

Figure 5.1. Four treatments replicated four times in a randomized complete block design.

When a productivity gradient is expected within the experimental area, blocks should be laid across the gradient and plots within a block laid parallel to the gradient as in Figure 5.1. Each treatment is assigned the same number of times, usually once, to experimental units within a block, but all or certain treatments can be replicated two or more times within a block. It is usually most efficient to have a single replicate of each treatment per block. To minimize experimental error, all precautions should be taken to treat the experimental units within a block as uniformly as possible.

RANDOMIZATION

After experimental units have been grouped into the desired blocks, the treatments are assigned at random to the units within each block, with a separate randomization being made for each block. For example, the four treatments of Figure 5.1 could be randomized in the following manner. Arbitrarily starting with row 15 of Table A.1, we proceed across this row until we have selected the digits 1 through 4, representing treatments A through D: 4, 1, 2, 3... is the order we will assign the treatments in block I. Then continuing across row 15 and back (from right to left) on row 16 we find 1, 4, 3, 2 and assign the treatments in that order in block II. Similarly, the randomization is completed for blocks III and IV.

ANALYSIS OF VARIANCE

The data we will analyze are the same we used in Chapter 4. The experiment was to determine the effect of implanting a hormone, stilbestrol, on the weight-gaining ability of male and female lambs. Thus the treatments were the factorial set of Table 5.1, the two factors being sex and stilbestrol, each factor having two levels. In this case, *blocks* were four different ranches. Thus the *replications* of Table 4.1 become *blocks* and the *treatments* become the factorial set of Table 5.1. The data are reorganized in Table 5.2. The analysis of variance is given in Table 5.3.

TABLE 5.1.
Treatments to determine the effect of stilbestrol ear implants on the gaining ability of wether and ewe lambs

	Stilbestrol	
Sex	0	3 mg/Animal
Female	FS_0	FS_3
Male	MS_0	MS_3

TABLE 5.2.
Weight gains of lambs grouped by treatment and block (pounds per lamb per 100 days)

Treatment		Block I	II	III	IV	Treatment total ($Y_{i.}$)	mean ($\bar{Y}_{i.}$)
A	FS_0	47	52	62	51	212	53
B	MS_0	50	54	67	57	228	57
C	FS_3	57	53	69	57	236	59
D	MS_3	54	65	74	59	252	63
Block total ($Y_{.j}$)		208	224	272	224	$928 = Y_{..}$	
Block mean ($\bar{Y}_{.j}$)		52	56	68	56		$58 = (\bar{Y}_{..})$

TABLE 5.3.
Analysis of variance

Source of variation	df	SS	MS	Observed F	Required F 5%	Required F 1%
Total	15	854				
Blocks	3	576	192.0	24.69	3.86	6.99
Treatments	3	208	69.3	8.91		
Error (BT)[a]	9	70	7.78			

[a]BT means the block by treatment interaction. It is the random failure of treatments to show the same effect in all blocks and not a true interaction that would imply that treatments respond differently in different blocks.

Sources of Variation and Degrees of Freedom

We now have an additional source of variation—that resulting from blocks. Since each treatment occurs the same number of times in each block, differences among blocks do not result from treatments but from other differences associated with the blocks. This component of the total sum of squares can be removed and the unaccounted error (experimental error) reduced accordingly.

Degrees of freedom are one less than the number observations associated with each source of variation. There are 16 experimental units (groups of lambs), therefore 15 df. There are four blocks and four treatments and therefore 3 df for each of these sources of variation. Error degrees of freedom can be found by subtraction, $15-3-3=9$ or by multiplying degrees of freedom for blocks by degrees of freedom for treatments, $3\times3=9$. In this design, when each treatment is replicated once in each block, degrees of freedom for error are always df blocks × df treatments.

Correction Term

$$C=\frac{Y_{..}^{2}}{rn}$$

where r is the number of replications and n is the number of treatments

$$C=\frac{928^2}{4(4)}=53824$$

Sums of Squares and Mean Squares

$$\text{BLOCKS. } SSB=\frac{\Sigma Y_{.j}^{2}}{n}-C$$

$$SSB=\frac{208^2+\ldots+224^2}{4}-53824=54400-53824=576$$

Note that the divisor n, in the term $\Sigma Y_{.j}^{2}/n$ is the number of variates making up each total in the numerator; in this case *the number of treatments.*

$$MSB=\frac{SSB}{df(B)}=\frac{576}{3}=192.0$$

Also, with a calculator programmed to compute a standard deviation: $MSB=s_B^{2}/n$, where s_B^{2} is the variance of the block totals, 208...224, and n is the number of variates in each block total.

$$MSB=\frac{678}{4}=192 \qquad \text{and} \qquad SSB=3(192)=576$$

$$\text{TREATMENTS. } SST=\frac{\Sigma Y_{i.}^{2}}{r}-C$$

$$SST=\frac{212^2+\ldots+252^2}{4}-53824=54032-53824=208$$

$$MST=\frac{SST}{df(T)}=\frac{208}{3}=69.3$$

Using a standard deviation key after entering treatment totals 212...252 gives 16.653 and $MST = 16.653^2/4 = 69.3$.

TOTAL. $SS = \Sigma Y_{ij}^2 - C$

$$SS = 47^2 + 52^2 + \ldots + 59^2 - C = 54678 - 53824 = 854$$

With a standard deviation key, enter 47...59 to get $s = 7.545$. $SS = 7.545^2(15) = 854$.

ERROR. $SSE = SS - SST - SSB$

$$SSE = 854 - 208 - 576 = 70$$

If the various sums of squares are calculated in the above order, SSE is readily obtained by subtraction, as soon as the total sum of squares is calculated.

$$MSE = \frac{SSE}{df(E)} = \frac{70}{9} = 7.78$$

THE WHAT AND WHY OF THE ANALYSIS

Before continuing with other aspects of the analysis of variance it will be helpful to look at what was done and why in calculating each mean square.

Mean Square for Blocks

Assuming a lack of real differences among the block means (the null hypothesis again), an estimate of the variability per experimental unit is calculated from the variance of *block* means. Thus $s^2 = MSB = ns_{\bar{y}_b}^2$, where n = number of treatments and $s_{\bar{y}_b}^2$ is the variance of block means. Note that this uses the relationship of a variance of means to the variance per experimental unit, $s^2 = ns_{\bar{y}}^2$. Since $s_{\bar{y}_b}^2 = \Sigma(\bar{Y}_{.j} - \bar{Y}_{..})^2/(r-1)$, the formula for MSB becomes:

$$MSB = n\left[\frac{\Sigma\left(\bar{Y}_{.j} - \bar{Y}_{..}\right)^2}{r-1}\right]$$

where $\bar{Y}_{.j}$ represents each block mean, $\bar{Y}_{..}$ is the general mean, and r is the number of block means. Calculating MSB gives:

$$\mathrm{MSB} = \frac{4\left[(52-58)^2 + (56-58)^2 + \ldots + (56-58)^2\right]}{4-1} = \frac{4(144)}{3} = 192.0$$

Mean Square for Treatments

Using the null hypothesis again and assuming no real differences among the treatment means, $s^2 = \mathrm{MST} = r s_{\bar{y}_t}^{\;2}$, where r = number of replications and $s_{\bar{y}_t}^{\;2}$ is the variance of treatment means. This is another estimate of the variance per experimental unit based on the variability among *treatment* means.

Again, the relationship between an estimated variance of means ($s_{\bar{y}}^2$) and the estimated variance of the individual variates of the parent population (s^2) is used. Expanding the formula gives

$$\mathrm{MST} = \frac{r\Sigma\left(\bar{Y}_{i.} - \bar{Y}_{..}\right)^2}{n-1}$$

where $\bar{Y}_{i.}$ = each of the treatment means, and n is the number of treatments. SST is the numerator, the denominator is degrees of freedom for treatment. The calculation gives

$$\mathrm{MST} = \frac{4\left[(53-58)^2 + (57-58)^2 + \ldots + (63-58)^2\right]}{4-1} = \frac{4(52)}{3} = 69.3$$

Mean Square for Error

MSE represents the variability among the experimental units that remain after the other sources of variation have been removed. It is informative to see what is involved in *removing* block and treatment effects.

The model for the randomized complete block design is $Y_{ij} = \bar{Y}_{..} + T_i + B_j + e_{ij}$. This says that any cell of a two-way table like Table 5.2 is made up of the mean of all the variates, $\bar{Y}_{..}$, a treatment effect, T_i, a block effect, B_j, and a residual component, e_{ij}, which is the unaccounted variability we call experimental error.

Each treatment and each block has its own effect defined as the difference between the treatment or block mean and the general mean. For example, the effect of treatment FS_0 is the same for all replications of this treatment and is $53 - 58 = -5$. Symbolically the T_i for $FS_0 = \bar{Y}_{1.} - \bar{Y}_{..}$, and all the treatment effects are collectively symbolized as $\bar{Y}_{i.} - \bar{Y}_{..}$, where $\bar{Y}_{i.}$ is any one of several treatment means. Similarly the block effects are defined as $B_j = \bar{Y}_{.j} - \bar{Y}_{..}$.

Replacing the T and B with their defined effects we have $Y_{ij} = \bar{Y}_{..} + (\bar{Y}_{i.} - \bar{Y}_{..}) + (\bar{Y}_{.j} - \bar{Y}_{..}) + e_{ij}$ Now we can rewrite the model to specify the error term for any cell of the two-way table as $e_{ij} = Y_{ij} - \bar{Y}_{..} - (\bar{Y}_{i.} - \bar{Y}_{..}) - (\bar{Y}_{.j} - \bar{Y}_{..})$. To determine e_{13} for example,

$$e_{13} = 62 - 58 - (53 - 58) - (68 - 58)$$

$$= 62 - 58 + 5 - 10$$

$$= -1$$

The definition of e_{ij} can be simplified further for this model by removing parentheses and canceling a $+\bar{Y}_{..}$ with the $-\bar{Y}_{..}$ to give

$$e_{ij} = Y_{ij} - \bar{Y}_{i.} - \bar{Y}_{.j} + \bar{Y}_{..}$$

and for $e_{13} = 62 - 68 - 53 + 58 = -1$ as before. When this is done for all cells of Table 5.2, we produce a table of error terms, Table 5.4.

The sum of squares of these error terms divided by the total degrees of freedom minus the degrees of freedom for the other identifiable sources of variation, blocks, and treatments is the MSE, which is s^2, the unaccounted variability per experimental unit; thus

$$s = \text{MSE} = \frac{0^2 + 1^2 + \ldots + -2^2}{15 - 3 - 3} = \frac{70}{9} = 7.78$$

TABLE 5.4.
A table of error terms. The variates of Table 5.2 with treatment and block effects removed

Treatment	Block I	II	III	IV
FS_0	0	1	−1	0
MS_0	−1	−1	0	2
FS_3	4	−4	0	0
MS_3	−3	4	1	−2

F VALUES

F ratios are used to evaluate the probabilities of obtaining treatment and block means that vary as much as those of our experiment if there are no real treatment or block differences. We have estimated σ^2, the population variance per experimental unit in three ways: (1) based on variation among treatment means (MST); (2) based on variation among block means (MSB); (3) based on variability among the experimental units with block and treatment effects removed (MSE). If there are no differences resulting from block and treatments, all three mean squares should be about equal.

$$F\ (\text{blocks}) = \frac{\text{MSB}}{\text{MSE}} = \frac{192.0}{7.78} = 24.69$$

$$F\ (\text{treatments}) = \frac{\text{MST}}{\text{MSE}} = \frac{69.3}{7.78} = 8.91$$

The required F values for statistical significance for degrees of freedom 3 (numerator) and 9 (denominator) are found in Table A.3 and recorded in the analysis of variance table (Table 5.3). Since our observed F value for blocks as well as for treatments exceed that required for significance at the 1% level, we can say that if the null hypotheses are true, the chances are less than 1 in 100 that our particular sample of blocks or treatments could have occurred by chance alone. We are willing to gamble that these chances did not occur, reject the null hypotheses, and conclude that there are real block and treatment differences. The next step is to determine which of the treatments are significantly different. This discussion is the subject of Chapter 6. Before leaving the randomized complete block design, we should comment on the improvement in efficiency over the completely randomized design. Because of the existence of sizable block differences and the removal of these block effects, the precision of our experiment was increased allowing us to detect treatment differences that could not be detected by the completely randomized design.

SUMMARY

In the randomized complete block design: Blocks are sets of experimental units that are arranged or selected prior to the allocation of treatments so that the existing variability is minimized within blocks and maximized between blocks. Treatments are randomly assigned the same number of times (usually once) to the experimental units within a block. An independent randomization is carried out for each block. Compared to the completely randomized design, the degrees of freedom for experimental error are reduced by the number of degrees of freedom for blocks. Block variability is removed from experimental error. Thus the greater the variability among blocks, the more efficient the design becomes in its ability to detect possible treatment differences.

6

MEAN SEPARATION

As we have seen, an experiment is conducted to answer certain questions the investigator poses in advance. These questions are important in determining the treatments to be included, the design of the experiment, and the appropriate method for comparison of treatment means.

Usually, treatments can be selected that make it possible to carry out planned F tests to answer important questions. For example, when two or more factors are to be studied, a factorial set of treatments makes it possible to answer questions as to how the factors may interact. And even if interactions are not present, inferences concerning the average effects of the factors are more widely applicable because each factor has been examined over a range of conditions. Levels of a treatment can be planned to determine not only whether there is a response to the treatment but also how best to characterize the response and to quantify the dose-response relationship. Treatments may be classified into groups with common characteristics and thus provide for meaningful F tests among the groups. Such planned F tests allow more precise mean separation than do multiple comparison tests. The latter should only be used where there are no logical relationships among the treatments.

A significant F value immediately raises the question: Which of the mean values are significantly different? Three widely used methods for mean separation are briefly described below.

LEAST SIGNIFICANT DIFFERENCE

This test should not be used unless the F test is significant. Strictly speaking, LSD should be used only to compare adjacent means in an array (means arranged in order of magnitude). When it is used indiscriminately to test all possible differences among several means, certain differences will be significant but not at the level of significance chosen. Instead of making comparisons at the 5% level, comparisons between means farther apart than two in an array will be made at lower levels of significance. LSD can be used for comparing adjacent means, and when it is used to make meaningful comparisons that are planned before the data are examined, it should not lead to many errors. The great advantage of LSD is that it is easy to calculate and provides a single figure for making comparisons.

As pointed out before, LSD is a form of the t test. Its formula is derived from the formula for the t test to test the statistical significance of the difference between two means: $t=(\bar{d}-\mu_{\bar{d}})/s_{\bar{d}}$. Let the difference between two means $(\bar{Y}_1-\bar{Y}_2=\bar{d})$ be the lower limit of the values we would expect 5% or more of the time by chance alone in drawing samples of mean differences from a population of mean differences where the mean is zero ($\mu_{\bar{d}}=0$). We replace $\bar{d}$ with LSD and $\mu_{\bar{d}}$ with zero, and the formula becomes $t=\text{LSD}/s_{\bar{d}}$. Solving for LSD gives $\text{LSD}=ts_{\bar{d}}$, where $s_{\bar{d}}=(s_1{}^2/r_1)+(s_2{}^2/r_2)$, $s_1{}^2$ and $s_2{}^2$ are the estimated variances of plots receiving treatments 1 and 2, respectively, and r_1 and r_2 are the number of experimental units receiving treatments 1 and 2, respectively. In an analysis of variance $s_1{}^2$ is assumed to estimate the same variance as $s_2{}^2$ and r_1 is usually equal to r_2, therefore $\text{LSD}=t\sqrt{2s^2/r}$, where s^2 is the mean square for error, r is the number of replications, and t is the tabular t value for degrees of freedom for error.

When comparing two treatments that are replicated a different number of times: $\text{LSD}=t\sqrt{(s^2/r_1)+(s^2/r_2)}$, where r_1 and r_2 are the number of replications for each treatment.

To illustrate the use of LSD, we will use it to separate the means of our lamb-stilbestrol experiment, Table 5.2 (Chapter 5). The mean effects are: $FS_0=53$; $MS_0=57$; $FS_3=59$; $MS_3=63$ lb gain per lamb per 100 days.

$$\text{LSD}_{.05}=t_{.05}\sqrt{\frac{2s^2}{r}}=2.262\sqrt{\frac{2(7.78)}{4}}$$

$$=2.262(1.972)=4.46 \text{ lb per animal per 100 days}$$

If we use LSD only to compare adjacent means, we conclude that there are no differences; but the F value tells us that there are differences. Using it to compare all means, we conclude that stilbestrol improved gaining ability in both female $(59-53=6)$ and male $(63-57=6)$ lambs. Differences in gaining ability associated with sex are not significant.

Testing differences by LSD is, in effect, making a t test for each difference and leads to the same statistical inference as F tests of the same differences. Researchers are often confused on this point and try all three tests to show a difference to be significant. Do not do it! They all give the same result. To illustrate, take the difference $FS_0-MS_0=53-57=-4$.

(1) LSD = 4.5. Therefore the difference is not significant.

(2) $t=(\bar{d}-\mu_{\bar{d}})/s_{\bar{d}}=4/1.972=2.028$. The tabular t for the 5% level and 9 df = 2.262. Therefore, again, the difference is not significant.

(3) $F=MS(FS_0-MS_0)/MSE$. Since $MS(FS_0-MS_0)$ is based on 1 df, it is also the $SS(MS_0-FS_0)$ and $SS(MS_0-FS_0)=(228^2+212^2)/4-(228+212)^2/8=24232-24200=32$. $F=32/7.78=4.11$. The tabular F for the 5% level and 1 and 9 df is

5.32. Therefore, once again, the difference is not significant. Note that $t^2 = F = 2.028^2 = 4.11$. There is always this relation between the two tests and they both lead to the same statistical conclusion.

LSD is a fixed-range test, since it provides one range for testing all differences. Other, more conservative fixed-range tests, are Tukey's and Scheffe's (see Bancroft reference at end of book).

MULTIPLE-RANGE TESTS

These tests are so named because they provide multiple ranges to make pairwise comparisons among several means. With means arrayed from the lowest to the highest, a multiple-range test gives significant ranges that become larger as the means to be compared are further apart in the array. A conservative multiple-range test that is considered to keep all mean separations at the level of significance specified is the Student-Newman-Keuls procedure (see Bancroft reference). In this book, only Duncan's multiple-range test is discussed, since it and/or the intelligent use of LSD following a significant F for treatments are adequate procedures for making logical pairwise comparisons.[1]

Duncan's Multiple-Range Test

This test is the most widely used of several multiple-range tests available. It provides protection against making mistakes inherent in the indiscriminate use of the LSD test. The test is identical to LSD for adjacent means in an array but requires progressively larger values for significance between means as they are more widely separated in the array. This test is used most appropriately when several unrelated treatments are included in an experiment, for example, for making all possible comparisons among the yielding abilities of several varieties. To illustrate the procedure, we will use the lamb implant experiment.

The test involves the calculation of shortest significant differences (D) for all possible relative positions between the treatment means when they are arrayed in order of magnitude. The D's are then used in an orderly procedure to determine statistical differences among the means. In most books the formula for D is given as $D = Qs_{\bar{y}}$, where Q is a tabularized value (Table A.7 of Steel and Torrie, 1960) depending upon the chosen level of significance, the degrees of freedom for error, and the relative separation of means in the array, and $s_{\bar{y}}$ is the standard error of a mean and is $\sqrt{MSE/r} = \sqrt{s^2/r}$. In this book $D = R(LSD)$, where R is a tabular value from Tables A.4 and A.5, chosen for the level of significance, degrees of

[1]For a discussion of various tests for random pairwise comparisons see S. G. Carmer, and M. R. Swanson, "An Evaluation of ten Pairwise Multiple Comparison Procedures by Monte Carlo Methods," *Journal of the American Statistical Association*, 68:66–74, 1973.

freedom for error, and the position of means in the array; and $LSD = t\sqrt{2s^2/r}$. In Tables A.4 and A.5, the R values are computed from Q values to facilitate the calculation of D from LSD.

Using our lamb experiment as an example, the procedure is as follows:

(1) Calculate the least significant difference.

$$LSD_{.05} = t\sqrt{\frac{2s^2}{r}} = 2.262\sqrt{\frac{2(7.78)}{4}} = 4.46$$

(2) Calculate D for relative position in the array of means. Since there are four means they can be 2, 3 or 4 apart. (Note: adjacent means are called 2 apart.)

Relative position in array (p of Table A.4)	2	3	4
Values of R, 5% level, Table A.4.	1.00	1.04	1.07
D = R(LSD)	4.5	4.6	4.8

(3) Arrange the means in order of magnitude and test for significant differences.

Treatment	FS_0	MS_0	FS_3	MS_3
Mean	53	57	59	63

Start by comparing the largest mean with the smallest, using the D for their positions relative to each other in the array (in this case $p = 4$, therefore $D = 4.8$). If the difference between these means equals or is larger than the D, the means are significantly different. ($63 - 53 = 10, D = 4.8$, therefore 63 is significantly larger than 53). Next compare the largest mean with the next smallest ($63 - 57 = 6, D = 4.6$; 63 is significantly larger than 57). Then the largest with the next smallest ($63 - 59 = 4, D = 4.5$; 63 is not significantly different from 59). *When a nonsignificant difference is found, a line can be drawn connecting these (and intervening) means.* Then repeat the process; start by comparing the second largest with the smallest, and so forth.

There is an *exception rule* used with Duncan's multiple-range test. It states that a difference between two means cannot be declared significant if the two means concerned are contained in a subset of means with a nonsignificant range. Thus, if among five means in an array, A has been found not significantly different from D, that is, $\underline{\text{A B C D}}$ E, and B is significantly different from E, it is not necessary to test B against D and C as they are in a subset with a nonsignificant range. The next step would be to test C against E, if this difference is not significant, C and E are connected, $\underline{\text{A B C D}}$ E (with a second line under C D E), and further testing is unnecessary. This procedure avoids making tests between means that are already connected by a line.

(4) Indicate statistical significance by lines or letters.

MS_3	FS_3	MS_0	FS_0
63	59	57	53

OR

MS_3	FS_3	MS_0	FS_0
63a	59ab	57bc	53c

Means connected by the same line or followed by a common letter are not significantly different at the 5% level. If letters are used, significant differences can be shown even if the means are not arrayed.

In our example, note that mean comparisons by Duncan's multiple-range test or LSD lead to the same conclusions ($MS_3 > MS_0$ and $FS_3 > FS_0$), but both tests lead us to conclude that there is no significant difference in gains between males and females ($MS_3 \not> FS_3$ and $MS_0 \not> FS_0$).

PLANNED F TESTS

In planning an experiment, we can often provide for F tests to answer pertinent questions. This involves partitioning the degrees of freedom and sum of squares for *treatments* into component comparisons. The components may be class comparisons or response trends. They can be tested by partitioning the degrees of freedom and sum of squares for treatment effects into meaningful single degrees of freedom and associated sums of squares. Skillfully selected treatments can answer as many independent questions as there are degrees of freedom. When the comparisons are independent, they are said to be orthogonal—a desirable characteristic, as the comparisons lead to clear-cut probability statements.

The power and simplicity of this method of mean separation is not appreciated among research workers as fully as it should be. The method involves the selection of *orthogonal coefficients*, and perhaps this term creates the impression that it is complicated and difficult. This is far from true. Actually, the method has three important advantages: (1) it enables one to answer specific, important questions about treatment effects; (2) the computations are simple; and (3) it provides a useful check on the treatment sum of squares.

Orthogonal Coefficients

The construction of a table of comparison coefficients is useful in checking for orthogonality and in the calculation of component sums of squares. Coefficients for trend comparisons come from tables of orthogonal polynomials such as Table A.11. Coefficients for class comparisons are constructed using the following simple rules.

1. If two groups of equal size are to be compared, simply assign coefficients of +1 to the members of one group and −1 to those of the other group. It is immaterial which group is assigned the positive coefficients.

2. In comparing groups containing different numbers of treatments, assign to the first group, coefficients equal to the number of treatments in the second group, and to the second group, coefficients of the opposite sign equal to the number of treatments in the first group. Thus, if among five treatments, the first two are to be compared to the last three, the coefficients would be +3, +3, −2, −2, −2.

3. Reduce coefficients to the smallest possible integers. For example, in comparing a group of two treatments with a group of four, by rule 2, we have coefficients +4, +4, −2, −2, −2, −2, but these can be reduced to +2, +2, −1, −1, −1, −1.

4. Interaction coefficients can always be found by multiplying the corresponding coefficients of the main effects.

Two rules are used to test independence of comparisons. Comparisons are independent and therefore orthogonal when (1) the sum of the coefficients for each comparison is zero and (2) the sum of the products of the corresponding coefficients of any two comparisons is zero.

An example of the construction of a table of orthogonal coefficients may be helpful. Suppose we are planning an experiment with a crop to test the efficiency of phosphorus fertilization by three methods: broadcast (B), shallow band placement (S), and deep band placement (D). For each one of these methods of placement, we will apply phosphorus at two rates (P_1 and P_2). A nonfertilized treatment (NT) is included also to establish a response to the phosphorus fertilizer. Across the top of the table (Table 6.1), we list the treatments. The comparison coefficients are written in as we list the comparisons we will make.

1. Is there a response to P? This can be decided by comparing NT with all the treatments receiving P. Since there are six of these, NT gets a coefficient of 6 and the others get −1, as they are being compared to a single group. Having made a comparison involving a single treatment with all the rest, we cannot use NT again if we want the comparisons to be orthogonal, and therefore NT gets a coefficient of 0 in the comparisons that follow.

2. Is the average response to P_2 greater than that to P_1? This means comparing $P_1B+P_1S+P_1D$ with $P_2B+P_2S+P_2D$. Since there are two groups, each of equal size, we assign +1 to one and −1 to the other.

3. Over both levels of P, is band placement superior to broadcast that is, $P_1S+P_1D+P_2S+P_2D$ versus P_1B+P_2B. Now we are comparing a group with four treatments with a group having two treatments and thus assign coefficients of 4 to the treatments in the group of two and -2 to the treatments in the group of four. Reducing these to the smallest possible integer gives coefficients of 2 and -1, respectively.

4. Considering band placement only, is there a difference between shallow and deep? That is, P_1S+P_2S versus P_1D+P_2D. The coefficients are 2 and -2 and reduce to 1 and -1.

5. Is the change in yield from P_1 to P_2 different for broadcast compared to band placement? This is the interaction of comparisons 2 and 3, and coefficients are found by multiplying the coefficients for these two comparisons for each treatment, that is, $0(0)=0, 1(2)=2, 1(-1)=-1, 1(-1)=-1, -1(2)=-2, -1(-1)=1, -1(-1)=1$.

6. And finally, is there a change in yield from P_1 to P_2 that is different for shallow compared to deep band placement? This is the interaction of comparisons 2 and 4, and coefficients are determined by multiplication of the coefficients for comparisons 2 and 4.

TABLE 6.1.
Coefficients for the partitioning of the sum of squares among six treatments into six independent (orthogonal) comparisons.

	Treatments						
Comparison	NT	P_1B	P_1S	P_1D	P_2B	P_2S	P_2D
1. Response to P	6	−1	−1	−1	−1	−1	−1
2. P_1 vs. P_2	0	1	1	1	−1	−1	−1
3. B vs. S+D	0	2	−1	−1	2	−1	−1
4. S vs. D	0	0	1	−1	0	1	−1
5. (P_1 vs. P_2)(B vs. S+D)	0	2	−1	−1	−2	1	1
6. (P_1 vs. P_2)(S vs. D)	0	0	1	−1	0	−1	1

NT=no treatment; P_1 and P_2=phosphorus fertilizer at rates 1 and 2, respectively; B, S, and D=broadcast, shallow band placement, and deep band placement, respectively.

In Table 6.1, note that coefficients of all rows sum to zero and that the sum of the products of the coefficients for the same treatments for any two comparisons sum to zero. For example comparisons 1 and 5: $6(0)+(-1)2+(-1)(-1)+(-1)(-1)+(-1)(-2)+(-1)1+(-1)1=0$. Thus we can be sure that the comparisons are orthogonal and that the sums of squares of the comparisons will add to the sums of squares for the six treatments.

For a simple example in the use of class comparison coefficients we will again use the lamb implant experiment.

Class Comparisons

In the selection of the treatments for this experiment, note that three specific questions were asked: (1) Considering all lambs, does implanting affect gaining ability? (2) Are there differences in gaining ability between male and female lambs? (3) Is the effect of implanting the same for both sexes? The answer to each of these questions involves a single degree of freedom. The coefficients for the three comparisons are given in Table 6.2.

In the implant comparison we are comparing lambs of both sexes implanted with stilbestrol with lambs of both sexes not implanted. This is a valid comparison, as equal groups of male and female lambs received each level of stilbestrol.

In comparing gains for each sex we are comparing the average rate of gain of all female lambs with that of all male lambs for both levels of stilbestrol. This also is a valid comparison, since equal groups of lamb of each sex were implanted.

If implanting caused a significantly greater rate of gain in one sex than in the other, we would say that there is a significant interaction between sex of lambs and implant. Coefficients for this comparison ($I \times S$) are determined by multiplying the coefficients for each treatment of the first two lines of Table 6.2.

TABLE 6.2.
Comparisons, treatments, treatment totals, and coefficients for partitioning for treatment sum of squares

Comparison	Treatments and Treatment Totals			
	FS_0 212	FS_3 236	MS_0 228	MS_3 252
Implant	+1	−1	+1	−1
Sex	+1	+1	−1	−1
$I \times S$	+1	−1	−1	+1

To compute sums of squares, mean squares, and to make F tests, we proceed as shown below and organize the results in Table 6.3.

In calculating the sums of squares for treatment components we will first use the correction term procedure and then illustrate the use of the comparison coefficients we constructed in Table 6.2. The latter procedure for calculating a sum of squares only works when the sum of squares involves a single degree of freedom.

SUM OF SQUARES FOR IMPLANT.

$$SSI = \frac{(212+228)^2+(236+252)^2}{8} - \frac{(928)^2}{16} = 144$$

In using comparison coefficients, we use the following rule to calculate a sum of squares:

$$SS = \frac{\left(\sum c_i Y_{i.}\right)^2}{r \sum c_i^2}$$

where c_i = comparison coefficients from Table 6.2, $Y_{i.}$ = treatment totals, and r = number of replicates.

$$SSI = \frac{\left[1(212)-1(236)+1(228)-1(252)\right]^2}{4\left[(1)^2+(-1)^2+(+1)^2+(-1)^2\right]} = \frac{(-48)^2}{4(4)} = 144$$

SUM OF SQUARES FOR SEX.

$$SSS = \frac{(212+236)^2+(228+252)^2}{8} - \frac{(928)^2}{16} = 64$$

TABLE 6.3.
Orthogonal partitioning of treatments of the lamb-implant experiment

Source of Variation	df	SS	MS	Observed F	Required F 5%	Required F 1%
Treatments	3	208	69.33	8.91	3.86	6.99
Implants	1	144	144	18.51	5.12	10.56
Sex	1	64	64	8.23		
I×S	1	0	0	0		
Error	9	70	7.78			

or by the coefficient method,

$$SSS = \frac{[1(212)+1(236)-1(228)-1(252)]^2}{4(4)} = \frac{(-32)^2}{16} = 64$$

SUM OF SQUARES FOR I×S.

$$SS(I \times S) = SST - SSI - SSS = 208 - 144 - 64 = 0$$

or with coefficients,

$$SS(I \times S) = \frac{[1(212)-1(236)-1(228)+1(252)]^2}{4(4)} = \frac{0^2}{16} = 0$$

Notice how much simpler the computations for the coefficient method are than for the correction term procedure. In each case, only one number needs to be squared instead of adding the squares of two large numbers, and a correction term is not needed. Note also that the sum of the three component sums of squares is exactly equal to the treatment sum of squares calculated in the usual manner, furnishing a check on the calculations.

Since each sum of squares has only a single degree of freedom, the mean square in each case is the same as the sum of squares.

F tests are made by dividing each mean square by MSE. *Note that by using these more sensitive F tests we have learned something that neither the LSD nor Duncan's multiple-range tests told us*. We now have good evidence that male lambs gain faster than female lambs.

For another example of partitioning a treatment sum of squares into subcomponents for planned F tests, see the section on Mean Separation in Chapter 7.

Trend Comparisons

It is often desirable to study a variable at several levels, for example, increments of a fertilizer, dates of harvest, or doses of pesticide or herbicide. In these cases, the experimenter is interested in the nature of the response of the experimental units to the varying levels of a treatment. The statistical analysis should be designed to evaluate the trend of the response.

Wherever possible, it is desirable to use an arithmetic series for levels of a factor. Equally spaced intervals for a treatment dose or for a time series estimate responses evenly throughout the range of the levels you choose and provide a better base for curve fitting than do series where the intervals between successive treatment levels are unequal. In addition, as you will see, there are great advantages in computing sums of squares and in fitting regression equations.

For a simple example we have chosen the lima bean seed treatment experiment illustrated in Figure 3.2. Note that the doses of insecticide are equally spaced: 0, 1/6, 1/3 oz of insecticide/100 lb seed. One objective was to determine the nature of the response to doses of insecticide with and without a fungicide treatment. A portion of the data from this experiment is given in Table 6.4, and the analysis of variance is in Table 6.5. Before proceeding with the trend comparisons, note how the factorial treatments have been partitioned in Table 6.5 into main effects and interaction. The computations for the sum of squares for treatments and the partitions follow.

$$SST = \frac{341^2 + 290^2 + \ldots + 460^2}{5} - C, \quad \text{where } C = \frac{2240^2}{6(5)}$$

$$= 176222.8 - 167253.33 = 8969.47$$

$$SSF = \frac{(314 + 290 + 244)^2 + (446 + 459 + 460)^2}{15} - C = 8003.33$$

$$SSI = \frac{(341 + 446)^2 + (290 + 459)^2 + (244 + 460)^2}{10} - C = 345.27$$

$$SS(F \times I) = SST - SSF - SSI$$

$$= 8969.47 - 8003.33 - 345.27 = 620.87$$

TABLE 6.4.
Lima bean seedlings emerged from 100 seeds planted per plot. F_0 and F_2 are 0 and 2 2/3 oz fungicide/100 lb seed respectively. I_0, I_1, and I_2 are 0, 1/6, and 1/3 oz insecticide/100 lb seed, respectively

Treatment		Blocks						
		I	II	III	IV	V	$Y_{i.}$	$\bar{Y}_{i.}$
F_0	I_0	55	69	71	78	68	341	68.2
F_0	I_1	65	47	55	64	59	290	58.0
F_0	I_2	47	37	58	48	54	244	48.8
F_2	I_0	91	76	92	92	95	446	89.2
F_2	I_1	85	93	97	88	96	459	91.8
F_2	I_2	84	94	94	96	92	460	92.0
$Y_{.j}$		427	416	467	466	464	$2240 = Y_{..}$	$74.7 = \bar{Y}_{..}$

TABLE 6.5.
Analysis of variance of lima bean seedling emergence

Source of Variation	df	SS	MS	Observed F	Required F 5%	Required F 1%
Total	29	10140.67				
Blocks	4	401.00	100.25			
Treatment	5	8969.47	1793.89			
Fungicide	1	8003.33	8003.33	207.82	4.35	8.10
Insecticide	2	345.27	172.64	4.48	3.49	5.85
F×I	2	620.87	310.44	8.05		
Error	20	770.20	38.51			

The significant interaction (F×I) indicates that the response to insecticide depends on whether or not seeds were also treated with the fungicide. The treatment means can be examined statistically for a significantly different linear trend in emergence as insecticide dose increases for the two fungicide treatments. Since there are two degrees of freedom for F×I, we can ask two independent questions. Two appropriate questions are: Is there a significant difference in *linear* response to insecticide for F_0 versus F_2, and is there a significant difference in some *nonlinear* response? To simplify the calculation of sums of squares, the first step is to set up a table of comparison coefficients (Table 6.6).

TABLE 6.6
Comparison coefficients for determining response functions of lima bean seedling emergence to dosage levels of insecticide.

Comparison	F_0I_0 341	F_0I_1 290	F_0I_2 244	F_2I_0 446	F_2I_1 459	F_2I_2 460
Fungicide	−1	−1	−1	1	1	1
Insecticide linear	1	0	−1	1	0	−1
Insecticide nonlinear	1	−2	1	1	−2	1
F×IL	−1	0	1	1	0	−1
F×INL	−1	2	−1	1	−2	1

(Column heading over treatments: Treatments and Treatment Totals)

The comparison "fungicide" is the same as in Table 6.5 and compares the mean of all plots of F_0 with the mean of all plots of F_2. It is a simple class comparison, and since both groups to be compared are of equal size, a -1 is assigned to the components of one group and a $+1$ to the components of the other. The coefficients for "insecticide linear" and "insecticide nonlinear" are taken from Table A.11 under n=3 for the three dosage levels of insecticide. We can use the coefficients of Table A.11 whenever treatment levels are equally spaced. Coefficients for F×IL and F×INL are obtained by multiplying the coefficients for fungicide with those for insecticide linear or insecticide nonlinear.

Each comparison of Table 6.6 involves a single degree of freedom, and so we can compute sums of squares from: $SS=(\Sigma c_i Y_{i.})^2/(r\Sigma c_i^2)$; thus,

$$SSF=\frac{(-341-290-244+446+459+460)^2}{5(6)}$$

$$=8003.33$$

$$SS(IL)=\frac{(341-244+446-460)^2}{5(4)}$$

$$=344.45$$

$$SS(INL)=\frac{[341-2(290)+244+446-2(459)+460]^2}{5(12)}$$

$$=0.82$$

$$SS(F\times IL)=\frac{(-341+244+446-460)^2}{5(4)}$$

$$=616.05$$

$$SS(F\times INL)=\frac{[-341+2(290)-244+446-2(459)+460]^2}{5(12)}$$

$$=4.82$$

Mean squares equal sums of squares, as each is based on a single degree of freedom, and F values are calculated by dividing each by MSE as in Table 6.7.

Note that the sums of squares for the five treatment components of Table 6.7 add to the sum of squares for treatments of Table 6.5. This is a check on our arithmetic; since the components are an orthogonal set, they must equal the sum of squares partitioned.

TABLE 6.7.
Mean squares and F values for testing the significance of responses to dosage levels of insecticide

Source of Variation	df	MS	Observed F	Required F 5%	Required F 1%
Fungicide	1	8003.33	207.8	4.35	8.10
Insecticide linear	1	344.45	8.9		
Insecticide nonlinear	1	0.82	0.02		
F×IL	1	616.05	16.0		
F×INL	1	4.82	0.1		
Error	20	38.51			

The F values for F×IL and F×INL show a highly significant interaction for the linear emergence of seedlings and no interaction to a response that differs from linear. Thus the experiment can be neatly and appropriately summarized, as in Figure 6.1, by linear regression lines that estimate the effect of increasing the dose of insecticide when seeds are or are not also treated with a fungicide. The calculation of the regression lines is left as practice after you learn regression in Chapters 13 and 14.

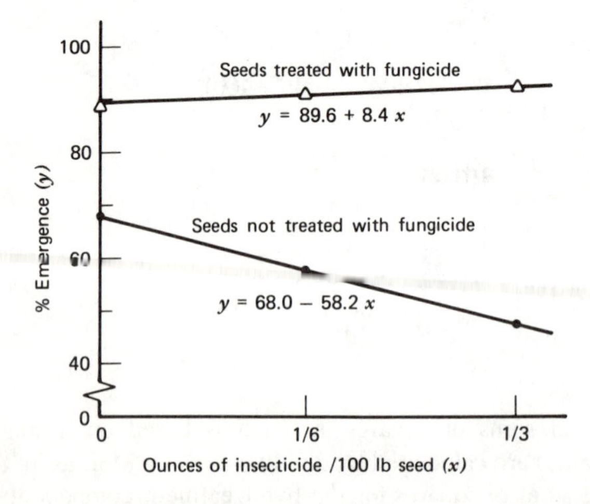

Figure 6.1. The effect of insecticide seed treatment, with and without a fungicide treatment, on the emergence of lima bean seedlings.

SUMMARY

The problem of deciding which treatment means are significantly different is called mean separation. There are three general approaches to mean separation: the use of least significant differences; the use of multiple-range tests; and through planned F tests.

Least significant difference is calculated as follows:

$$LSD = t\sqrt{\frac{2(MSE)}{r}}$$

where t is a tabulated value chosen for the degrees of freedom for error and the level of significance desired, MSE is the mean square for error, and r is the number of variates on which the means to be separated are based. To separate two means based on unequal numbers of variates,

$$LSD = t\sqrt{\frac{MSE}{r_1} + \frac{MSE}{r_2}}$$

Duncan's multiple-range test is the most popular of a number of range tests available; it is calculated as $D = R(LSD)$ where R is a tabular value for degrees of freedom for error, level of significance, and distance apart of two means in an array of treatment means. LSD is the least significant difference.

Planned F tests usually offer the most precise procedure for mean separation. As many independent questions can be asked and answered by F tests as there are degrees of freedom for treatments. The questions should be planned before the experiment is conducted.

The sum of squares for a single degree of freedom can be calculated from a set of coefficients whose sum is zero by the equation:

$$SS = \frac{\left(\sum c_i Y_{i.}\right)^2}{r \sum c_i^2}$$

where c_i is the set of coefficients, $Y_{i.}$ is a set of treatment totals, and r is the number of variates making up each total. Two comparisons are orthogonal (independent) if their coefficients and the products of corresponding coefficients add to zero. If as many orthogonal comparisons are made as there are degrees of freedom for treatments, their sums of squares will add to the sum of squares for treatments.

Coefficients for measuring trends can be obtained from Table A.11 if treatment levels are equally spaced and from Table A.11a for some sets of unequally spaced treatments.

7

THE LATIN SQUARE DESIGN

In this design the randomization of treatments is restricted further by grouping them into columns as well as rows. Thus it is possible to remove variability from experimental error associated with both these effects. Each treatment occurs the same number of times (usually once) in each row and column. The design will afford a more precise comparison of treatment effects than the randomized block design only if there is appreciable variation associated with the columns.

Rows and columns may refer to the spacial distribution of experimental units or to the order in which treatments are performed. In Figure 7.1, the treatments, A, B, and C, are three different makes of desk calculators to be tested; *columns* are three different operators and *rows* are the six different times the three operators test the machine. Each operator tests each machine two times and all three machines are tested in each time period. Thus the effects of time period and operators are measurable sources of variation that are independent of the machines and can be removed from the total variability of the experiment, reducing experimental error. This is an example of a double latin square. When

Row (time periods)	Column (operators) I	II	III
I	B	A	C
II	C	B	A
III	A	C	B
IV	B	C	A
V	C	A	B
VI	A	B	C

Figure 7.1. Three treatments in a double latin square. Sources of variation and degrees of freedom are: rows = 5; columns = 2; treatments = 2; error = 8. Treatments (A, B, C), are three different desk calculators.

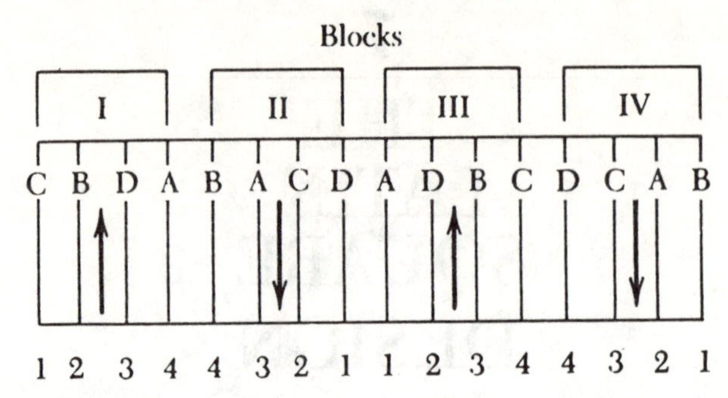

Figure 7.2. A latin square with four seed treatments (A, B, C and D) assigned to seeder units 1, 2, 3, and 4. The arrows indicate direction of planter travel. Sources of variation and degrees of freedom are: blocks=3; planter units=3; seed treatment=3; error=6.

the number of treatments is small and there is good reason to believe that there will be appreciable effects of columns and rows, variation can be removed in two directions by using two latin squares (each independently randomized).

There are times when a latin square may be advantageous when plots form a continuous line. Consider, for example, an experiment designed to test four seed treatments where individual plots are to be single rows throughout the experimental area. A seeder with four planter units is to be used. Planter units may differ in seeding rate. To remove the planter effect, each seed treatment can be assigned to a different seeder unit in each of four blocks so that each treatment is seeded the same number of times by each seeding unit as in Figure 7.2.

A latin square requires at least as many replications as there are treatments and therefore is not practical for experiments with a large number of treatments. Most commonly used latin squares are those having from four to eight treatments, with a single experimental unit per treatment in each column and row.

RANDOMIZATION

Start with any latin square (systematic or randomized) with the number of treatments required for your experiment. For example, suppose we wish to randomize six treatments, A, B, C, D, E, and F. We start with latin square below (Fig. 7.3); go to a table of random numbers (Table A.1); pick an arbitrary starting place, for example, row 5; and proceed across and back on row 6 assigning the numbers 1, 3, 5, 4, 2, 6 to rows 1 through 6. Continuing along row 6 of the table of random numbers and back (left to right) on row 7, assign the numbers 4, 2, 5, 1, 3, 6 to the columns. The new latin square is now completed as in Figure 7.4 by rearranging the rows and columns of the old square as indicated by the random numbers.

Rows	Columns					
	4	2	5	1	3	6
1	B	D	E	F	A	C
3	C	E	A	D	F	B
5	A	F	C	B	E	D
4	D	A	F	C	B	E
2	F	B	D	E	C	A
6	E	C	B	A	D	F

Figure 7.3. Procedure for rerandomization of a 6×6 latin square. Rows and columns are to be rerandomized in the order indicated by a table of random numbers. This results in the latin square of Figure 7.4.

Row	Column						Row totals, $Y_{i..}$
	I	II	III	IV	V	VI	
I	F 28.2	D 29.1	A 32.1	B 33.1	E 31.1	C 32.4	186.0
II	E 31.0	B 29.5	C 29.4	F 24.8	D 33.0	A 30.6	178.3
III	D 30.6	E 28.8	F 21.7	C 30.8	A 31.9	B 30.1	173.9
IV	C 33.1	A 30.4	B 28.8	D 31.4	F 26.7	E 31.9	182.3
V	B 29.9	F 25.8	E 30.3	A 30.3	C 33.5	D 32.3	182.1
VI	A 30.8	C 29.7	D 27.4	E 29.1	B 30.7	F 21.4	169.1
Column totals $Y_{.j.}$	183.6	173.3	169.7	179.5	186.9	178.7	$1071.7 = Y_{...}$
	Treatments						
	A(1) $(NH_4)_2SO_4$	B(2) NH_4NO_3	C(3) $CO(NH_2)_2$	D(4) $Ca(NO_3)_2$	E(5) Na NO_3	F(6) No N	
Totals, $Y_{..k}$	186.1	182.1	188.9	183.8	182.2	148.6	
Means, $\bar{Y}_{..k}$	31.0	30.4	31.5	30.6	30.4	24.8	

Figure 7.4. A 6×6 latin square. Each treatment appears once in each row and in each column. The treatments are five nitrogen source materials, all applied to give 100 lb of nitrogen per acre, and a nonfertilized control. The values are sugar beet root yields in tons per acre.

ANALYSIS OF VARIANCE

We will analyze the data of Figure 7.4 where the variates can be classified in three ways: rows, columns, and treatments. Rows are the i's and go from 1 to r. The columns are the j's and go from 1 to c. Treatments are indicated by the k subscript, and k goes from 1 to n. In the usual latin square, $r=c=n$.

We start by completing the first two columns of Table 7.1.

Sources of Variation and Degrees of Freedom

Degrees of freedom are, as usual, one less than the number of observations associated with each source of variation: df total $= rc-1=6(6)-1=35$; df rows $=r-1=6-1=5$; df columns $=c-1=6-1=5$; df treatments $=n-1=6-1=5$. Degrees of freedom for error can be obtained by subtraction: $35-5-5-5=20$, or by $(r-1)(c-1)-(n-1)=5(5)-5=20$.

Correction Term

$$C=\frac{Y_{...}^{2}}{rc}=\frac{1071.7^2}{6(6)}=31903.91$$

Sums of Squares and Mean Squares

ROWS

$$SSR=\frac{\sum Y_{i..}^{2}}{c}-C=\frac{186.0^2+\ldots+169.1^2}{6}-31903.91=32.19$$

where c is the number of plots in each row.

$$MSR=\frac{SSR}{df(R)}=\frac{32.19}{5}=6.438$$

COLUMNS

$$SSC=\frac{\sum Y_{.j.}^{2}}{r}-c$$

where r is the number of plots in each column.

$$SSC=\frac{183.6^2+\ldots+178.8^2}{6}-31903.91=33.67$$

$$MSC=\frac{SSC}{df(C)}=\frac{33.67}{5}=6.734$$

TABLE 7.1.
Analysis of variance, sugar beet nitrogen source trial

Source of Variation	Degrees of Freedom df	Sums of Squares SS	Mean Squares MS	Observed F	Required F	
					5%	1%
Total	35	281.88				
Rows	5	32.19	6.438	4.26	2.71	4.10
Columns	5	33.67	6.734	4.45		
Treatments	5	185.77	37.154	24.56		
Error (RC−T)	20	30.25	1.513			

TREATMENTS

$$SST = \frac{\sum Y_{..k}^{2}}{r} - C$$

where r is the number of replicates of each treatment.

$$SST = \frac{148.6^2 + \ldots + 182.2^2}{6} - 31903.91 = 185.77$$

$$MST = \frac{SST}{df(T)} = \frac{185.77}{5} = 37.154$$

TOTAL

$$SS = \sum Y_{ijk}^{2} - C = 28.2^2 + 32.1^2 + \ldots + 27.4^2 + 29.1^2 - 31903.91$$

$$= 32185.79 - 31903.91 = 281.88$$

ERROR

$$SSE = SS - SSR - SSC - SST = 281.88 - 32.19 - 33.67 - 185.77 = 30.25$$

$$MSE = \frac{SSE}{df(E)} = \frac{30.25}{20} = 1.513$$

CALCULATORS PROGRAMMED TO COMPUTE STANDARD DEVIATION. The sums of squares and mean squares can also be computed from a standard deviation of totals. For example, to compute SSR, enter each row total

(186.0...169.1) with the appropriate entry key of your calculator and obtain the standard deviation of the row totals = 6.21496. Square the standard deviation and divide by the number of variates in each total you entered: $(6.21496)^2/6 = \text{MSR} = 6.4376$. Multiplying by df(R) = SSR = 6.4376(5) = 31.19, as before.

F Values

$$F(\text{rows}) = \frac{\text{MSR}}{\text{MSE}} = \frac{6.438}{1.513} = 4.26$$

$$F(\text{columns}) = \frac{\text{MSC}}{\text{MSE}} = \frac{6.734}{1.513} = 4.45$$

$$F(\text{treatments}) = \frac{\text{MST}}{\text{MSE}} = \frac{37.154}{1.513} = 24.56$$

All three F ratios are based on 5 and 20 degrees of freedom. The required values for statistical significance are obtained from Table A.3 and entered in the analysis of variance table. All three sources of variation are classified highly significant. From this we conclude that there are real differences among rows and columns as well as treatments.

MEAN SEPARATION

In planning the sugar beet experiment to evaluate the effects of different sources of nitrogen, the investigator posed several questions that were to be answered by partitioning the sum of squares for treatments into the orthogonal set of comparisons indicated in Table 7.2.

TABLE 7.2.
An orthogonal partitioning of the treatments of Figure 7.4.

Source of Variation	df	SS	Observed MS	F	Required F 5%	Required F 1%
Treatments	5	185.77	37.154	24.56	2.71	4.10
No N vs. N	1	180.200	180.200	119.10	4.35	8.10
Organic N vs. inorganic N	1	3.816	3.816	2.52		
Ammonium N vs. nitrate N	1	0.202	0.202	0.13		
$(NH_4)_2SO_4$ vs. NH_4NO_3	1	1.334	1.334	0.88		
$NaNO_3$ vs. $Ca(NO_3)_2$	1	0.213	0.213	0.14		
Error	20	30.25	1.513			

TABLE 7.3.
Treatment coefficients to check for orthogonality of comparisons and to facilitate the computation of sums of squares

Comparison	Treatments and Treatment Totals					
	No N 148.6	$(NH_4)_2SO_4$ 186.1	NH_4NO_3 182.1	$CO(NH_2)_2$ 188.9	$Ca(NO_3)_2$ 183.8	$NaNO_3$ 182.2
No N vs. N	+5	−1	−1	−1	−1	−1
Organic N vs. inorganic N	0	−1	−1	+4	−1	−1
NH_4-N vs. NO_3-N	0	+1	+1	0	−1	−1
$(NH_4)_2SO_4$ vs. NH_4NO_3	0	+1	−1	0	0	0
$Ca(NO_3)_2$ vs. $NaNO_3$	0	0	0	0	+1	−1

Note that all rows sum to zero and that the sum of the products of the corresponding coefficients of any two comparisons is zero, and therefore the treatment comparisons are orthogonal.

The coefficients for testing the orthogonality of the comparisons and for completing Table 7.2 are shown in Table 7.3.

Sums of squares can be calculated as follows from the treatment totals:

$$\text{SS(no N vs. N)} = \frac{148.6^2}{6} + \frac{(186.1 + \ldots + 182.2)^2}{30} - \frac{1071.7^2}{36}$$

$$= 3680.327 + 28403.787 - 31903.914 = 180.200$$

When the comparison involves a single degree of freedom, the shorter method of calculation using the orthogonal polynomials of Table 7.3 is: $SS = (\Sigma c_i Y_{..k})^2 / (r\Sigma c_i^2)$, where the c_i are the coefficients of Table 7.3, the $Y_{..k}$ are the treatment totals, and r is the number of replicates in each treatment total. Thus

$$\text{SS(no N vs. N)} = \frac{[5(148.6) - 186.1 - 182.1 - 188.9 - 183.8 - 182.2]^2}{6(30)}$$

$$= \frac{-180.1^2}{180} = 180.200$$

The denominator, 6(30), is found by summing the squares of the coefficients of the terms in the numerator and multiplying this by the number of variates making up each term of the numerator; thus

$$6\left[(5^2)+(-1)^2+(-1)^2+(-1)^2+(-1)^2+(-1)^2\right]=6(30)$$

$$\text{SS (organic N vs. inorganic N)} = \frac{188.9^2}{6} + \frac{(186.1+182.1+183.8+182.2)^2}{24}$$

$$-\frac{(188.9+186.1+182.1+183.8+182.2)^2}{30} = \frac{188.9^2}{6} + \frac{734.2^2}{24} - \frac{923.1^2}{30}$$

Note that the third term is a new correction term.

$$=5947.202+22460.402-28403.787=3.816$$

The shorter calculation is

$$=\frac{\left[4(188.9)-186.1-182.1-183.8-182.2\right]^2}{6(20)}=3.816$$

$$\text{SS}(\text{NH}_4-\text{N vs. NO}_3-\text{N}) = \frac{(186.1+182.1)^2+(183.8+182.2)^2}{12}$$

$$-\frac{(186.1+182.1+183.8+182.2)^2}{24}$$

$$=\frac{368.2^2+366.0^2}{12}-\frac{734.2^2}{24}$$

$$=22460.603-22460.402=0.201$$

Or

$$=\frac{(186.1+182.1-183.8-182.2)^2}{6(4)}=\frac{(2.2)^2}{24}=0.202$$

$$\text{SS}\left[(\text{NH}_4)_2\text{SO}_4 \text{ vs NH}_4\text{NO}_3\right] = \frac{186.1^2+182.1^2}{6} - \frac{(186.1+182.1)^2}{12}$$

$$=11298.937-11297.603=1.334$$

Or

$$= \frac{(186.1 - 182.1)^2}{2(6)} = \frac{4.0}{12} = 1.333$$

$$SS[Ca(NO_3)_2 \text{ vs } NaNO_3] = \frac{183.8^2 + 182.2^2}{6} - \frac{(183.8 + 182.2)^2}{12}$$

$$= 11163.213 - 11163.000 = 0.213$$

Or

$$= \frac{(183.8 - 182.2)^2}{2(6)} = \frac{1.6^2}{12} = 0.213$$

Mean squares are obtained by dividing the sums of squares by their associated degrees of freedom; since, in this case, each comparison involves a single degree of freedom, SS = MS.

F values are calculated by dividing each MS by MS for error. Required F values are tabular values from Table A.3 for 1 and 20 df. We now have an F test to answer each of the questions posed when the experiment was planned. The only significant F value is for the comparison no N vs. N; all others are quite low, leading to the conclusion that there was a response to nitrogen but that beets responded similarly to all N sources.

SUMMARY

In a latin square:

Experimental units are organized into two categories other than treatments. These two categories are usually referred to as *rows* and *columns* with regard to the organization of data in a two way table.

Each treatment is assigned the same number of times (usually once) within each category so that differences between categories are not due to treatment effects.

At least as many replications are required as there are treatments. Latin squares are usually not practical with more than eight treatments.

Only when both categories (rows and columns) vary appreciably will the latin square design improve the detection of treatment differences over the randomized complete block.

8

THE SPLIT-PLOT DESIGN

Split-plot designs, and a variation, the split-block, are frequently used for factorial experiments in which the nature of the experimental material or the operations involved make it difficult to handle all factor combinations in the same manner or when the investigator wishes to increase precision in estimating certain effects and is willing to sacrifice precision in estimating certain others. The basic split-plot design involves assigning the treatments of one factor to main plots arranged in a completely random, randomized complete block or a latin square design. The treatments of the second factor are assigned to subplots within each main plot. The design usually sacrifices precision in estimating the average effects of the treatments assigned to main plots. It often improves the precision for comparing the average effects of treatments assigned to subplots and, when interactions exist, for comparing the effects of subplot treatments for a given main plot treatment. This arises from the fact that experimental error for main plots is usually larger than the experimental error used to compare subplot treatments. Usually, the error term for subplot treatments is smaller than would be obtained if all treatment combinations were arranged in a randomized complete block design.

Note the experiment of Figure 8.1. It involves two factors, nitrogen fertilizer (N) at two levels ($n=2$) and green manures (G) of four types ($g=4$). The total number of treatments for this trial are $n \times g = 8$. Note that all eight treatments occur once in each of the three blocks but that within a block all treatments of a common nitrogen level occur together. Also note that with respect to the nitrogen levels, we have a randomized complete block with two treatments in three blocks. The degrees of freedom for these six main plots are partitioned as for a randomized complete block in the "split-plot" column of Table 8.1 where the two designs are compared.

The restriction on the randomization of the treatments within a block results in two error terms for the split-plot design. The main plot error is usually larger, as it involves variability among the larger more widely spaced main plots, and the subplot error is usually smaller, as it involves variability among closely spaced subplots within the main plots.

The split-plot design can be used with more than two treatment factors, as it is not necessary to have an additional split for each factor. For example, to test two varieties at two levels of nitrogen fertilizer applied at two different times, main plots could be the four combinations of variety and nitrogen levels that could be split for time of applying the nitrogen. One replication of this trial could be:

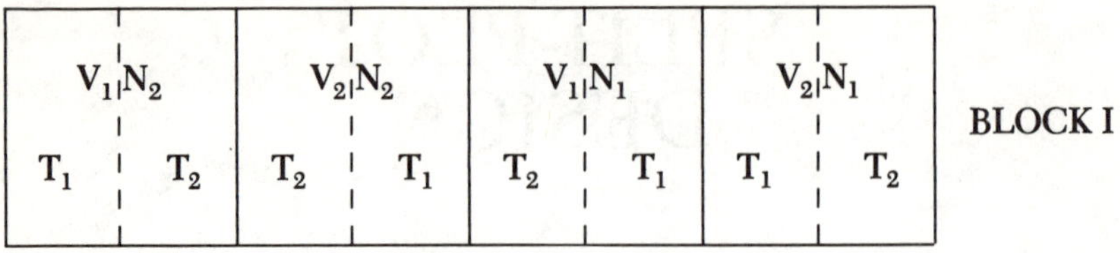

V_1N_2		V_2N_2		V_1N_1		V_2N_1		
T_1	T_2	T_2	T_1	T_2	T_1	T_1	T_2	BLOCK I

With three factors (A, B, C) each at two or more levels, there are six different possibilities for main plots: A, B, C, AB, AC, BC.

Each variation of the split-plot design imposes certain restrictions as to the error term that may be used to test treatment effects. It is important, therefore, to assign factors in a manner that gives the greatest precision for comparing the interactions and average treatment effects in which you are most interested. Some skill and experience are required in laying out split-plot experiments, and you are urged to consult someone with experience in the use of this design.

TABLE 8.1.
ANOVA outline for the two factor experiment of Figure 8.1 as a split-plot and a randomized complete block design. The brackets and arrows indicate appropriate error terms for testing treatment effects in the two designs.

Source of Variation	Degrees of Freedom	Split-Plot	RCB[a]
Subplots	$ngb-1$	23	23
Main plots	$nb-1$	5	—
Blocks	$b-1$	2	2
Nitrogen	$n-1$	1 ←	1 ←
MP error	$(b-1)(n-1)$	2 ┘	—
Green manures	$(g-1)$	3 ←	3 ←
N×G	$(n-1)(g-1)$	3 ←	3 ←
SP error	$(b-1)[(g-1)+(n-1)(g-1)]$	12 ┘	—
Error (RCB)[a]	$(b-1)[(n-1)+(g-1)+(n-1)(g-1)]$		14 ┘

[a] Randomized complete block.

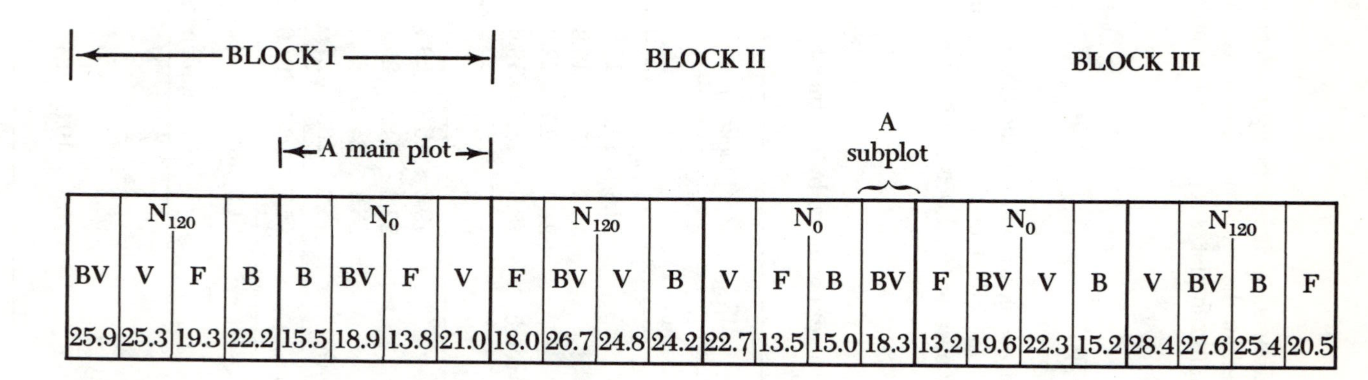

Figure 8.1. Split-plot design. Main plots (N_{120}, N_0) are nitrogen fertility levels. Subplots BV, V, F, B, are green manure treatments. All plots are laid out in strips through the field in three blocks. Plot yields of the sugar beet crop following the green manure treatments are given in tons of roots per acre.

RANDOMIZATION

The randomization of the treatments assigned to main plots is carried out as prescribed for the design selected for the main plot treatments. Subplot treatments are then randomized within each main plot, a separate randomization being made for each main plot.

ANALYSIS OF VARIANCE

To illustrate computational procedure, we will use the experiment of Figure 8.1. The trial was designed to test the effect of three green manure crops on the subsequent production of sugar beets at two levels of nitrogen fertilization. At the

TABLE 8.2.
Sugar beet root yields (tons per acre) organized by treatments, main plots and blocks.

Treatments		Blocks (j)				
Pounds N/acre (i)	Green Manure (k)	I	II	III	Totals	Means
0	Fallow	13.8	13.5	13.2	40.5	13.5
	Barley	15.5	15.0	15.2	45.7	15.2
	Vetch	21.0	22.7	22.3	66.0	22.0
	Barley-vetch	18.9	18.3	19.6	56.8	18.9
Main plot totals ($Y_{1j.}$)		69.2	69.5	70.3	$209.0 = Y_{1..}$	17.4
120	Fallow	19.3	18.0	20.5	57.8	19.3
	Barley	22.2	24.2	25.4	71.8	23.9
	Vetch	25.3	24.8	28.4	78.5	26.2
	Barley-vetch	25.9	26.7	27.6	80.2	26.7
Main plot totals ($Y_{2j.}$)		92.7	93.7	101.9	$288.3 = Y_{2..}$	24.0
Block totals ($Y_{.j.}$)		161.9	163.2	172.2	$497.3 = Y_{...}$	20.7

	Green Manures			
	F	B	V	BV
Totals ($Y_{..k}$)	98.3	117.5	144.5	137.0
Means ($\bar{Y}_{..k}$)	16.4	19.6	24.1	22.8

Symbols for treatment factors and levels:
N = nitrogen, n = 2; G = green manure, g = 4; B = block, b = 3.

TABLE 8.3.
Analysis of variance. Sugar beet, nitrogen × green manure experiment.

Source of Variation	df	SS	MS	Observed F	Required F 5%	Required F 1%
Subplots	23	516.12				
Main plots	5	274.92				
Blocks	2	7.87	3.935			
Nitrogen	1	262.02	262.020	104.18	18.51	98.49
Main plot error, BN	2	5.03	2.515			
Green manures	3	215.26	71.753	118.99	3.49	5.95
N×G	3	18.70	6.233	10.34		
Subplot error, BG+B(N×G)	12	7.24	0.603			

outset it was assumed that sugar beets would respond differently to the green manures, depending on the level of nitrogen fertility, and thus the objective was to compare the effect of the green manures as precisely as possible at each level of fertility. Therefore the main plots were to be two levels of nitrogen fertilization applied to the sugar beets at thinning time and replicated three times in a randomized complete block design. Subplots were to be green manures, grown during the fall and winter preceding the planting of sugar beets. The green manure treatments were barley (B), vetch (V), barley and vetch grown together (BV), and fallow (F). Nothing was allowed to grow in fallow plots prior to planting sugar beets. The plots were laid out as in Figure 8.1. Yields of sugar beet following the green manures are given for each subplot and organized for analysis in Table 8.2.

The first step is to determine sources of variation and associated degrees of freedom—the first two columns of Table 8.3.

Sources of Variation and Degrees of Freedom

Total degrees of freedom in the experiment are one less than the number of subplots, $ngb-1=(2)(4)(3)-1=23$. Main plots are listed as a source of variation as their partition leads to df for main plot error; $df(MP)=nb-1=2(3)-1=5$. Degrees of freedom for main plots are partitioned according to the design in which they are arranged, in this case the randomized complete block: blocks $=b-1=3-1$; nitrogen $=n-1=2-1$; main plot error, often called error a, $=(b-1)(n-1)=(3-1)(2-1)$. Degrees of freedom for green manures are $(g-1)=4-1$; and for treatment factor interaction, N×G, $=(n-1)(g-1)=(2-1)(4-1)$. Degrees of freedom for subplot error, usually called error b, can be obtained by subtraction, paying attention to the indentation of the sources of variation, that is, $23-5-3-3=12$; or by the addition of degrees of freedom for the interaction of blocks with green manures and nitrogen × green manures: $(b-1)(g-1)+(b-1)(n-1)(g-1)=(3-1)(4-1)+(3-1)(2-1)(4-1)=12$.

Correction Term

$$C=\frac{Y_{...}^{\ 2}}{ngb}=\frac{497.3^2}{2(4)(3)}=10{,}304.47$$

Sums of Squares and Mean Squares

Blocks:

$$SSB=\frac{\Sigma Y_{.j.}^{\ 2}}{ng}-C=\frac{161.9^2+163.2^2+172.2^2}{2(4)}-C=7.87.$$

Note that the denominator (8) is the number of variates making up each term in the numerator.

$$MSB=\frac{SSB}{df(B)}=\frac{7.87}{2}=3.935$$

Nitrogen:

$$SSN=\frac{\Sigma Y_{i..}^{\ 2}}{gb}-C=\frac{209.0^2+288.3^2}{3(4)}-C=262.02$$

$$MSN=\frac{SSN}{df(N)}=\frac{262.02}{1}=262.02$$

Main plots:

$$SS(MP)=\frac{\Sigma Y_{ij.}^{\ 2}}{g}-C=\frac{69.2^2+\ldots+101.9^2}{4}-C=274.92$$

Main plot error:

$$SS(MPE)=SS(MP)-SSB-SSN=274.92-7.87-262.02=5.03$$

$$MS(MPE)=\frac{SS(MPE)}{df(MPE)}=\frac{5.03}{2}=2.515$$

Green manure treatments:

$$SSG=\frac{\Sigma Y_{..k}^{2}}{nb}-C=\frac{98.3^2+\ldots+137.0^2}{2(3)}-C=215.26$$

$$MSG=\frac{SSG}{df(G)}=\frac{215.26}{3}=71.753$$

N×GM:

$$SS(N\times G)=\frac{\Sigma Y_{i.k}^{2}}{b}-C-SSN-SSG=\frac{40.5^2+\ldots+80.2^2}{3}-C$$

$$-262.02-215.26=18.70$$

$$MS(N\times G)=\frac{SS(N\times G)}{df(N\times G)}=\frac{18.70}{3}=6.233$$

Subplots:

$$SS(SP)=\Sigma Y_{ijk}^{2}-C=13.8^2+15.5^2+\ldots+27.6^2-C=516.12$$

Subplot error:

$$SS(SPE)=SS(SP)-SS(MP)-SSG-SS(N\times G)$$

$$=516.12-274.92-215.26-18.70=7.24$$

$$MS(SPE)=\frac{SS(SPE)}{df(SPE)}=\frac{7.24}{12}=0.603$$

THE STANDARD DEVIATION KEY. With a calculator programmed to compute $s=\sqrt{\Sigma(Y_i-\bar{Y})^2/(r-1)}$, sums of squares are obtained by the following rule. Enter each total with the appropriate key. When all are entered, depress the standard deviation key, square the displayed value to obtain s^2, divide by the number of variates making up each total entered, and multiply by df. For example, SS(MP): Enter $69.2, 69.5, \ldots, 101.9$; depress standard deviation key $=14.830$; square $=219.9377$; divide by $4=54.9844=$ MS(MP); multiply by $5=274.92=$ SS(MP).

F Values

Nitrogen effects are tested using MS(MPE); green manures and the interaction of nitrogen and green manures are tested using MS(SPE). F for N is $262.02/2.515=104.18$. F for N×G is $6.233/0.603=10.34$. The highly significant F value for

N×G indicates a difference in the comparative response of the sugar beet crop to the green manures at the different fertility levels. The crux of the experiment is to isolate and understand the interaction—a problem in mean separation and agronomic interpretation.

MEAN SEPARATION

Pertinent F Tests

By partitioning the sum of squares for the N×G interaction, we gain insight into the nature of the interaction. There are several ways this can be done, but partitioning to answer the following three questions appears logical. Did the sugar beets respond differently at the two nitrogen levels to: vetch versus no vetch; fallow versus barley; vetch versus barley and vetch? Table 8.4 gives the treatment totals and a set of orthogonal coefficients for use in calculating the interaction components as well as other single degree of freedom comparisons.

To partition the 3 df for N×G we must first set down coefficients for partitioning the main effects into single degree of freedom components. Coefficients for N are simple, as there are only two groups. The four green manure treatments are partitioned to answer the three questions posed for interaction, but this is done for green manure effects over both levels of N: V+BV vs. F+B, F vs.

TABLE 8.4.
Orthogonal coefficients for the indicated comparisons.

		Treatments, Treatment Totals and Means							
		N_0				N_{120}			
		F	B	V	BV	F	B	V	BV
	$Y_{i.k}$	40.5	45.7	66.0	56.8	57.8	71.8	78.5	80.2
Comparison	$\bar{Y}_{i.k}$	13.5	15.2	22.0	18.9	19.3	23.9	26.2	26.7
1. N		−1	−1	−1	−1	1	1	1	1
2. V vs. no V		−1	−1	1	1	−1	−1	1	1
3. F vs. B		−1	1	0	0	−1	1	0	0
4. V vs. BV		0	0	−1	1	0	0	−1	1
5. N×(V vs. no V)		1	1	−1	−1	−1	−1	1	1
6. N×(F vs. B)		1	−1	0	0	−1	1	0	0
7. N×(V vs. BV)		0	0	1	−1	0	0	−1	1

B, and V vs. BV. The calculation of sums of squares for these three comparisons would add to the sum of squares for green manures in Table 8.3. These comparisons would not mean much, however, as we already know that the green manures have a differential effect, depending on the N level, and our aim is to look at this interaction in more detail by calculating F tests for comparisons 5, 6, and 7 of Table 8.4. The coefficients for these comparisons are obtained by multiplying coefficients for comparisons 1 and 2, 1 and 3, and 1 and 4.

Sums of squares for the three interaction components are computed as below and entered in Table 8.5.

$$SS[N\times(V \text{ vs. no } V)] = \frac{(40.5+45.7-66.0-56.8-57.8-71.8+78.5+80.2)^2}{3(8)}$$

$$= \frac{7.5^2}{24} = 2.344$$

Again, note the use of the formula for calculating a sum of squares with a single degree of freedom: $SS=(\Sigma c_i Y_{i.k})^2/(r\Sigma c_i^2)$. In these cases all the coefficients (c_i's) are ± 1, and it is not necessary to write them in the numerators.

$$SS[N\times(F \text{ vs. } B)] = \frac{(40.5-45.7-57.8+71.8)^2}{3(4)} = 6.453$$

$$SS[N\times(V \text{ vs. } BV)] = \frac{(66.0-56.8-78.5+80.2)^2}{3(4)} = 9.901$$

The three single degree of freedom F tests of Table 8.5 provide answers to the three questions posed above.

TABLE 8.5.
Variance components for interaction.

Source of Variation	df	SS	MS	Observed *F*	Required *F* 5%	Required *F* 1%
N×G	3	18.70	6.233	10.34	3.49	5.95
N×(V vs. no V)	1	2.344	2.344	3.88	4.75	9.33
N×(F vs. B)	1	6.453	6.453	10.70		
N×(V vs. BV)	1	9.901	9.901	16.42		
SPE	12	7.24	0.603			

N×(V VERSUS NO V). The differences in the response to vetch at N_0 compared to N_{120} are not significantly different. Referring to the means of Table 8.4, the change in mean plot yield for plots receiving vetch versus no vetch at N_0 is not significantly different from the change in mean plot yield for plots receiving vetch versus no vetch at N_{120}; that is $(22.0+18.9-13.5-15.2)/2=6.1$ compared to $(26.2+26.7-19.3-23.9)/2=4.85$ gives a difference of $6.1-4.85=1.25$, which is not significantly different from zero. The divisor 2 is to keep the comparison on a per plot basis.

A t test can also be used to make this comparison but will lead to the same statistical conclusion. We show it here to point out the equivalence of the two tests and to illustrate the greater ease of the F test. We are examining a difference of differences: $6.1-4.85=1.25$. The appropriate t test is $t=(\bar{d}_1-\bar{d}_2)/s_{\bar{d}_1-\bar{d}_2}$ where $\bar{d}_1=6.1$, $\bar{d}_2=4.85$, and $s_{\bar{d}_1-\bar{d}_2}$ is the standard error of a difference of differences and is computed as

$$s_{\bar{d}_1-\bar{d}_2}=\sqrt{s_{\bar{d}_1}{}^2+s_{\bar{d}_2}{}^2}=\sqrt{\frac{2s_1^2}{r_1}+\frac{2s_2^2}{r_2}}$$

When s_1^2 and s_2^2 estimate a common variance and $r_1=r_2$, then $s_{\bar{d}_1-\bar{d}_2}=\sqrt{4s^2/r}=\sqrt{4s^2/6}=\sqrt{4(0.603)/6}=0.634$. The divisor 6 is the number of variates in the mean differences being compared—in this case each mean difference is the average of two means each based on three replicates.

Substituting into the above t formula gives $t=(6.1-4.85)/0.634=1.25/0.634=1.97$, a nonsignificant t value, since the required $t_{.05}$ for 12 df is 2.179. Note $t^2=F=(1.97)^2=3.88$, the F value of Table 8.5.

N×(F VERSUS B). The difference between fallow and barley at N_0 is significantly less than at N_{120}. That is, $15.2-13.5=1.7$ is significantly less than $23.9-19.3=4.6$. Compared to fallow, the response to barley was $4.6-1.7=2.9$ tons /acre more with than without fertilizer N. Confidence limits for this difference of differences can be calculated from $CL_{95}=\bar{d}_1-\bar{d}_2\pm ts_{\bar{d}_1-\bar{d}_1}$ where t is the tabular value for 12 df and the 5% level; $s_{\bar{d}_1-\bar{d}_2}=\sqrt{4s^2/r}=\sqrt{4(0.603)/3}=0.897$; and $CL_{95}=2.9\pm2.179(.897)=2.9\pm2.0=0.9$ to 4.9 tons/acre. That is, with a confidence of 95% we can say that, under these conditions, the beneficial effect of barley green manure was between 0.9 and 4.9 tons/acre more when the sugar beets were fertilized with nitrogen than when they were not.

N×(V VERSUS BV). There is a significant loss in root yield, $23.0-18.9=3.1$ tons/acre, from the barley-vetch compared to the straight vetch green manure that does not occur when the sugar beets are given N fertilizer, $26.2-26.7=-0.5$. Confidence limits for the difference of differences are: $CL_{95}=3.1-(-0.5)\pm$

TABLE 8.6.
The effect of green manures and nitrogen fertilization on sugar beet root yield.

Pounds of N per Acre	Green Manure Treatments			
	Fallow	Barley	Vetch	Barley-Vetch
		Roots, Tons/Acre		
0	13.5	15.2	22.0	18.9
120	19.3	23.9	26.2	26.7

LSD, 5%: between green manures at the same N level, 1.4; between green manures at different N levels, 2.9.

2.179(0.897) = 3.6 ± 2.0 = 1.6 to 5.6 tons/acre.

This experiment might be summarized as in Tables 8.6 and 8.7. Table 8.6 presents the relevant effects of the experiment, and Table 8.7 gives the statistical information germane to a discussion of the significant interaction. A common procedure is to use single, double or triple asterisks to denote statistical significance at the 5, 1, and 0.1% level, respectively. Means for the average effects of nitrogen or green manures are not presented, as the strong interaction makes them rather meaningless. The LSDs of Table 8.6 are not really necessary but do provide approximate guides for interpreting the results.

TABLE 8.7.
Mean squares for interaction and interaction components of the effect of nitrogen and green manure treatments on sugar beet root yield.

Source of Variation	df	Mean Square
N×G	3	6.233**
N×(V vs. no V)	1	2.344
N×(F vs. B)	1	6.453**
N×(V vs. BV)	1	9.901**
Subplot error	12	0.603
Main plot error	2	2.515

Standard Errors and LSDs

At times, LSDs or multiple-range tests may be desirable. For these tests, standard errors are calculated based on variability among experimental units to which treatments are applied. With the split-plot design the calculation of standard errors for certain kinds of treatment comparisons becomes more complicated, as can be seen in Table 8.8, because we have two sources of experimental error—that involving main plots and that involving subplots.

Note that the standard error for comparing subplot treatment means within a main plot involves only the subplot error, but when comparisons are made between subplot treatment means for different main plots, the standard error involves both main plot and subplot errors. Skipping much tedious algebra, it turns out that the latter standard error is a weighted average of Ea and Eb, the weighting factor for Ea is 1, and that for Eb is $b-1$. As $1+b-1=b$, the denominator turns out to be br, where b is the number of subplot treatments and r is the number of replications.

To illustrate computation, LSDs for all possible comparisons of the means of the sugar beet green manure × nitrogen fertility trial of Table 8.2 are given below.

TABLE 8.8.
Standard errors for a split-plot design.

Means Compared	Standard Error of a Mean[a] ($s_{\bar{y}}$)
Main plot treatments: $A_1 - A_2$	$\sqrt{\frac{Ea}{rb}}$
Subplot treatments: $B_1 - B_2$	$\sqrt{\frac{Eb}{ra}}$
Subplot treatments for the same main-plot treatment: $B_1A_1 - B_2A_1$	$\sqrt{\frac{Eb}{r}}$
Subplot treatments for different main plot treatments: $B_1A_1 - B_1A_2$ or $B_1A_1 - B_2A_2$	$\sqrt{\frac{(b-1)Eb+Ea}{rb}}$

[a] Note the use of $s_{\bar{y}}$ in the determination of LSD or D: $LSD = t\sqrt{2}\, s_{\bar{y}}$; $D = R(LSD)$.
Ea = MS(MPE), Eb = MS(SPE), a = number of main plot treatments,
b = number of subplot treatments, r = number of replications.
A = treatments applied to main plots, B = treatments applied to subplots.

LSD FOR DIFFERENCES BETWEEN MAIN PLOT TREATMENTS. (between nitrogen means)

$$LSD_{.05} = t_a\sqrt{\frac{2(Ea)}{rb}}$$

where t_a is the tabular t value for df for Ea.

$$LSD_{.05} = 4.303\sqrt{\frac{2(2.515)}{3(4)}} = 4.303(0.647) = 2.8 \text{ tons/acre}$$

LSD FOR DIFFERENCES BETWEEN SUBPLOT TREATMENTS. (among green manure means)

$$LSD_{.05} = t_b\sqrt{\frac{2(Eb)}{ra}}$$

where t_b = tabular t value for df for Eb.

$$LSD_{.05} = 2.179\sqrt{\frac{2(0.603)}{3(2)}} = 2.179(0.448) = 1.0 \text{ ton/acre}$$

LSD FOR DIFFERENCES BETWEEN SUBPLOT TREATMENTS FOR THE SAME MAIN PLOT TREATMENT. (among green manure means for the same nitrogen level)

$$LSD_{.05} = t_b\sqrt{\frac{2Eb}{r}} = 2.179\sqrt{\frac{2(0.603)}{3}} = 2.179(0.634) = 1.4 \text{ tons/acre}$$

LSD FOR DIFFERENCES BETWEEN SUBPLOT TREATMENTS FOR DIFFERENT MAIN PLOT TREATMENTS. (to compare different green manure means at different nitrogen levels or to compare means for the same green manure treatment at different nitrogen levels)

$$LSD_{.05} = t_{ab}\sqrt{\frac{2[(b-1)Eb + Ea]}{rb}}$$

where t_{ab} is a weighted t value somewhere between the tabular values for t_a and t_b

and is calculated as follows:

$$t_{ab} = \frac{(b-1)(Eb)(t_b)+Ea(t_a)}{(b-1)Eb+Ea} = \frac{(4-1)(0.603)(2.179)+2.515(4.303)}{(4-1)(0.603)+2.515}$$

$$= \frac{14.764}{4.324} = 3.414$$

$$LSD_{.05} = 3.414\sqrt{\frac{2\left[(4-1)(0.603)+2.515\right]}{3(4)}} = 3.414(0.849) = 2.9 \text{ tons/acre}$$

If the eight treatment combinations had been randomized within each block, the design would have been the randomized complete block. The error mean square would then be

$$EMS = \frac{SS(MP)+SS(SP)}{df(MP)+df(SP)}$$

$$= \frac{5.03+7.24}{2+12}$$

$$= 0.876$$

and the LSD for all treatment comparisons would be

$$LSD = t\sqrt{\frac{2(0.876)}{3}} = 2.145(0.764) = 1.6 \text{ tons/acre}$$

(Note, t is the tabular value for 14 df at the 5% level.)

A comparison of the LSDs indicate the relative efficiencies of the two designs in separating treatment effects. Note the improved power (smaller LSDs) of the split-plot in separating the means of the subplot treatments and the comparison of subplot treatments *within* a main plot treatment and the loss of precision (larger LSDs) in comparing main plot treatments and subplot treatments *across* main plot treatments.

SUMMARY

The split-plot design is often useful for a factorial set of treatments. The design involves the random assignment of one treatment factor or combination of factors to main plots which are then split for the random assignment of another factor or combination of factors. Compared to the randomized complete block design, precision is lost in making comparisons among main plot treatments and subplot treatments for different main plot treatments, but precision is often improved for comparisons among subplot treatments and for subplot treatments within main plot treatments.

9

THE SPLIT-SPLIT PLOT

The addition of a third factor by splitting subplots of a split-plot design results in a split-split plot. This technique is often quite useful for a three-factor experiment to facilitate field operations or when it is desirable to keep treatment combinations together. However, the additional restriction on randomization makes it necessary to compute a third error term that is used to test for main effects of the factor applied to the second split and for all interactions involving this factor. The arrangement may have certain advantages in physical operations with the experimental units, but the necessity for the third error term can make mean separation quite complicated. You are urged to consult a biometrician before employing this scheme.

Randomization procedure is the same as for the split-plot design, with the subplots being split into sub-subplots, equal in number to the levels of factor three, to which the third factor is randomly assigned—a new randomization for each set of sub-subplots. Figure 9.1 illustrates the partial layout of a split-split plot to evaluate the effects of dates of planting, aphid control, and date of harvest on the control of aphid-borne sugar beet viruses. The procedure for the stepwise handling of data from such an experiment will be illustrated with the effect of these treatments on root yield.

ORGANIZATION OF DATA

Data are organized by treatments and blocks in Table 9.1. Table 9.2 is formed to provide totals for the two-way interactions and main effects.

ANALYSIS OF VARIANCE

The completed analysis of variance is given in Table 9.3. The stepwise procedure for completing the table is as follows.

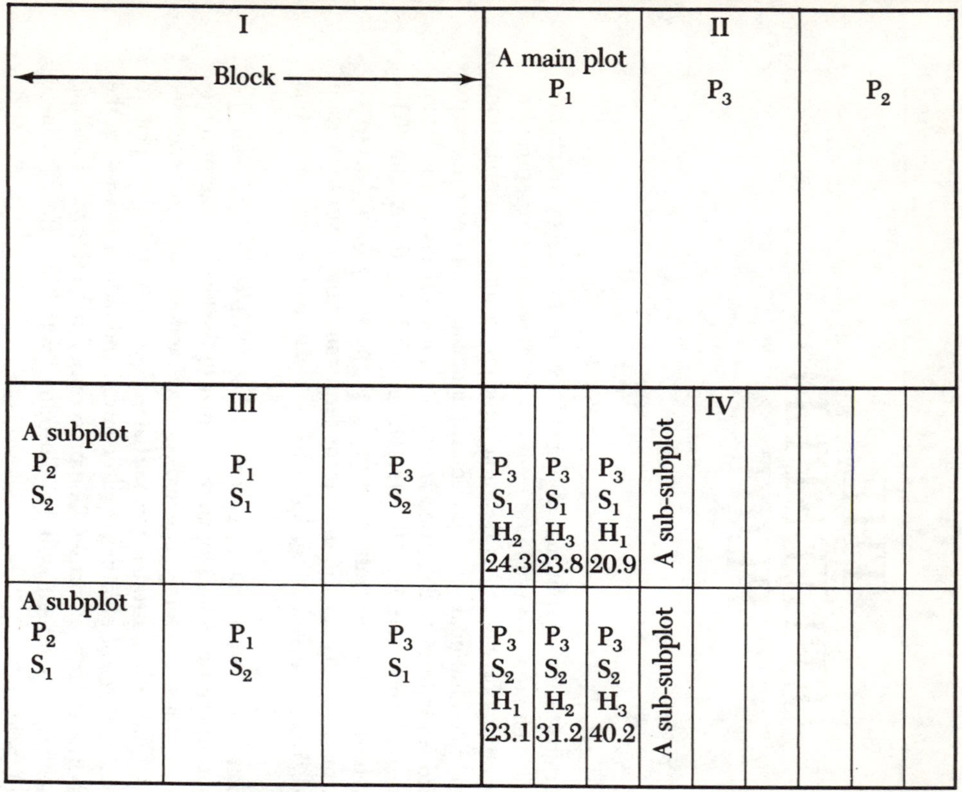

I

Block

A main plot P_1

II

P_3

P_2

III

IV

A subplot P_2 S_2

P_1 S_1

P_3 S_2

P_3 S_1 H_2 24.3

P_3 S_1 H_3 23.8

P_3 S_1 H_1 20.9

A sub-subplot

A subplot P_2 S_1

P_1 S_2

P_3 S_1

P_3 S_2 H_1 23.1

P_3 S_2 H_2 31.2

P_3 S_2 H_3 40.2

A sub-subplot

Figure 9.1. Features of a split-split plot for a sugar beet virus control experiment. Main plots are dates of planting (P_1, P_2, P_3) arranged in randomized complete blocks (I, II, III, IV). Subplots are not sprayed (S_1) and sprayed (S_2) for aphid control. Sub-subplots are dates of harvest at 4 week intervals (H_1, H_2, H_3). Sugarbeet root yields are shown for the sub-subplots of the P_3 main plot in block IV. Complete data from this experiment are organized in Table 9.1.

TABLE 9.1.
Sugar beet root yields (tons per acre), split-split plot, organized by treatment and block.

Treatments			Blocks (j)					
P(i)	S(k)	H(l)	I	II	III	IV	Totals	Means
1	1	1	25.7	25.4	23.8	22.0	96.9	24.2
		2	31.8	29.5	28.7	26.4	116.4	29.1
		3	34.6	37.2	29.1	23.7	124.6	31.2
		SP Totals $Y_{1j1.}$	92.1	92.1	81.6	72.1	337.9 = $Y_{1.1.}$	28.2
	2	1	27.7	30.3	30.2	33.2	121.4	30.4
		2	38.0	40.6	34.6	31.0	144.2	36.0
		3	42.1	43.6	44.6	42.7	173.0	43.2
		SP Totals $Y_{1j2.}$	107.8	114.5	109.4	106.9	438.6 = $Y_{1.2.}$	36.6
		MP Totals $Y_{1j..}$	199.9	206.6	191.0	179.0	776.5 = $Y_{1...}$	
2	1	1	28.9	24.7	27.8	23.4	104.8	26.2
		2	37.5	31.5	31.0	27.8	127.8	32.0
		3	38.4	32.5	31.2	29.8	131.9	33.0
		SP Totals $Y_{2j1.}$	104.8	88.7	90.0	81.0	364.5 = $Y_{2.1.}$	30.4
	2	1	38.0	31.0	29.5	30.7	129.2	32.3
		2	36.9	31.9	31.5	35.9	136.2	34.0
		3	44.2	41.6	38.9	37.6	162.3	40.6
		SP Totals $Y_{2j2.}$	119.1	104.5	99.9	104.2	427.7 = $Y_{2.2.}$	35.6
		MP Totals $Y_{2j..}$	223.9	193.2	189.9	185.2	792.2 = $Y_{2...}$	
3	1	1	23.4	24.2	21.2	20.9	89.7	22.4
		2	25.3	27.7	23.7	24.3	101.0	25.2
		3	29.8	29.9	24.3	23.8	107.8	27.0
		SP Totals $Y_{3j1.}$	78.5	81.8	69.2	69.0	298.5 = $Y_{3.1.}$	24.9

TABLE 9.1.
Continued.

Treatments			Blocks (j)					
P(i)	S(k)	H(l)	I	II	III	IV	Totals	Means
	2	1	20.8	23.0	25.2	23.1	92.1	23.0
		2	29.0	32.0	26.5	31.2	118.7	29.7
		3	36.6	37.8	34.8	40.2	149.4	37.4
	SP Totals $Y_{3j2.}$		86.4	92.8	86.5	94.5	$360.2 = Y_{3.2.}$	30.0
	MP Totals $Y_{3j..}$		164.9	174.6	155.7	163.5	$658.7 = Y_{3...}$	
	B Totals $Y_{.j..}$		588.7	574.4	536.6	527.7	$2227.4 = Y_{....}$	

$C = (2227.4)^2/72 = 68907.0939$, $\Sigma Y_{ijkl}^2 = 71747.70$

Symbols for treatment factors and levels: P = date of plant, p = 3 dates; S = sprays for aphid control, s = 2; H = harvest date, h = 3; B = blocks, b = 4.

TABLE 9.2.
Totals for two-way interactions and main effects.

Totals for Two-Way Interactions

	$P \times S (Y_{i.k.})$		$P \times H (Y_{i..l})$				$S \times H (Y_{..kl})$	
	S_1	S_2	H_1	H_2	H_3		S_1	S_2
P_1	337.9[a]	438.6	218.3[b]	260.6	297.6	H_1	291.4[c]	342.7
P_2	364.5	427.7	234.0	264.0	294.2	H_2	345.2	399.1
P_3	298.5	360.2	181.8	219.7	257.2	H_3	364.3	484.7

Totals for Main Effects

Plant Date ($Y_{i...}$)			Spray Treatment ($Y_{..k.}$)		Harvest Date ($Y_{...l}$)		
P_1	P_2	P_3	S_1	S_2	H_1	H_2	H_3
776.5	792.2	658.7	1000.9	1226.5	634.1	744.3	849.0

[a] From Table 9.1: total for P_1S_1 over all harvests and blocks.
[b] Total for P_1H_1 over all sprays and blocks = 96.9 + 121.4 = 218.3.
[c] Total for S_1H_1 over all plant dates and blocks = 96.9 + 104.8 + 89.7 = 291.4.

TABLE 9.3.
Analysis of variance, split-split plot.

Source of Variation	df	SS	MS[a]	Observed F	Required F 5%	Required F 1%
Sub-subplots	71	2840.6061				
Subplots	23	1542.8128				
Main plots	11	698.9028				
Blocks, B	3	143.4561	47.8187			
Plant dates, P	2	443.6886	221.8443←	11.91	5.14	10.92
Main plot error, BP	6	111.7581	18.6264			
Spray treatment, S	1	706.8800	706.8800←	81.21	5.12	10.56
P×S	2	40.6875	20.3438←	2.34	4.26	8.02
Subplot error, BS+B(P×S)	9	78.3425	8.7047			
Harvest dates, H	2	962.3353	481.1676←	102.80	3.26	5.25
P×H	4	13.1097	3.2774←	0.70	2.63	3.89
S×H	2	127.8308	63.9154←	13.66	3.26	5.25
P×S×H	4	44.0192	11.0048←	2.35	2.63	3.89
Sub-subplot error, BH+B(P×H)+B(S×H)+B(P×S×H)	36	168.4983	4.6805			

[a] Brackets indicate formation of F ratios.

Sources of Variation and Degrees of Freedom

Degrees of freedom for the sources of variation listed in Table 9.3 are:

Sub-subplots = pshb − 1 = 3(2)(3)(4) − 1 = 71
Subplots = psb − 1 = 23
Main plots = pb − 1 = 11
Blocks = b − 1 = 3
Plant dates = p − 1 = 2
Main plot error = (b − 1)(p − 1) = 6, or 11 − 3 − 2 = 6
Spray treatment = s − 1 = 1
P×S = (p − 1)(s − 1) = 2
Subplot error = (b − 1)(s − 1) + (b − 1)(p − 1)(s − 1) = 3 + 6 = 9,
or 23 − 11 − 2 − 1 = 9
Harvest date = h − 1 = 2
P×H = (p − 1)(h − 1) = 4

$S \times H = (s-1)(h-1) = 2$
$P \times S \times H = (p-1)(s-1)(h-1) = 4$
Sub-subplot error $= (b-1)(h-1) + (b-1)(p-1)(h-1) + (b-1)(s-1)(h-1)$
$+ (b-1)(p-1)(s-1)(h-1) = 6 + 12 + 6 + 12 = 36,$
or $71 - 23 - 2 - 4 - 2 - 4 = 36$

If the 18 treatments of this experiment had been laid out in randomized complete blocks, there would be a single error term with $df = (b-1)(t-1) = 3(17) = 51$, which is the sum of the degrees of freedom for the three error terms of Table 9.3 ($6 + 9 + 36 = 51$). Thus the splitting of plots partitions degrees of freedom and sums of squares for error into components having fewer degrees of freedom but usually with each successive term having a smaller mean square. Compare the mean squares for the three error terms in Table 9.3.

Correction Term

$$C = \frac{Y_{....}^2}{pshb} = \frac{2227.4^2}{3(2)(3)(4)} = 68907.0939.$$

Sums of Squares and Mean Squares

$$SSB = \frac{\Sigma Y_{.j..}^2}{psh} - C = \frac{588.7^2 + \ldots + 527.7^2}{3(2)(3)} - C = 143.4561$$

$$SSP = \frac{\Sigma Y_{i...}^2}{shb} - C = \frac{776.5^2 + \ldots + 658.7^2}{2(3)(4)} - C = 443.6886$$

$$SS(MP) = \frac{\Sigma Y_{ij..}^2}{sh} - C = \frac{199.9^2 + \ldots + 163.5^2}{2(3)} - C = 698.9028$$

$$SS(MPE) = SS(MP) - SSB - SSP = 111.7581$$

$$SSS = \frac{\Sigma Y_{..k.}^2}{phb} - C = \frac{1000.9^2 + 1226.5^2}{3(3)(4)} - C = 706.8800$$

$$SS(P \times S) = \frac{\Sigma Y_{i.k.}^2}{hb} - C - SSP - SSS = \frac{337.9^2 + \ldots + 360.2^2}{3(4)} - C - SSP - SSS$$
$$= 40.6875$$

$$SS(SP) = \frac{\Sigma Y_{ijk.}^2}{h} - C = \frac{92.1^2 + \ldots + 94.5^2}{3} - C = 1524.8128$$

$$SS(SPE) = SS(SP) - SS(MP) - SSS - SS(P \times S) = 78.3425$$

$$SSH = \frac{\Sigma Y_{...l}^2}{psb} - C = \frac{634.1^2 + \ldots + 849.0^2}{3(2)(4)} - C = 962.3353$$

$$SS(P\times H)=\frac{\Sigma Y_{i..l}^{2}}{sb}-C-SSP-SSH=\frac{218.3^2+\ldots+257.2^2}{2(4)}-C-SSP-SSH$$
$$=13.1097$$

$$SS(S\times H)=\frac{\Sigma Y_{..kl}^{2}}{pb}-C-SSS-SSH=\frac{291.4^2+\ldots+484.7^2}{3(4)}-C-SSS-SSH$$
$$=127.8308$$

$$SS(P\times S\times H)=\frac{\Sigma Y_{i.kl}^{2}}{b}-C-SSP-SSS-SSH-SS(P\times S)-SS(P\times H)-SS(S\times H)$$

$$=\frac{96.9^2+\ldots+149.4^2}{4}-C-SSP-SSS-SSH-SS(P\times S)$$
$$-SS(P\times H)-SS(S\times H)=44.0192$$

$$SS(SSP)=\Sigma Y_{ijkl}^{2}-C=25.7^2+\ldots+40.2^2-C=2840.6061$$

$$SS(SSPE)=SS(SSP)-SS(SP)-SSH-SS(P\times H)-SS(S\times H)-SS(P\times S\times H)$$
$$=168.4983$$

Mean squares are formed as usual by dividing SS's by appropriate degrees of freedom, for example, $MS(SSPE)=168.4983/36=4.6805$.

The Standard Deviation Key

With a calculator programmed to calculate $s=\sqrt{(Y_i-\bar{Y})^2/(r-1)}$, totals can be entered to compute mean squares and sum of squares. For example, for SSB, enter the block totals, 588.7, 574.4, 536.6, and 527.7; depress the standard deviation key, $s=29.3383$; square s to give $s^2=860.7367$; divide by the number of experimental units in each of the totals squared (18) to give $47.8187=MSB$; multiply by df B(3) to give $143.4561=SSB$.

F Values

The main plot error mean square is used to test the effects of plant date; subplot error MS to test the effects of spray treatment and the interaction $P\times S$; and sub-subplot error MS to test the remaining sources of variation—those associated with the sub-subplot treatments.

MEAN SEPARATION

The actual procedure used for mean separation will depend on the nature of the treatments, the questions the experimenter set out to answer, and the results of the initial analysis. For our example, the analysis tells us that the effects of the spray treatments and harvest dates were similar for all dates of planting (nonsignificant F values for P×S, P×H, and P×S×H) but that the plants that were sprayed for aphid control behaved quite differently with respect to harvest date than did plants that were not sprayed (highly significant F value for S×H).

Partitioning Interaction

Table 9.4 is set up to examine the S×H interaction in more detail. The means of Table 9.4 show increasing root yield as the harvest season progresses, with an indication of a more rapid rate of yield increase for the S_2 compared to the S_1 treatment. Since the harvest dates were at four-week intervals, we can use the coefficients of Table A.11 under $n=3$ to make it easy to partition the sum of squares for harvest date into a component to account for a linear increase with advancing harvest date and a residual component to show the portion of the sum of squares not accounted for by a linear trend. With the 2 df for harvest date partitioned, we can partition the 2 df for S×H into a linear and residual effect. Using the mean square for SSP error (Table 9.3), we calculate the F values of Table 9.4 and find a highly significant difference in the linear response of the S_1 compared to the S_2 treatment with respect to date of harvest. There is also a significant S×H residual component due to the small increase in root yield from

TABLE 9.4.
Coefficients for partitioning sums of squares due to spray treatment, harvest date, and the S×H interaction; the resulting mean squares; and F ratios.

		S×H Treatments						
		S_1H_1	S_1H_2	S_1H_3	S_2H_1	S_2H_2	S_2H_3	
	Totals	291.4	345.2	364.3	342.7	399.1	484.7	Mean	
Comparison	Means	24.3	28.8	30.4	28.6	33.3	40.4	Squares	F[a]
S		−1	−1	−1	1	1	1	706.8800	
H Linear		−1	0	1	−1	0	1	962.1252	205.6
H Residual		1	−2	1	1	−2	1	0.2101	<1
S×HL		1	0	−1	−1	0	1	99.4752	21.25
S×HR		−1	2	−1	1	−2	1	28.3556	6.06

[a] F values are calculated by dividing the mean squares by the mean square for the SSP error of Table 9.3. The tabular F required for statistical significance is for 1 and 36 df and for 5%=4.11 and for 1%=7.39.

harvest two to harvest three for S_1 (28.8 to 30.4 tons/acre) compared to the much larger increase for S_2 (33.3 to 40.4 tons/acre). A biological interpretation that makes sense is that the sugar beets not sprayed for virus suppression show a progressively lower rate of growth as the harvest season advances, while the plants with less virus show a more or less constant rate of growth over the time interval of the three harvests. This interpretation can be illustrated as in Figure 9.2 by showing the increase in root yield over the harvest periods as linear for the S_2 treatment and quadratic for the S_1 treatment. It would not be appropriate to extrapolate beyond the harvest dates involved, as both trend lines would level off as winter approached and not continue upward as indicated for S_2 or decline as the quadratic equation predicts for S_1. Within the limits of the harvest dates, however, both equations graphically illustrate the effect of the more severe level of virus infestation on the root yield of sugar beets and provide an objective procedure for estimating yield for the two treatments over the fall harvest period.

The calculation of the regression equations of Figure 9.2 is left as an exercise after you have learned the shortcut regression methods of Chapter 15. The procedure for computing the single degree of freedom mean squares of Table 9.4 are given below.

$$\text{SS Sprays} = \frac{(-291.4-345.2-364.3+342.7+399.1+484.7)^2}{4(3)6} = 706.8800$$

Note the use of the formula for calculating a single degree of freedom sum of squares: $SS = (\Sigma c_i Y_i)^2/(r\Sigma c_i^2)$. The c_i's are class comparison coefficients of Table 9.4, and r is the number of variates in each term of the numerator. Here $r = bp = 4(3)$.

$$\text{SS(H Linear)} = \frac{(-291.4+364.3-342.7+484.7)^2}{4(3)4} = 962.1252$$

$$\text{SS(H Residual)} = \frac{[291.4-2(345.2)+364.3+342.7-2(399.1)+484.7]^2}{4(3)12}$$
$$= 0.2101$$

$$\text{SS(S} \times \text{HL)} = \frac{(291.4-364.3-342.7+484.7)^2}{4(3)4} = 99.4752$$

$$\text{SS(S} \times \text{HR)} = \frac{[-291.4+2(345.2)-364.3+342.7-2(399.1)+484.7]^2}{4(3)12}$$
$$= 28.3556$$

As checks on arithmetic, note that SS(H Linear) + SS(H Residual) = SSH of Table 9.3 and that SS(S × HL) + SS(S × HR) = SS(S × H) of Table 9.3.

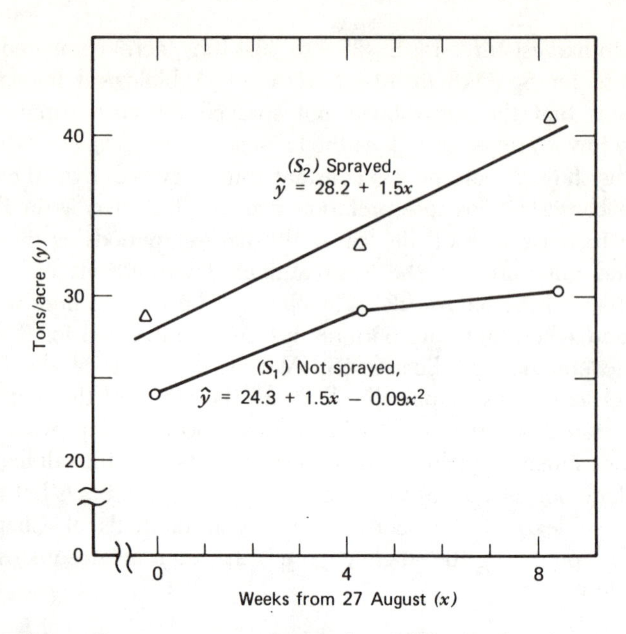

Figure 9.2. Effect of vector control on the fall growth of sugar beets. The difference in the two lines shows the nature of the S×H interaction. The equations can be used to estimate root yields produced by the two treatments over the fall harvest period.

Standard Errors and LSD's

For some experiments involving split-split plots it may be desirable to separate certain means by LSD or multiple-range tests, and thus it is necessary to know the appropriate standard errors for use in these tests. Standard errors for testing main effects of the factors applied to main plots and subplots and for their interactions are the same as given in Table 8.8 except that c(number of factor C treatments) is a multiplier in each denominator. Standard errors for separating means of the factor applied to the sub-subplots and for interactions with the other two factors are given in Table 9.5 along with t values that must be calculated for certain comparisons.

For a complete summary of the results of an experiment, it is usually good practice to give the means of the highest order of treatment factor combinations and the means of the factor combinations that appear particularly relevant to the conclusions to be made along with some procedure for approximate mean separation. For our example, Table 9.6 gives the means of plant date × spray treatment × date of harvest, the plant date means, and the means of the highly significant interaction, S×H. Footnotes to the table give LSD's for approximate mean separation. The computation of these LSD's, all at the 5% level, using the standard errors of Tables 8.8 and 9.5, are shown below.

TABLE 9.5.
Standard errors and t computations for the separation of means involving C treatments.

Means Compared	Standard Error ($s_{\bar{y}}$)	t Values[a]
C means	$\sqrt{\frac{Ec}{rab}}$	t_c
C means for same A	$\sqrt{\frac{Ec}{rb}}$	t_c
C means for same B	$\sqrt{\frac{Ec}{ra}}$	t_c
B means for same or different C	$\sqrt{\frac{(c-1)Ec+Eb}{rac}}$	$t_{bc}=\frac{(c-1)Ect_c+Ebt_b}{(c-1)Ec+Eb}$
A means for same or different C	$\sqrt{\frac{(c-1)Ec+Ea}{rbc}}$	$t_{ac}=\frac{(c-1)Ect_c+Eat_a}{(c-1)Ec+Ea}$
C means for same A and B	$\sqrt{\frac{Ec}{r}}$	t_c
B means for same A and same or different C	$\sqrt{\frac{(c-1)Ec+Eb}{rc}}$	$t_{bc}=\frac{(c-1)Ect_c+Ebt_b}{(c-1)Ec+Eb}$
A means for same or different B and C	$\sqrt{\frac{b(c-1)Ec+(b-1)Eb+Ea}{rbc}}$	$t_{abc}=\frac{b(c-1)t_c+(b-1)Ebt_b+Eat_a}{b(c-1)+(b-1)Eb+Ea}$

[a] t_a, t_b, t_c indicate tabular t values from Table A.2 for degrees of freedom for Ea, Eb, and Ec, respectively.
Key: A, B, and C are treatments applied to mainplots, subplots, and sub-subplots at levels a, b, and c, respectively; r is the number of replications. Ea, Eb, and Ec are main, subplot, and sub-subplot error mean squares, respectively. To compute LSD and D, note that $LSD=t\sqrt{2}\,s_{\bar{y}}$ and $D=R(LSD)$.

LSD, PLANT DATE MEANS. $LSD=t_a\sqrt{(2Ea)/rbc}$. Note that c has been included in the denominator of this formula from Table 8.8 to keep the standard error on a sub-subplot basis.

$$LSD=2.447\sqrt{[2(18.6264)]/4(2)3}=2.447(1.246)=3.0 \text{ tons/acre.}$$

LSD, H MEANS FOR SAME P AND S TREATMENTS. For example, $P_1S_1H_1-P_1S_1H_2$. $LSD = t_c\sqrt{(2Ec)/r} = 2.028\sqrt{[2(4.6805)]/4} = 2.028(1.530) = 3.1$ tons /acre. Note that t_c is based on 36 df and is determined by linear interpolation between tabular t's from Table A.2 for 35 and 40 df.

TABLE 9.6.
Effect of plant date, spray treatment and date of harvest on sugar beet root production.

Plant Date	Spray Treatment	Harvest Date 8/27	9/24	10/22	Plant Date Means[a]
			(Roots, tons/acre) $P \times S \times H$ means[b]		
3/2	No	24.2	29.1	31.2	32.3
	Yes	30.4	36.0	43.2	
4/2	No	26.2	32.0	33.0	33.0
	Yes	32.3	34.0	40.6	
5/2	No	22.4	25.2	27.0	27.4
	Yes	23.0	29.7	37.4	
		Spray treatment × harvest date means[c]			
Not sprayed		24.3	28.8	30.4	
Sprayed		28.6	33.3	40.4	

[a] LSD, 5%: 3.0
[b] LSD, 5% between harvest dates for same plant date and spray treatment: 3.1; between spray treatments for the same plant date and same or different harvest date: 3.7; between plant date means for the same or different spray treatment or harvest date: 4.4. The $P \times S \times H$ interaction is not significant at the 5% level.
[c] LSD, 5% between H dates for the same spray treatment: 1.8; between spray treatments for the same or different H date: 2.1. The $S \times H$ interaction is significant at the 0.1% level.

LSD, SPRAY TREATMENT MEANS FOR THE SAME P AND THE SAME OR DIFFERENT H. For example, $P_1S_1H_1 - P_1S_2H_1$ or $P_1S_1H_1 - P_1S_2H_2$.

$$LSD = t_{bc}\sqrt{\frac{2[(c-1)Ec + Eb]}{rc}} \qquad t_{bc} = \frac{(c-1)Ect_c + Ebt_c}{(c-1)Ec + Eb}$$

$$t_{bc} = \frac{(3-1)(4.6805)2.028 + 8.7047(2.262)}{(3-1)4.6805 + 8.7047} = 2.141$$

$$LSD = 2.141\sqrt{\frac{2[(3-1)4.6805+8.7047]}{4(3)}} = 2.141(1.735) = 3.7 \text{ tons/acre}$$

LSD, PLANT DATE MEANS FOR SAME OR DIFFERENT S AND H. For example, $P_1S_1H_1 - P_2S_1H_1$ or $P_1S_2H_1 - P_2S_1H_1$.

$$LSD = t_{abc} = \sqrt{\frac{2[b(c-1)Ec+(b-1)Eb+Ea]}{rbc}} \qquad t_{abc} = 2.242$$

(see Table 9.5 for formula)

$$LSD = 2.242\sqrt{\frac{2[2(3-1)4.6805+(2-1)7.7047+18.6264]}{4(2)3}}$$

$$= 2.242(1.959) = 4.4 \text{ tons/acre}$$

LSD, H DATE MEANS FOR SAME S. For example, $S_1H_1 - S_1H_2$.

$$LSD = t_c\sqrt{\frac{2Ec}{ra}} = 2.028\sqrt{\frac{2(4.6805)}{4(3)}} = 2.028(0.883) = 1.8 \text{ tons/acre}$$

LSD, S MEANS FOR SAME OR DIFFERENT H. For example, $S_1H_1 - S_2H_1$ or $S_1H_1 - S_2H_2$.

$$LSD = t_{bc}\sqrt{\frac{2[(c-1)Ec+Eb]}{rac}}$$

$$= 2.141\sqrt{\frac{2[(3-1)4.6805+8.7047]}{4(3)3}} = 2.141(1.002)$$

$$= 2.1 \text{ tons/acre}$$

SUMMARY

The split-split plot is an extension of the split-plot principle with subplots being split into sub-subplots to which a third treatment factor is assigned. The analysis of variance is more complicated in that there are three error terms for testing treatment effects. Usually, the factor assigned to sub-subplots and the interactions involving this factor are more precisely evaluated than are the other treatment components. Mean separation is complicated by the three error terms.

10

THE SPLIT BLOCK

In this variation of the split-plot design, the subunit treatments are applied in strips across an entire replication of main plot treatments. If the main plots are in a latin square, the subunit treatments can be in strips across an entire row or column of main plots. This arrangement often facilitates physical operations concerning the subunits but sacrifices precision in comparing the main effects of factor B. It often improves precision in comparing the AB interaction, especially in comparing B means for a given A treatment. When this is the primary effect in which you are interested, the design is quite useful. Before employing it, however, it is wise to consult with someone experienced in its use.

Figure 10.1 illustrates a single replicate of split-plots compared to a split-block. In the latter, note that the subunit treatments are continuous across the entire block of main plots, and thus each subunit treatment splits the block. Another term applicable to this layout is strip-plot, as both A and B treatments are in strips. The A and B treatments are independently randomized in each replication.

Table 10.1 shows the partitioning of degrees of freedom for the two layouts of Figure 10.1, assuming four replications for each layout. Note that the split-block arrangement necessitates the division of the split-plot error b into two error terms and provides fewer degrees of freedom for testing B treatment main effects. But

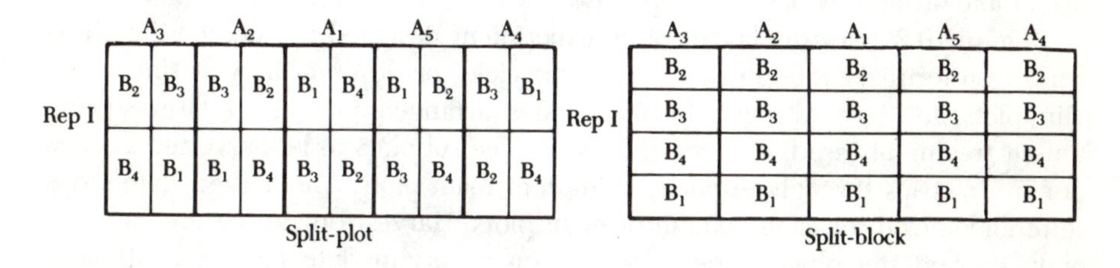

Figure 10.1. A single replicate of split plots compared to a split block. The experiment involves five treatments of factor A and four treatments of factor B. In a split-plot layout, B treatments are independently randomized within each A treatment plot, while in a split block, the B treatments are in strips across the entire block of A plots.

TABLE 10.1.
Degrees of freedom for the split-plot design and the split-block variation of Figure 10.1.

Source of Variation		Degrees of Freedom: Split-Plot		Degrees of Freedom: Split-Block	
Subplots	rab − 1	79		79	
Main plots	ra − 1	19		19	
Blocks	r − 1	3		3	
A	a − 1	4		4	
MP error	(r − 1)(a − 1)	12	Error a	12	Error a
B	b − 1	3		3	
AXB	(a − 1)(b − 1)	12		12	
Strip-plot error	(r − 1)(b − 1)			9	Error b
Subplot error	(r − 1)(a − 1)(b − 1)	45	Error b*	36	Error c

r = 4 replications, a = 5 factor A treatments, b = 4 factor B treatments. Brackets and arrows indicate the use of appropriate error terms for F tests.
*Combines df for error b and c of the split-block.

since variability associated with the strips across the main plots is now removed from the split-plot error b to give error c of the split-block layout, the latter is smaller and often provides a more precise F test for testing for interactions.

Figure 10.2 gives the layout of an experiment designed to examine the effect of nitrogen fertilizer rate on sugar beet root yield for different harvest times. The main plots are four nitrogen fertilizer rates arranged in a 4×4 latin square. Subunit treatments are five dates of harvest. The subplots to be harvested at each date are in strips through an entire column of main plots. The harvest date strips are rerandomized for each column of main plots. Harvest operations are easier to conduct when the plots to be harvested on a certain date form a continuous column. This arrangement, however, necessitates the calculation of a separate error term to test for the main effect of harvest dates. The root yield for each subplot is given in Figure 10.2 along with totals for main plots, rows, columns, and harvest date strip-plots. These data, along with the treatment totals of Table 10.2, are required to compute the sums of squares for the ANOVA of Table 10.3. The procedure for these calculations follows the table.

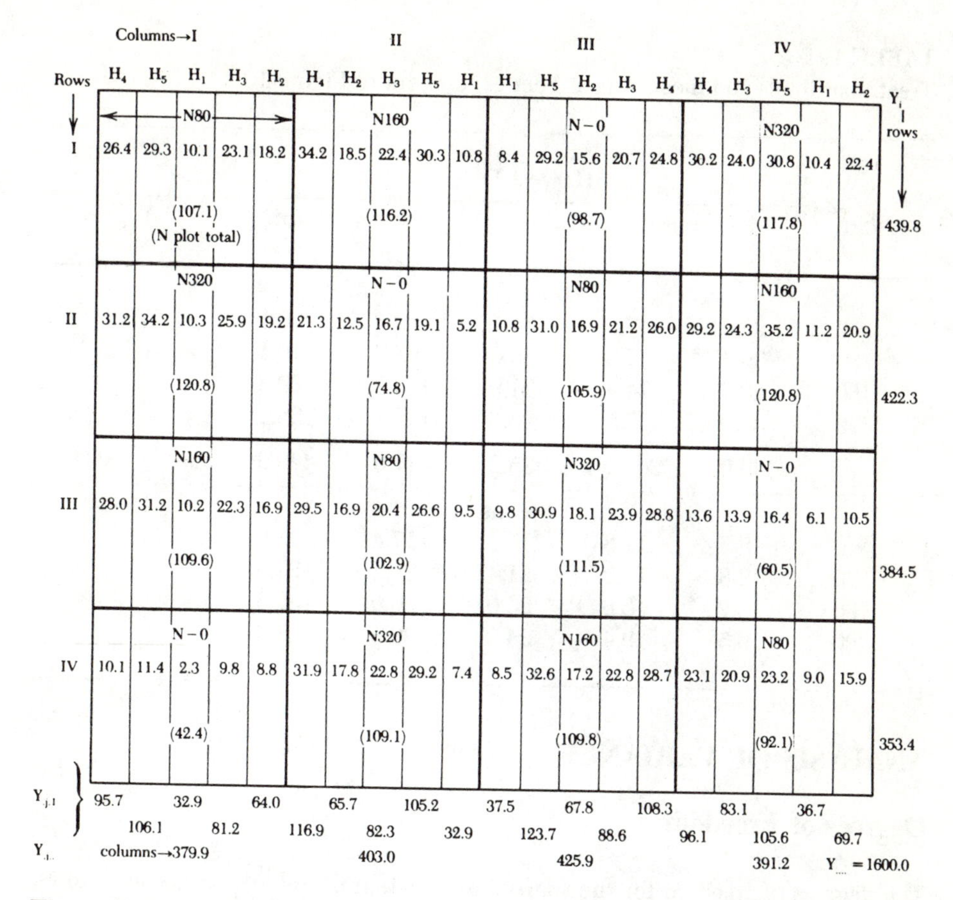

Figure 10.2. Layout of a sugar beet experiment, plot yields (tons of roots per acre) and totals. Main plot treatments are pounds of fertilizer N per acre arranged in a 4×4 latin square. Subplot treatments are five dates of harvest at three-week intervals. Note that the same harvest date continues through all N plots in a column; thus each column of main plots becomes a "split-block." The number of weeks from planting to harvest for H_1 through H_5 are, respectively, 20, 23, 26, 29, and 32. Note that any subplot can be identified as Y_{ijkl}, where i = row (r = 4), j = column (c = 4), k = nitrogen rate (n = 4), and l = harvest (h = 5).

TABLE 10.2.
Treatment totals and means, sugar beet experiment of Figure 10.2.

N rate	Harvest Date 1	2	3	4	5	$Y_{..k.}$
			Totals ($Y_{..kl}$)			
0	22.0	47.4	61.1	69.8	76.1	276.4
80	39.4	67.9	85.6	105.0	110.1	408.0
160	40.7	74.4	91.9	120.1	129.3	456.4
320	37.9	77.5	96.6	122.1	125.1	459.2
$Y_{...l}$	140.0	267.2	335.2	417.0	440.6	$Y_{....}$ = 1600.0
			Means			
0	5.5	11.8	15.3	17.4	19.0	
80	9.8	17.0	21.4	26.2	27.5	
160	10.2	18.6	23.0	30.0	32.3	
320	9.5	19.4	24.2	30.5	31.3	

ANALYSIS OF VARIANCE

Degrees of Freedom

The degrees of freedom for the sources of variation of Table 10.3 are as follows. Note rows, $r=4$; columns, $c=4$; nitrogen rate, $n=4$; harvest dates, $h=5$.

Subplots	$rch-1=4(4)(5)-1=79$	Error a	$(r-1)(c-1)-(n-1)=3(3)-3=6$
Main plots	$rc-1=4(4)-1=15$	H dates	$(h-1)=(5-1)=4$
Rows	$r-1=4-1=3$	Error b	$(c-1)(h-1)=3(4)=12$
Columns	$c-1=4-1=3$	NXH	$(n-1)(h-1)=3(4)=12$
N rates	$(n-1)=4-1=3$	Error c	$(c-1)(n-1)(h-1)=3(3)(4)=36$

Correction Term

$$C=\frac{Y_{....}^{2}}{rnh}$$

where r is the number of replications, n is the number of N levels, and h is the number of harvest dates.

$$C=\frac{1600^2}{4(4)(5)}=32000.00$$

TABLE 10.3.
Analysis of variance. Split-block design

Source of Variation	df	SS	MS	Observed F	Required F 5%	Required F 1%
Subplots	79	5542.680				
Main plots	15	1503.720				
Rows	3	224.657	74.886			
Columns	3	58.063	19.354			
N levels	3	1101.328	367.109	18.41	4.76	9.78
Error a, RC-N	6	119.672	19.945			
H dates	4	3710.765	927.691	111.92	3.26	5.41
Error b, CH	12	99.467	8.289			
N×H	12	157.147	13.096	6.59	2.03	2.72
Error c, C(NXH)	36	71.581	1.988			

Sums of Squares

$$SSR=\frac{Y_{i...}^{2}}{nh}-C$$

$$=\frac{439.8^2+\ldots+353.4^2}{4(5)}-C=32224.657-C=224.657$$

$$SSC=\frac{Y_{.j..}^{2}}{nh}-C$$

$$=\frac{379.9^2+\ldots+391.2^2}{4(5)}-C=32058.063-C=58.063$$

$$SSN=\frac{\Sigma Y_{..k.}^{2}}{rh}-C$$

$$=\frac{276.4^2+\ldots+459.2^2}{4(5)}-C=33101.328-C=1101.328$$

$$SS(\text{main plots})=\frac{\Sigma Y_{ijk.}^{2}}{h}-C$$

$$=\frac{107.1^2+\ldots+92.1^2}{5}-C=33503.720-C=1503.720$$

$$SS(Ea) = SS(\text{main plots}) - SSR - SSC - SSN$$
$$= 1503.720 - 224.657 - 58.063 - 1101.328 = 119.672$$

$$SSH = \frac{\Sigma Y_{...l}^{2}}{rn} - C$$

$$= \frac{140.0^2 + \ldots + 440.6^2}{4(4)} - C = 35710.765 - C = 3610.765$$

$$SS(Eb) = \frac{\Sigma Y_{.j.l}^{2}}{r} - C - SSC - SSH$$

$$= \frac{95.7^2 + \ldots + 69.7^2}{4} - C - SSC - SSH$$

$$= 35868.295 - C - SSC - SSH = 99.467$$

$$SS(N \times H) = \frac{\Sigma Y_{..kl}^{2}}{r} - C - SSN - SSH$$

$$= \frac{22.0^2 + \ldots + 125.1^2}{4} - C - SSN - SSH$$

$$= 36969.240 - C - SSN - SSH = 157.147$$

$$SS(\text{subplots}) = \Sigma Y_{ijkl}^{2} - C$$

$$= 26.4^2 + 29.3^2 + \ldots + 15.9^2 - C = 37542.68 - C$$
$$= 5542.680$$

$$SS(Ec) = SS(\text{subplots}) - SS(\text{M plots}) - SSH - SS(Eb) - SS(N \times H)$$
$$= 5542.680 - 1503.720 - 3710.765 - 99.467 - 157.147$$
$$= 71.581$$

Mean Squares

Mean squares are obtained by dividing sums of squares by the degrees of freedom associated with each. For example:

$$MS(Ec) = \frac{SS(Ec)}{df(Ec)} = \frac{71.581}{36} = 1.988$$

The Standard Deviation Key

The use of a correction term can be avoided with a calculator programmed to compute $s=\sqrt{\Sigma(Y_i-\bar{Y})^2/(r-1)}$. To compute any mean square, enter the appropriate totals, obtain s^2, and divide by the number of variates in each total entered. For example, to compute SSR, enter 439.8, 422.3, 384.5, 353.4; depress the key to give s=38.7003; square 38.7003 to give 1497.7133; divide by 20 to give 74.886=MSR; multiply by 3 (i.e., df R) to obtain 224.657=SSR.

F Values and Mean Separation

F values are determined by dividing mean squares by appropriate error terms; Ea for nitrogen, Eb for harvest date, and Ec for the N×H interaction.

The highly significant F value for N×H indicates a different response to N depending on harvest date. An understanding of this interaction is crucial to the interpretation of the results. By partitioning N rates and H dates into polynomial trend comparisions, we can also partition the sum of squares for N×H to learn the nature of the interaction. Polynomial coefficients for partitioning unequally spaced nitrogen rates are not easy to come by, which is one of the reasons for stressing equally spaced treatment rates. Some of these coefficients are given in Table A.11a, in which coefficients for our four N rates are under the series 0, 1, 2, 4. Coefficients for partitioning the five equally spaced harvest dates are in Table A.11 under n=5. The coefficients for N rates and harvest dates are assigned to the treatments of our experiment in Table 10.4. Interaction coefficients are obtained by multiplication. For example, the coefficients for NL×HL are: $-7(-2)=14$, $-7(-1)=7$, $-7(0)=0$, and so forth. Note that coefficients for all the comparisons meet the two rules for orthogonality given in Chapter 6 and thus the sums of squares for the single degree of freedom comparisons will add to each multiple degree of freedom sum of squares partitioned, thus providing a test of computational accuracy.

The single degree of freedom sums of squares are computed using the rule $SS=\Sigma(c_iY_i)^2/(r\Sigma c_i^2)$, where c_i are the comparison coefficients, Y_i are treatment totals, and r is the number of variates in each total. For example, the sum of squares associated with the linear response to nitrogen is

$$SS(NL)=\frac{[-7(22.0)-7(47.4)+\ldots+9(125.1)]^2}{4(700)}=730.7301$$

The other single df sums of squares are computed similarly and entered in Table 10.4.

F tests are made using the appropriate error mean squares from Table 10.3: error a for the N comparisons, error b for the H comparisons, and error c for the

TABLE 10.4.
Orthogonal coefficients for partitioning the treatment sum of squares into single degree of freedom trend comparisons, the resulting sums of squares, and their statistical significance.

	N Rates, Harvest Dates, and Treatment Totals																						
	N_0					N_{80}					N_{160}					N_{320}							Statistical Significance
	1	2	3	4	5	1	2	3	4	5	1	2	3	4	5	1	2	3	4	5		Sum of	of 1 df
Comparisons	22.0	47.4	61.1	69.8	76.1	39.4	67.9	85.6	105.0	110.1	40.7	74.4	91.9	120.1	129.3	37.9	77.5	96.6	122.1	125.1	df	Squares	Comparisons[a]
N rates																					3	1101.328	
Linear	−7	−7	−7	−7	−7	−3	−3	−3	−3	−3	1	1	1	1	1	9	9	9	9	9	1	730.730	**
Quadratic	7	7	7	7	7	−4	−4	−4	−4	−4	−8	−8	−8	−8	−8	5	5	5	5	5	1	359.593	**
Residual	−3	−3	−3	−3	−3	8	8	8	8	8	−6	−6	−6	−6	−6	1	1	1	1	1	1	11.005	ns
H dates																					4	3710.765	
Linear	−2	−1	0	1	2	−2	−1	0	1	2	−2	−1	0	1	2	−2	−1	0	1	2	1	3525.006	**
Quadratic	2	−1	−2	−1	2	2	−1	−2	−1	2	2	−1	−2	−1	2	2	−1	−2	−1	2	1	166.980	**
Cubic	−1	2	0	−2	1	−1	2	0	−2	1	−1	2	0	−2	1	−1	2	0	−2	1	1	0.006	ns
Residual	1	−4	6	−4	1	1	−4	6	−4	1	1	−4	6	−4	1	1	−4	6	−4	1	1	18.772	ns
N×H																					12	157.147	
NL×HL	14	7	0	−7	−14	6	3	0	−3	−6	−2	−1	0	1	2	−18	−9	0	9	18	1	98.899	**
NL×HQ	−14	7	14	7	−14	−6	3	6	3	−6	2	−1	−2	−1	2	18	−9	−18	−9	18	1	5.194	ns
NL×HC	7	−14	0	14	−7	3	−6	0	6	−3	−1	2	0	−2	1	−9	18	0	−18	9	1	1.015	ns
NL×HR	−7	28	−42	28	−7	−3	12	−18	12	−3	1	−4	6	−4	1	9	−36	54	−36	9	1	5.043	ns
NQ×HL	−14	−7	0	7	14	8	4	0	−4	−8	16	8	0	−8	−16	−10	−5	0	5	10	1	38.660	**
NQ×HQ	14	−7	−14	−7	14	−8	4	8	4	−8	−16	8	16	8	−16	10	−5	−10	−5	10	1	2.595	ns
NQ×HC	−7	14	0	−14	7	4	−8	0	8	−4	8	−16	0	16	−8	−5	10	0	−10	5	1	1.359	ns
NQ×HR	7	−28	42	−28	7	−4	16	−24	16	−4	−8	32	−48	32	−8	5	−20	30	−20	5	1	1.557	ns
NR×HL	6	3	0	−3	6	−16	−8	0	8	16	12	6	0	−6	−12	−2	−1	0	1	2	1	1.536	ns
NR×HQ	−6	3	6	3	−6	16	−8	−16	−8	16	−12	6	12	6	−12	2	−1	−2	−1	2	1	0.755	ns
NR×HC	3	−6	0	6	−3	−8	16	0	−16	8	6	−12	0	−12	6	−1	2	0	−2	1	1	0.384	ns
NR×HR	−3	12	−18	12	−3	8	−32	48	−32	8	−6	24	−36	24	−6	1	−4	6	−4	1	1	0.151	ns

[a]To determine F ratios, use appropriate error MS's from Table 10.3: Ea for N rates, Eb for H dates, and Ec for N×H comparisons.

$N \times H$ comparisons. For example, F for the interaction comparison $NL \times HL = 98.899/1.988 = 49.75$, which far exceeds the tabular value from Table A.3 of 7.39 for significance at the 1% level (df=1 and 36). Note that $NL \times HL$ and $NQ \times HL$ are the only two components of the $N \times H$ interaction that are statistically significant.

To proceed further and show an appropriate way to present the results of this experiment requires some understanding of polynomial and multiple regression, and so we will delay this until Chapter 16. (If you wish to see how it comes out, look at page 262.) For now, note that the total of the six significant single degree of freedom sums of squares account for 99% of the sum of squares due to all 19 treatment components, that is, $(730.730 + 359.593 + 3525.006 + 166.980 + 98.899 + 38.659)/(1101.328 + 3710.765 + 157.147) = 0.990$.

STANDARD ERRORS

Table 10.5 gives standard errors to use for mean separation by LSD and multiple-range tests.

TABLE 10.5.
Standard Errors for a Split-Block.

Means Compared	Standard Error ($s_{\bar{y}}$)	t Values
A means	$\sqrt{\frac{Ea}{rb}}$	t_a
B means	$\sqrt{\frac{Eb}{ra}}$	t_b
A means for the same or different B	$\sqrt{\frac{(b-1)Ec+Ea}{rb}}$	$t_{ac} = \frac{(b-1)Ec(t_c)+Ea(t_a)}{(b-1)Ec+Ea}$
B means for the same A	$\sqrt{\frac{(a-1)Ec+Eb}{ra}}$	$t_{bc} = \frac{(a-1)Ec(t_c)+Eb(t_b)}{(a-1)Ec+Eb}$

A = treatments applied to the main plots, B = treatments applied to the subplots; a, b, and r are the number of main plot treatments, subplot treatments, and replications, respectively; Ea, Eb, and Ec are error mean squares; t_a, t_b, t_c are tabular t values for df for Ea, Eb, and Ec, respectively. To compute LSD and D, note that $LSD = t\sqrt{2}\ s_{\bar{y}}$, and to compute Duncan's multiple range, $D = R(LSD)$.

To illustrate the use of Table 10.5, we will compute interaction LSD's tc compare the means of Table 10.2. Neither LSD's nor multiple-range tests would, however, be appropriate for separating the means of this experiment, as a great deal more can be learned by the more powerful method of Chapter 16.

LSD (5%) between N means for the same or different H.

$$LSD = t_{ac}\sqrt{\frac{2[(b-1)Ec+Ea]}{rb}}$$

$$t_{ac} = \frac{(b-1)Ec(t_c)+Ea(t_a)}{(b-1)Ec+Ea} = \frac{(5-1)1.988(2.028)+19.945(2.447)}{(5-1)1.988+19.945}$$

$$= \frac{64.9321}{27.897} = 2.328$$

$$LSD = 2.328\sqrt{\frac{2[(5-1)1.988+19.945]}{4(5)}} = 2.328(1.670) = 3.4 \text{ tons/acre}$$

LSD (5%) between H means for the same N.

$$LSD_{0.05} = t_{bc}\sqrt{\frac{2[(a-1)Ec+Eb]}{ra}}$$

$$t_{bc} = \frac{(a-1)Ec(t_c)+Eb(tb)}{(a-1)Ec+Eb} = \frac{(4-1)1.988(2.028)+8.289(2.179)}{(4-1)1.988+8.289}$$

$$= \frac{30.1567}{14.253} = 2.116$$

$$LSD = 2.116\sqrt{\frac{2[(4-1)1.988+8.289]}{4(4)}} = 2.116(1.335) = 2.8 \text{ tons/acre}$$

SUMMARY

In the split-block arrangement:

A block of plots receiving the treatments of factor A is split, so that each treatment of factor B occurs in a continuous strip across the block. An independent randomization of the treatments of factor B is made for each block of plots of factor A.

The advantages of the layout are the facilitation of physical operations and the possibility of greater precision in estimation of the A×B interaction.

The disadvantages are a loss in precision in determining the effects of factor B, more complex computations, and complications in mean separation.

11

SUBPLOTS AS REPEATED OBSERVATIONS

The split-plot principle can be applied to experiments where successive observations are made on the same whole units over a period of time. For example, a fertilizer trial or variety trial with a perennial crop might be harvested several times during a year and/or for two or more years. The plots to which the treatments are assigned can be called main plots, and the several harvests can be called subplots. A subplot in this case, however, differs from the usual subplot in that it consists of data taken from the entire main plot rather than from a designated portion as is the case with the usual split-plot.

There are no unusual problems in analyzing data on a main plot basis for a single observation date or for the totals over several dates of observation. But F values arising from testing the effects of successive observations and the interaction of main plot treatments with successive observations may not be distributed as F, and too many significant effects may result.

A stepwise procedure and suggestions for handling data from such experiments are given in the following example. The data are dry matter forage yields from an alfalfa variety trial. There are four varieties randomized in five complete blocks. To simplify matters, we will consider data from only four harvests, two early and two late, and only for two years.

ANALYSIS FOR EACH SET OF OBSERVATIONS

An ANOVA should be carried out for each harvest. The organization of data as in Table 11.1 generates the necessary totals to complete an ANOVA for each harvest as well as for an annual analysis.

The ANOVA for each harvest is given in Table 11.2. The degrees of freedom and sums of squares for "varieties" are partitioned as shown because varieties 1 and 2 are closely related, variety 2 being a selection from variety 1. The procedure for completing an ANOVA for a single harvest date is given below Table 11.2 for harvest 1.

TABLE 11.1.

First-year data from an alfalfa variety trial laid out as a randomized complete block with four varieties ($v=4$), five blocks ($b=5$), and four harvests ($h=4$). Data are tons per acre of dry alfalfa.

Variety (i)	Harvest (k)	Blocks (j) 1	2	3	4	5	$Y_{i.k}$	$\bar{Y}_{i.k}$
1	1	2.69	2.40	3.23	2.87	3.27	14.46	2.89[a]
2	1	2.87	3.05	3.09	2.90	2.98	14.89	2.98
3	1	3.12	3.27	3.41	3.48	3.19	16.47	3.29
4	1	3.23	3.23	3.16	3.01	3.05	15.68	3.14
	$Y_{.j1}$	11.91	11.95	12.89	12.26	12.49	$61.50 = Y_{..1}$	
1	2	2.74	1.91	3.47	2.87	3.43	14.42	2.88
2	2	2.50	2.90	3.23	2.98	3.05	14.66	2.93
3	2	2.92	2.63	3.67	2.90	3.25	15.37	3.07
4	2	3.50	2.89	3.39	2.90	3.16	15.84	3.17
	$Y_{.j2}$	11.66	10.33	13.76	11.65	12.89	$60.29 = Y_{..2}$	
1	3	1.67	1.22	2.29	2.18	2.30	9.66	1.93
2	3	1.47	1.85	2.03	1.82	1.51	8.68	1.74
3	3	1.67	1.42	2.81	1.51	1.76	9.17	1.83
4	3	2.60	1.92	2.36	1.92	2.14	10.94	2.19
	$Y_{.j3}$	7.41	6.41	9.49	7.43	7.71	$38.45 = Y_{..3}$	
1	4	1.92	1.45	1.63	1.60	1.96	8.56	1.71
2	4	2.00	2.03	1.71	1.60	1.96	9.30	1.86
3	4	2.03	1.96	1.85	1.82	2.40	10.06	2.01
4	4	2.07	1.89	1.92	1.82	1.78	9.48	1.90
	$Y_{.j4}$	8.02	7.33	7.11	6.84	8.10	$37.40 = Y_{..4}$	
		Variety × block totals (main plots, $Y_{ij.}$)					Y_i	$\bar{Y}_{i..}$
1		9.02	6.98	10.62	9.52	10.96	47.10	9.42[b]
2		8.84	9.83	10.06	9.30	9.50	47.53	9.51
3		9.74	9.28	11.74	9.71	10.60	51.07	10.21
4		11.40	9.93	10.83	9.65	10.13	51.94	10.39
	$Y_{.j.}$	39.00	36.02	43.25	38.18	41.19	$197.64 = Y_{...}$	

[a]In tons per acre per harvest. [b]In tons per acre per year.

TABLE 11.2.
Analysis of variance for each harvest of the first year

Source of Variation	df	Har 1 SS	Har 1 MS	Har 2 SS	Har 2 MS	Har 3 SS	Har 3 MS	Har 4 SS	Har 4 MS
Total	19	1.1801		3.1045		3.3016		0.8376	
Blocks	4	0.1651	0.0413	1.7249	0.4312	1.2562	0.3140	0.3112	0.0778
Varieties	3	0.4729	0.1576	0.2547	0.0849	0.5660	0.1887	0.2295	0.0765
1+2 vs. 3+4	1	0.3920	0.3920*	0.2268	0.2268	0.1567	0.1567	0.1411	0.1411*
1 vs. 2	1	0.0185	0.0185	0.0058	0.0058	0.0960	0.0960	0.0548	0.0548
3 vs. 4	1	0.0624	0.0624	0.0221	0.0221	0.3133	0.3133	0.0336	0.0336
Error	12	0.5421	0.0452	1.1249	0.0937	1.4794	0.1233	0.2969	0.0247

*Ratio of MS to error MS exceeds tabular F required for significance at the 5% level. Tabular $F_{0.05}$ (1 and 12 df) = 4.75.

V = varieties, v = 4; H = harvests, h = 4; B = blocks, b = 5.

$$C = \frac{Y_{..1}^2}{vb} = \frac{61.50^2}{(4)5} = 189.1125$$

$$SSB = \frac{\Sigma Y_{.j1}^2}{v} - C = \frac{11.91^2 + \ldots + 12.49^2}{4} - C = 0.1651$$

$$SSV = \frac{\Sigma Y_{i.1}^2}{b} - C = \frac{14.46^2 + \ldots + 15.68^2}{5} - C = 0.4729$$

$$SS(V1+2 \text{ vs. } 3+4) = (14.46 + 14.89 - 16.47 - 15.68)^2/5(4) = 0.3920.$$

Note that this single degree of freedom computation and the two that follow involve the use of the rule: $SS = (\Sigma c_i Y_i)^2/(r\Sigma c_i^2)$. For these three computations the c_i's are all + or −1.

$$SS(V1 \text{ vs. } V2) = \frac{(14.46 - 14.89)^2}{5(2)} = 0.0185$$

$$SS(V3 \text{ vs. } V4) = \frac{(16.47 - 15.68)^2}{5(2)} = 0.0624$$

$$SS(\text{total}) = 2.69^2 + \ldots + 3.05^2 - C = 1.1801$$

$$SS(\text{error}) = SS(\text{total}) - SSB - SSV = 0.5421$$

Mean squares are obtained by dividing SS's by appropriate degrees of freedom, for example, MSV = 0.4729/3 = 0.1576. F values for testing variety effects are found by dividing MS's for variety components by the mean square for error for that particular harvest, for example, for harvest 1, F for V1+2 vs. V3+4 = 0.3920/0.0452 = 8.67. Varieties 3 and 4 average 0.28, 0.21, 0.18, and 0.17 tons/acre more than varieties 1 and 2 for harvests 1 through 4, respectively. Since the difference (V1+2)−(V3+4) is statistically significant for harvests 1 and 4, it appears logical to assume real differences for this comparison for harvests 2 and 3 also, even though F values are not significant at the 5% level.

Annual Analysis

An annual analysis is carried out and organized as in Table 11.3. A stepwise procedure for the computations, identical to those used in a split-plot, are given below.

TABLE 11.3.
Analysis of variance, first year, alfalfa variety trial

Source of Variation	df	SS	MS	F	Tabular F 5%	Tabular F 1%
Subplots	79	34.8690				
Main plots	19	5.0769				
Blocks, B	4	1.9386	0.4846			
Varieties, V	3	0.9014	0.3005	1.61	3.49	5.95
1+2 vs. 3+4	1	0.8778	0.8778	4.71	4.75	9.33
1 vs. 2	1	0.0046	0.0046			
3 vs. 4	1	0.0189	0.0189			
MP error, BV	12	2.2369	0.1864			
Harvests, H	3(1)	26.4452	8.8151	155.2	4.49[a]	8.53[a]
V×H	9(3)	0.6217	0.0690	1.21	3.24[a]	5.29[a]
Subplot error, BH+B(V×H)	48(16)	2.7252	0.0568			

[a]Tabular F values are for degrees of freedom in parentheses.

$$C = \frac{Y_{...}^2}{vhb} = \frac{197.64^2}{4(4)5} = 488.2696$$

Note that putting h in the denominator keeps the observations on a per-harvest basis.

$$SSB = \frac{\Sigma Y_{.j.}^2}{vh} - C = \frac{39.00^2 + \ldots + 41.19^2}{4(4)} - C$$

$$= 490.2082 - C = 1.9386$$

If you have a calculator preprogrammed to compute $s = \sqrt{(Y_i - \bar{Y})^2/(r-1)}$, SSB and the other sums of squares can be found by the following procedure: enter the appropriate totals, get s, square s, and divide s^2 by the number of variates in each total entered. The result is MS which, multiplied by the appropriate degrees of freedom equals SS. For example, s^2 of block totals = 7.75437. Divide by 16 = 0.48465 = MSB. Multiply by 4 = 1.9386 = SSB.

$$SSV = \frac{\Sigma Y_{i..}^2}{hb} - C = \frac{47.10^2 + \cdots + 51.94^2}{4(5)} - C$$

$$= 489.1710 - C = 0.9014$$

$$SS(V1+2 \text{ vs. } 3+4) = \frac{(\Sigma c_i Y_{i..})^2}{bh\Sigma c_i^2}$$

Note that the comparison coefficients, the c_i's, are all + or −1.

$$SS(V1+2 \text{ vs. } 3+4) = \frac{(47.10 + 47.53 - 51.07 - 51.94)^2}{5(4)4}$$

$$= 0.8778$$

$$SS(MP) = \left(\frac{\Sigma Y_{ij.}^2}{h}\right) - C = \frac{9.02^2 + \cdots + 10.13^2}{4} - C = 5.0769$$

$$SS(MP\ error) = SS(MP) - SSB - SSV = 2.2369$$

$$SSH = \frac{\Sigma Y_{..k}^2}{bv} - C = \frac{61.50^2 + \cdots + 37.40^2}{5(4)} - C$$

$$= 26.4452$$

$$SS(V \times H) = \left(\frac{\Sigma Y_{i.k}{}^2}{b} \right) - C - SSV - SSH$$

$$= \frac{14.46^2 + \cdots + 9.48^2}{5} - C - SSV - SSH$$

$$= 0.6217$$

$$SS(\text{subplots}) = \Sigma Y_{ijk}{}^2 - C = (2.69^2 + \cdots + 1.78^2) - C$$

$$= 523.1386 - C = 34.8690$$

Note also that

$$SS(SP) = SS(\text{total}) \text{ for } H_1 + \cdots + SS(\text{total})H_4 + SSH$$

$$= 1.1801 + \cdots + 0.8376 + 26.4452 = 34.8690$$

$$SS(SP\ \text{error}) = SS(SP) - SS(MP) - SSH - SS(V \times H)$$

$$= 34.8690 - 5.0769 - 26.4452 - 0.6217 = 2.7252$$

Mean squares are obtained by dividing sums of squares by degrees of freedom, for example, $MSV = 0.9014/3 = 0.3005$.

F Values and Mean Separation

The brackets connecting mean squares of Table 11.3 indicate the error terms used in calculating F ratios. F values for harvests and V×H should be large before concluding the existence of real differences. A conservative approach recommended by many statisticians is to require larger F values for significance. It is suggested that degrees of freedom for harvest date be used to divide degrees of freedom for H, V×H, and subplot error (values in parentheses in Table 11.3) and to select tabular F values on the basis of the resulting degrees of freedom (those in Table 11.3 with the superscript a). Considering the larger F value for harvests, there is little doubt that there are real differences among the mean effects of harvest dates. There is no evidence for a real interaction of V×H.

Note that most of the variability among varieties is due to V1+2 vs. V3+4 and that the F value for this comparison is nearly significant at the 5% level.

STANDARD ERRORS. Standard errors used in LSD and multiple-range tests are the same as for the normal split-plot design with respect to the mean effects of the factor applied to the main plots (in this case varieties), but they differ from the split-plot for means of the repeated observation (harvest dates) and the interaction of main plot treatment × repeated observation (V×H). In the following discussion, we more or less follow the procedure given by Steel and Torrie, (1960).

For a review of the use of a standard error in the calculation of LSD and Duncan's multiple-range test, see Chapter 6. Briefly, $LSD = ts_{\bar{d}}$ and $D = R(LSD)$.

1. Comparing two A means, $V_1 - V_2$

a. On a per-harvest basis: $s_{\bar{d}} = \sqrt{\dfrac{2(\text{MP error})}{bh}}$

$$s_{\bar{d}} = \sqrt{\frac{2(0.1864)}{(5)4}} = 0.1365 \qquad LSD_{0.05} = 2.179(0.1365) = 0.30 \text{ tons/acre}$$

b. On an annual basis: $s_{\bar{d}} = \sqrt{\dfrac{2h(\text{MP error})}{b}}$

$$s_{\bar{d}} = \sqrt{\frac{2(4)(0.1864)}{5}} = 0.5461 \qquad LSD = 2.179(0.5461) = 1.19 \text{ tons/acre}$$

2. Comparing two B means, $H_1 - H_2$

$$s_{\bar{d}} = \sqrt{\frac{2(\text{SP error})}{bv}}$$

$$s_{\bar{d}} = \sqrt{\frac{2(0.0568)}{5(4)}} = 0.0754 \qquad LSD = 2.120(0.0754) = 0.16 \text{ tons/acre}$$

where $t_{.05}$ is for 16 df.

3. Comparing two A means at the same level of B, $V_1H_1 - V_2H_1$

$$s_{\bar{d}} = \sqrt{\frac{2(E_i)}{b}}.$$

where E_i is the error for the analysis of the harvest under consideration. For H_1

$$s_{\bar{d}} = \sqrt{\frac{2(0.0452)}{5}}, \qquad s_{\bar{d}} = 0.1345,$$

$$LSD = 2.179(0.1345) = 0.29 \text{ tons/acre}$$

4. Comparing two B means for the same or different A, $V_1H_1 - V_1H_2$ or

$V_1H_1 - V_2H_2$

$$s_{\bar{d}} = \sqrt{\frac{2(E_1 + E_2)}{2b}}$$

where E_1 and E_2 are error MS's for the two harvests and are averaged.

$$s_{\bar{d}} = \sqrt{\frac{2(0.0452 + 0.0937)}{2(5)}} = 0.1667$$

$$LSD = 2.179(0.1667) = 0.36 \text{ tons/acre}$$

COMBINING TWO OR MORE YEARS

In addition to analyzing the performance of varieties for each year, the researcher usually is interested in variety performance over a series of years and the possible interaction of varieties with years. Several years' results, involving several harvests each year, may be combined as a split-split-plot analysis with varieties as main plots, years as split-plots, and harvests as split-split plots. However, the interaction of varieties × years × harvests usually is not of primary importance. Annual analyses plus an analysis of yearly whole plot totals over a sêries of years is usually all that is required in making decisions as to varietal suitability.

To illustrate the procedure for combining yearly total variety plot yields over a period of years, we will use data from two years only. The procedure is the same as for the analysis of harvests within a year. Table 11.4 provides the necessary data. Note that the variety × block totals of year 1, Table 11.1, are the data for year 1 in Table 11.4.

The Analysis for Each Year

The annual analysis needed is the main plot analysis of Table 11.3 for each year. Since we now want the data on a per-plot per-year basis rather than on a per-plot per-harvest basis as for Table 11.3, we multiply the sums of squares of Table 11.3 by the number of harvests to complete Table 11.5. Thus, for year 1 the ANOVA of Table 11.5 is completed by

$$SSB = (\text{SSB for year 1})4 = (1.9386)4 = 7.7544$$

$$SSV = (0.9014)4 = 3.6056$$

$$\text{SS error} = (2.2369)4 = 8.9476$$

TABLE 11.4.
Tons of dry forage per main plot per year for years 1 and 2, alfalfa variety trial. (Note that data for year 1 are the same as for the bottom portion of Table 11.1).

Variety (i)	Year (k)	Blocks (j) 1	2	3	4	5	$Y_{i.k}$	$\bar{Y}_{i.k}$
1	1	9.02	6.98	10.62	9.52	10.96	47.10	9.42
2	1	8.84	9.83	10.06	9.30	9.50	47.53	9.51
3	1	9.74	9.28	11.74	9.71	10.60	51.07	10.21
4	1	11.40	9.93	10.83	9.65	10.13	51.94	10.39
	$Y_{.j1}$	39.00	36.02	43.24	38.18	41.19	$197.64 = Y_{..1}$	
1	2	11.88	11.33	11.81	12.22	10.65	57.89	11.58
2	2	12.15	10.98	12.20	11.30	12.54	59.15	11.83
3	2	12.92	11.95	12.05	11.88	13.19	61.99	12.40
4	2	11.74	11.62	11.54	12.00	11.74	58.64	11.73
	$Y_{.j2}$	48.69	45.86	47.60	47.40	48.12	$237.67 = Y_{..2}$	
		Variety × block totals (main plots, $Y_{ij.}$)					$Y_{i..}$	$\bar{Y}_{i..}$
1		20.90	18.31	22.43	21.74	21.64	104.99	10.50
2		20.99	20.79	22.26	20.60	22.04	106.68	10.67
3		22.66	21.23	23.79	21.59	23.79	113.06	11.31
4		23.14	21.55	22.37	21.65	21.87	110.58	11.06
	$Y_{.j.}$	86.69	81.88	90.85	85.58	89.31	$435.31 = Y_{...}$	

TABLE 11.5.
Analyses of variance of total yield per plot for each year.

Source of Variation	df	Year 1 SS	Year 1 MS	Year 2 SS	Year 2 MS
Blocks	4	7.7544	1.9386	1.1261	0.2815
Varieties	3	3.6054	2.2018	1.9254	0.6418
1+2 vs. 3+4	1	3.5112	3.5112	0.6444	0.6444
1 vs. 2	1	0.0184	0.0184	0.1588	0.1588
3 vs. 4	1	0.0756	0.0756	0.1122	0.1122
Error	12	8.9476	0.7456	3.7462	0.3120

The same procedure is used to complete Table 11.5 for year 2. Note that in the second year there were no statistically significant variety effects but that the major portion of the variability among "varieties" was due to the comparison, variety 1+2 vs. 3+4.

Putting the Years Together

The ANOVA of Table 11.6 is completed from the data of Table 11.4 and combines the yearly totals over the two years in a manner analogous to combining harvests within a year (Table 11.3). The sums of squares are obtained as shown below. Mean squares are sums of squares divided by their own degrees of freedom.

TABLE 11.6.
ANOVA of annual yields over two years.

Source of Variation	df	SS	MS	F	Tabular F 5%	Tabular F 1%
Subplots	39	67.1654				
Main plots	19	14.0138				
Blocks, B	4	6.1058	1.5264			
Varieties, V	3	4.0323	1.3441	4.16	3.49	5.95
1+2 vs 3+4	1	3.5820	3.5820	11.09	4.75	9.33
1 vs 2	1	0.1428	0.1428			
3 vs 4	1	0.3075	0.3075			
MP error, VB	12	3.8757	0.3230			
Years	1	40.0600	40.0600	55.29	4.49	8.53
V×Y	3	1.4985	0.4995			
(V1+2 vs. 3+4)×Y	1	0.5736	0.5736			
(V1 vs. 2)×Y	1	0.0344	0.0344			
(V3 vs. 4)×Y	1	0.8904	0.8904			
Subplot error, BY+B (V×Y)	16	11.5931	0.7246			

V=varieties, v=4; B=blocks, b=5; Y=years, y=2.

$$C=\frac{\Sigma Y_{ijk}^{\;2}}{vby}=\frac{435.31^2}{4(5)2}=4737.3699$$

$$SSB=\frac{\Sigma Y_{.j.}^{\;2}}{vy}-C=\frac{86.69^2+\ldots+89.31^2}{4(2)}-C=6.1058$$

$$SSV=\frac{\Sigma Y_{i..}^{2}}{by}-C=\frac{104.99^2+\ldots+110.58^2}{5(2)}-C=4.0323$$

$$SS(V1+2\text{ vs. }3+4)=\frac{(\Sigma c_i Y_{i..})^2}{by\Sigma c_i^{2}}$$

where c_i are class comparison coefficients, in this case + and −1.

$$SS(V1+2\text{ vs. }3+4)=\frac{(104.99+106.68-113.06-110.58)^2}{5(2)4}=3.5820$$

$$SS(V1\text{ vs. }V2)=\frac{(104.99-106.68)^2}{5(2)2}=0.1428$$

$$SS(V3\text{ vs. }V4)=\frac{113.06-110.58}{5(2)2}=0.3075$$

$$SS(MP)=\frac{\Sigma Y_{ij.}^{2}}{vb}-C=\frac{20.90^2+\ldots+21.87^2}{4(5)}-C=14.0138$$

$$SS(MP\text{ error})=SS(MP)-SSB-SSV=3.8757$$

$$SSY=\frac{\Sigma Y_{..k}^{2}}{vb}-C=\frac{197.64^2+237.67^2}{4(5)}-C=40.0600$$

$$SS(V\times Y)=\frac{\Sigma Y_{i.k}^{2}}{b}-C-SSV-SSY$$

$$=\frac{47.10^2+\ldots+58.64^2}{5}-C-SSV-SSY=1.4985$$

This interaction sum of squares is partitioned by attention to the class comparison coefficients given in Table 11.7.

TABLE 11.7.
Orthogonal coefficients for partitioning year and variety comparisons.

Comparison	Annual Variety Totals							
	V_1Y_1 47.10	V_2Y_1 47.53	V_3Y_1 51.07	V_4Y_1 51.94	V_1Y_2 57.89	V_2Y_2 59.15	V_3Y_2 61.99	V_4Y_2 58.64
Y	+	+	+	+	−	−	−	−
V1+2 vs. 3+4	+	+	−	−	+	+	−	−
V1 vs. 2	+	−	0	0	+	−	0	0
V vs. 4	0	0	+	−	0	0	+	−
(V1+2 vs. 3+4)×Y	+	+	−	−	−	−	+	+
(V1 vs. 2)×Y	+	−	0	0	−	+	0	0
(V3 vs. 4)×Y	0	0	+	−	0	0	−	+

SS(V1+2 vs. 3+4)×Y

$$= \frac{(47.10+47.53-51.07-51.94-57.89-59.15+61.99+58.64)^2}{5(8)}$$

$$= 0.5736$$

$$SS(V1\text{ vs. }2)\times Y = \frac{(47.10-47.53-57.89+59.15)^2}{5(4)} = 0.0344$$

$$SS(V3\text{ vs. }4)\times Y = \frac{(51.07-51.94-61.99+58.64)^2}{5(4)} = 0.8904$$

$$SS(\text{Subplots}) = \frac{\Sigma Y^2_{ijk}}{vby} - C = \frac{9.02^2 + \ldots + 11.74^2}{4(5)2} - C$$

$$= 67.1654$$

$$SS(\text{SP error}) = SS(SP) - SS(MP) - SSY - SS(V\times Y) = 11.5931$$

F values are determined by dividing MS's by the error term indicated by the brackets and arrows of Table 11.6. There are no indications for interactions of varieties with years, but the large F value for years indicates a real year effect despite the doubtful wisdom of using the subplot error MS to make the F test. The fact that MS subplot error is larger than the MS MP error lends justification to this conclusion.

Note that there is little doubt that varieties 3 and 4 are superior to 1 and 2, since the F value exceeds the tabular 1% value. There is no evidence that variety 2 is really better than 1 or that variety 3 is better than 4.

STANDARD ERRORS. The calculation of standard errors is analogous to that for the annual analysis. Standard errors and LSDs pertinent to the significant effects of this analysis are given below.

1. Comparing two variety means: $s_{\bar{d}} = \sqrt{\dfrac{2(\text{MP error})}{by}}$

$$s_{\bar{d}} = \sqrt{\frac{2(0.3230)}{5(2)}} = 0.2542 \qquad \text{LSD} = 2.179(0.2542) = 0.55$$

2. Comparing variety 1 and 2 vs 3 and 4: $s_{\bar{d}} = \sqrt{\dfrac{2(\text{MP error})}{by\,2}}$

$$s_{\bar{d}} = \sqrt{\frac{2(0.3230)}{5(2)2}} = 0.1797 \qquad \text{LSD} = 2.179(0.1797) = 0.39$$

In calculating standard errors, a rule to follow is that the denominator should equal the number of variates going into the means to be compared. Thus, 2 is placed in the denominator because we are comparing the mean of variates 1 and 2 with the mean of variates 3 and 4.

SUMMARY

Periodic sampling of main plots for yield, as repeated harvests of perennial variety plots, repeated picking of fruit from the same trees, or repeated sampling of soil plots over time for nutrient content are most properly analyzed as the split-plot design. Data are analyzed as for split-plots, but caution should be used in concluding that there are real effects for the repeated observation and its interactions with main plot treatments unless F values are large.

12

TRANSFORMATIONS (WHAT TO DO WHEN DATA BREAK THE RULES)

Research workers who are content to learn the "recipes" for carrying out an analysis of variance, without attempting to learn and understand the underlying principles, may be headed for serious trouble. Whether they realize it or not, they are making certain assumptions about the data when they perform an analysis of variance. If the data do not conform to these assumptions, such an analysis may cause workers to reach conclusions that are not justified. They may also overlook important conclusions that would be reached if the data were properly analyzed.

ASSUMPTIONS OF THE ANALYSIS OF VARIANCE

The assumptions on which an analysis of variance is based are briefly as follows:

1. The error terms are randomly, independently, and normally distributed.
2. The variances of different samples are homogeneous.
3. Variances and means of different samples are not correlated.
4. The main effects are additive.

We now discuss these four assumptions in more detail.

Normality

Fortunately, deviations from the assumption of normality do not affect the validity of the analysis of variance too seriously. There are tests for normality, but it is rather pointless to apply them unless the number of samples we are dealing with is fairly large. Independence implies that there is no relation between the size of the error terms and the experimental grouping to which they belong. Since adjacent plots in a field tend to be more closely related to each other than randomly scattered plots, it is important to avoid having all plots receiving a given treatment occupying adjacent positions in the field. This is one of the main reasons for the insistence on not dividing a plot receiving a certain treatment into subplots and referring to these as replicates. The best insurance against seriously violating the first assumption of the analysis of variance is to carry out the randomization appropriate to the particular experimental design you are using.

Homogeneity of Variances

The first reference in this book to analysis of variance (Chapter 3), dealt with a simple example with two treatments each replicated five times. You will note that we assumed that the variances within each treatment both estimated a common variance. We therefore felt justified in using the average of these two variances as a better estimate of σ^2 than either one alone. Similarly, in Chapter 4 we used a "pooled error mean square," or an average of four variances to give us the best estimate of the common variance.

If the variances within different treatments were, in fact, different, we would not be justified in pooling them. Suppose, for example, that the replicates in two of the treatments were actually samples from populations with large variances, while those of the other two treatments were from populations with much smaller variances. It should be obvious that the difference required for significance would be greater for the two highly variable treatments than for the two less variable ones. Averaging the large and small variances could give very misleading results. The difference between the two treatments with large variances might be declared significant when, in reality, it could easily have occurred by chance. On the other hand, the difference between the two treatments with small variances might be declared nonsignificant when, in fact, it was real. The following data from a hypothetical experiment with four treatments, each replicated five times, will illustrate this situation:

	Replicate							
Treatment	1	2	3	4	5	Total	Mean	s^2
A	3	1	5	4	2	15	3	2.5
B	6	8	7	4	5	30	6	2.5
C	12	6	9	3	15	45	9	22.5
D	20	14	11	17	8	70	14	22.5

Carrying out the analysis of variance in the usual way, we get:

Source of variation	df	SS	MS	F
Treatments	3	330	110	8.8**
Error	16	200	12.5	

Note that the error mean square is the average of the four individual variances within the treatments. The F value is highly significant. Let us now calculate an LSD:

$$LSD_{.05} = t\sqrt{2EMS/r} = 2.12\sqrt{5} = 4.74$$

Since the mean difference between treatments A and B is only 3, we would conclude that this was not significant. The mean difference between C and D is 5, and this would be called significant at the 5% level. We note, however, that the variances of C and D are nine times as large as those of A and B. The assumption that the variances are homogeneous is open to considerable doubt. It would, therefore, be more reasonable to analyze A and B separately from C and D.

The analysis for A and B is:

Source of variation	df	SS	MS	F
Treatments	1	22.5	22.5	9*
Error	8	20.0	2.5	

For C and D:

Source of variation	df	SS	MS	F
Treatments	1	62.5	62.5	2.78ns
Error	8	180	22.5	

We are now led to just the opposite conclusions regarding the differences between A and B and between C and D. Later we will show how to test data for homogeneity of variances. As to what we can do when we encounter data in which the variances are not homogeneous, there are several courses we can follow. First, we can separate the data into groups such that the variances within each group are homogeneous. Then each group can be analyzed separately as we did in the example above. Second, we can use a method described in more advanced statistics texts, which involves a rather complicated procedure of weighting means according to their variances. Third, we might be able to transform the data in such a way that they will be homogeneous. We discuss this method further on in this chapter.

Independence of Means and Variances

In some data, there is a definite relation between the means of samples and their variances. This is a special case and the most common cause of heterogeneity of variance. A positive correlation between means and variances is often encountered when there is a wide range of sample means.

Suppose, for example, that an experimenter was testing the effects of several insecticides on aphids and measuring the effectiveness by counting the number of aphids per leaf after application. If the means of two rather ineffective treatments were 305 and 315, he would naturally hesitate to attach much importance to this difference. On the other hand, if the means of two other treatments were 5 and 15, he might be inclined to feel that this difference was appreciable, impressed with the fact that one of these was three times as large as the other. Under the assumption that the variances are homogeneous and unrelated to the means, he would have to attach as much importance to the difference between 305 and 315 as that between 5 and 15, for the actual differences are the same in both cases. He probably would have an uneasy feeling that something was wrong. An examination of the various samples would almost certainly reveal that, in general, the samples with high means would also have large variances and those with low means would have small variances. Thus the assumption that the means and variances are not correlated would be false, and an ordinary analysis of variance of the raw data would not be valid.

Let us take a more extreme example. Some experimenters want to test the effect of a new vitamin on the weights of animals. They wish to include a wide range of animals in their tests, so they choose mice, chickens, and sheep. Common sense would tell us that a difference of a half pound in the mean weights of two lots of sheep would be considered negligible and easily attributed to chance. A difference of a half pound in the mean weights of two lots of chickens would be considered very large, but not beyond the realm of possibility. A difference of a half pound in the mean weights of two lots of mice would be looked upon as utterly fantastic. Admittedly this is an extreme and almost absurd example, but it serves to emphasize the point that the assumption of the independence of variances and means should not be accepted blindly. We should examine the data and, if necessary, test the validity of the assumption before we proceed with an analysis of variance.

Other types of data that often show a relation between variances and means are data based on counts and data consisting of proportions or percentages. Now, suppose that we find that there is a relation between variances and means. Does this mean we are forced to abandon the analysis of variance as a method for analyzing the data? Fortunately, it is often not the case. We can frequently transform the data in such a way that the assumption of independence between variances and means will be valid. Then we can proceed with an analysis of variance on the transformed data.

Additivity

For each experimental design there is a mathematical model called a *linear additive model*. For a completely randomized design, this model is $Y_i = \bar{Y} + t_i + e_i$, which says that the value of any experimental unit is made up of the general mean plus the treatment effect plus an error term. The corresponding model for a randomized complete block design is $Y_{ij} = \bar{Y} + t_i + b_j + e_{ij}$, which says that any experimental unit is made up of the general mean plus a treatment effect plus a block effect plus an error term. The important thing to note in these models is that the terms are *added*, hence the term additivity.

The model for a randomized complete block, for example, implies that a treatment effect is the same for all blocks and that the block effect is the same for all treatments. In other words, if a treatment is found to increase the yield a certain average amount above the general mean, it is assumed that it has this same effect in the high-yielding blocks as in the low-yielding blocks.

One can conceive of many situations where this assumption would *not* be correct. For example, in an experiment to test the effect of N on yield, some blocks might yield less than others because of a low natural nitrogen level in the soil. We might expect the plots in such blocks to benefit more from the addition of nitrogen than plots in blocks where the natural supply of nitrogen was already adequate. On the other hand, suppose that the low yield was due to an inadequate moisture supply. We might then expect the addition of nitrogen to do very little good in these low-yielding blocks but produce an appreciable increase in yield in blocks in which there was sufficient water. Another situation might be one in which the effect of a treatment is to increase the yield by a certain percentage or proportion. This is referred to as a *multiplicative treatment effect*.

In any of the above cases, the assumption of additivity would be incorrect; this fact must be recognized in analyzing the data. In the case of multiplicative treatment effects, there are again transformations that will change the data to fit the additive model.

TESTS FOR VIOLATIONS OF THE ASSUMPTIONS

We are now ready to give some specific examples of data that fail to meet one or more of the assumptions of the analysis of variance. We show how to test these assumptions and the ways in which the data may be transformed so that they will conform. Table 12.1 gives some hypothetical data that might be obtained from an experiment such as that discussed earlier, dealing with the effects of a new vitamin on mice, chickens, and sheep.

TABLE 12.1.
Weights, in pounds, of vitamin-treated and control animals, in a randomized complete block experiment

	Block					
Treatment	I	II	III	IV	Total	Mean
Mice—control	0.18	0.30	0.28	0.44	1.2	0.3
Mice—vitamin	0.32	0.40	0.42	0.46	1.6	0.4
Subtotals	0.50	0.70	0.70	0.90	2.8	0.35
Chickens—control	2.0	3.0	1.8	2.8	9.6	2.40
Chickens—vitamin	2.5	3.3	2.5	3.3	11.6	2.90
Subtotals	4.5	6.3	4.3	6.1	21.2	2.65
Sheep—control	108.0	140.0	135.0	165.0	548.0	137.0
Sheep—vitamin	127.0	153.0	148.0	176.0	604.0	151.0
Subtotals	235.0	293.0	283.0	341.0	1152.0	144.0
Grand totals	240.0	300.0	288.0	348.0	1176.0	49.0

Analyzing the data by the methods used in Chapters 5 and 6 results in the following analysis of variance:

Source of variation	df	SS	MS	F
Blocks	3	984.00	328.00	2.63
Treatments	5	108,713.68	21,742.74	174.43**
Species	2	108,321.16	54,160.58	434.51**
Vitamins	1	142.11	142.11	1.14
Species X Vitamins	2	250.41	125.20	1.00
Error	15	1,869.72	124.65	

The highly significant difference among species does not surprise us at all. It does seem very strange that we did not find a significant difference due to vitamins, especially since every animal in every replicate receiving the vitamin showed a greater weight than the corresponding control animal. It also seems strange that we find no evidence of interaction between vitamin effects and species, since the apparent response to vitamins is so different in the different species. If we accept this analysis at its face value, we would have to conclude that the experiment was virtually a total failure. All we seemed to learn was that mice, chickens, and sheep differ in weight. Even here, if we partition the species effect into two comparisons, one comparing sheep with chickens and mice; the other comparing chickens with mice, we find we cannot even show a significant difference between chickens and mice.

Let us look at the data with the assumptions of the analysis of variance in mind and see what can be done if some of the assumptions prove false. First, we can look at the error terms to see whether they are randomly, independently, and normally distributed. To do this we remove the general mean, the treatment effects, and the block effects from each cell of the table as we did in Chapter 5. This gives a table of error terms, Table 12.2.

TABLE 12.2.
Error components in vitamin experiment

	Block				
Treatment	I	II	III	IV	Total
Mice—control	8.88	−1.00	0.98	−8.86	0
Mice—vitamin	8.92	−1.00	1.02	−8.94	0
Chickens—control	8.60	−0.40	0.40	−8.60	0
Chickens—vitamin	8.60	−0.60	0.60	−8.60	0
Sheep—control	−20.00	2.00	−1.00	19.00	0
Sheep—vitamin	−15.00	1.00	−2.00	16.00	0
Totals	0	0	0	0	

These error terms certainly do not appear to be randomly distributed. They are apparently not independent, because in each block the error terms for the two members of each species are closely related. Finally, their distribution looks as though it deviates from normal considerably, since there are two modal classes, one between 8.5 and 9.0 and the other between −8.5 and −9.0. The first assumption of an analysis of variance did not stand up very well under close scrutiny.

TABLE 12.3.
Variances and their logs for groups in vitamin experiment

Treatment	df	s_i^2	Coded s_i^2	Log coded s_i^2
Mice—control	3	0.0115	11.5	1.06
Mice—vitamin	3	0.0035	3.5	0.54
Chickens—control	3	0.3467	346.7	2.54
Chickens—vitamin	3	0.2133	213.3	2.33
Sheep—control	3	546.0	546,000.	5.74
Sheep—vitamin	3	425.3	425,300.	5.63
Totals	18		971,875.	17.84
Mean			161,979.	
Log of mean			5.209	

Next, we examine the assumption of the homogeneity of variances. To do this, we need to learn a test known as *Bartlett's Test for Homogeneity of Variances*.

First, we need to calculate the variance among the four replicates of each treatment combination. For the mouse controls this will be

$$\frac{0.18^2+0.30^2+0.28^2+0.44^2-(1.2^2/4)}{\text{number of replicates}-1}=0.0115$$

After each such variance is computed, they are entered in a table as shown in Table 12.3.

The purpose of coding the variances is to avoid negative logarithms. We can multiply the variances by any constant we choose without altering the test. It is desirable to have all the coded values be 1 or greater, so we have coded by multiplying each s_i^2 by 1000. It is easiest to use common logarithms; two digits in the mantissa are usually sufficient. The mean of the coded variances is found by dividing their total by the number of samples, and the log of this mean is entered. We are now ready to calculate what is called the unadjusted chi-square.

The general formula for samples of unequal size is

$$\chi^2=2.3026(\log\bar{s}^2\times\Sigma\text{df})-\Sigma(\text{df}\times\log s_i^2)$$

When the samples are all of the same size, as in our example, this reduces to

$$\begin{aligned}\chi^2&=2.3026\,\text{df}(n\log\bar{s}^2-\Sigma\log s_i^2)\\&=2.3026(3)[6(5.209)-17.84]\\&=92.66\end{aligned}$$

The factor 2.3026 in these formulas is the factor for converting common logs to natural logs, n is the number of samples, and df is the degrees of freedom per sample.

The unadjusted chi-square must be adjusted by dividing by a correction factor, C. When the sample sizes are unequal, the required formula is

$$C = 1 + \frac{1}{3(n-1)}\left(\Sigma\frac{1}{df} - \frac{1}{\Sigma df}\right)$$

With equal sample sizes, this formula reduces to

$$C = 1 + \frac{(n+1)}{3n(df)}$$

In our example,

$$C = 1 + \frac{7}{3(6)(3)} = 1.13$$

Then χ^2 adjusted $= \chi^2$ unadjusted$/C = 92.66/1.13 = 82.00$.

We now refer to chi-square Table A.6 at 5 degrees of freedom (one less than the number of samples) and find that 82 far exceeds the tabular value at the 0.1% level of significance (20.517). The evidence that the variances are heterogeneous is therefore very convincing.

The next assumption to examine is that of independence between the means and variances. A quick glance at the data is sufficient to convince us that this assumption is certainly incorrect because the high means have very large variances and the low means have very small variances.

An important question to answer, in order to decide which transformation to use, is whether it is the variances or the standard deviations that are more nearly proportional to the means. We construct a table of ratios as shown in Table 12.4.

TABLE 12.4.
Ratios of variances and standard deviations to means in vitamin experiment

Treatment	$\bar{Y}$	s_i^2	s_i	$s_i^2/\bar{Y}$	$s_i/\bar{Y}$
M-C	0.3	0.01147	0.107	0.04	0.36
M-V	0.4	0.00347	0.059	0.01	0.15
C-C	2.4	0.3467	0.589	0.14	0.24
C-V	2.9	0.2133	0.462	0.07	0.16
S-C	137.0	546.0	23.367	3.98	0.17
S-V	151.0	425.3	20.624	2.82	0.14

We see that the ratio of variances to means increases markedly with the means, while the ratio of standard deviations to means remains fairly constant. (In other words, the standard deviations are roughly proportional to the means.) Incidentally, if the variances and means were unrelated, both of these ratios would be expected to decrease as the means increase.

The final assumption to examine is that of additivity. Under this assumption we would expect the block effects to be approximately the same for all treatments. From Table 12.1 we see that the average difference between block 1 and block 4 was 18 lb. However, the average differences between these two blocks in the case of mice, chickens, and sheep were 0.2, 0.8, and 53.0 lb, respectively.

The formal test for additivity is called *Tukey's test*. This test is applicable to any two-way classification such as a randomized complete block experiment in which the data are classified by blocks and treatments.

We need a table such as Table 12.5, that contains the raw data from Table 12.1 with the block and treatment effects calculated in the margins.

Note that the sums of both block effects and treatment effects add to zero. To carry out the additivity test we need to calculate

$$Q=\Sigma Y_{ij}\left(\bar{Y}_{i.}-\bar{Y}_{..}\right)\left(\bar{Y}_{.j}-\bar{Y}_{..}\right)$$

which says that we multiply each cell in the table by the corresponding treatment and block effects and sum all the products.

In our example

$$Q=0.18(-48.7)(-9.0)+\ldots+176.0(102.0)(9.0)=90{,}140.56$$

TABLE 12.5.
Calculation of block and treatment effects

Treatment	Block I	Block II	Block III	Block IV	Mean $(\bar{Y}_{i.})$	Treatment Effect $\bar{Y}_{i.}-\bar{Y}_{..}$
M-C	0.18	0.30	0.28	0.44	0.3	−48.7
M-V	0.32	0.40	0.42	0.46	0.4	−48.6
C-C	2.00	3.00	1.80	2.80	2.4	−46.6
C-V	2.50	3.30	2.50	3.30	2.9	−46.1
S-C	108.0	140.0	135.0	165.0	137.0	88.0
S-V	127.0	153.0	148.0	176.0	151.0	102.0
Mean	40.0	50.0	48.0	58.0	49.0	
$\bar{Y}_{.j}-\bar{Y}_{..}$	−9.0	1.0	−1.0	9.0		

The sum of squares for nonadditivity is then found as follows:

$$\text{SS nonadditivity} = \frac{(Q^2 \times \text{total experimental units})}{(\text{SSTr} \times \text{SSB})}$$

Applying this equation to our example gives

$$\text{SS nonadditivity} = \frac{(90{,}140.56^2 \times 24)}{(108{,}713.68 \times 984.0)}$$

$$= 1822.94$$

This is a portion of the block × treatment or error sum of squares, which can be partitioned as follows:

Source of Variation	df	SS	MS	F
Error (B × Tr)	15	1869.72		
Nonadditivity	1	1822.94	1822.94	545.79
Residual	14	46.78	3.34	

The F value observed far exceeds the required F value of 8.86 at the 1% level for 1 and 14 degrees of freedom (from Table A.3), so there is strong evidence that the assumption of additivity is incorrect.

We have now checked all the assumptions of the analysis of variance and found that our data does not satisfy any of them. It is no wonder that the analysis of variance gave disappointing results.

Perhaps the most sensible way of analyzing these data is to handle each species separately. The analyses are as follows:

Species	Source of variation	df	SS	MS	F
Mice	Blocks	3	0.0400	0.0133	8.31
	Vitamins	1	0.0200	0.0200	12.50*
	Error	3	0.0048	0.0016	
Chickens	Blocks	3	1.64	0.547	41.00**
	Vitamins	1	0.50	0.500	37.5**
	Error	3	0.04	0.013	
Sheep	Blocks	3	2834.0	944.7	157.4**
	Vitamins	1	392.0	392.0	66.3**
	Error	3	18.0	6.0	

These results are certainly much more satisfactory than the original overall analysis of variance. These analyses are valid, because within any one species the data conform to the basic assumptions quite well. The only shortcoming of these analyses is that they tell us little about whether the different species react similarly to the vitamins. This is perhaps not a very important question, and in practice the research worker would no doubt be content to stop at this point. However, we will follow the other procedure of transforming the data to show the remarkable results that can be achieved.

THE LOG TRANSFORMATION

We must now answer the question of how to transform the data. Whenever we have data where the standard deviations (not the variances) of samples are roughly proportional to the means, the most effective transformation is a log transformation. Another criterion for deciding on this transformation is the evidence of multiplicative rather than additive main effects. Both of these criteria are met in the data we are dealing with, so we will try transforming the data to logs and see what happens.

TABLE 12.6.
Data of vitamin experiment transformed to log 10X

	Block					
Species—Treatment	I	II	III	IV	Total	Mean
Mice—control	0.26	0.48	0.45	0.64	1.83	0.4575
Mice—vitamin	0.51	0.60	0.62	0.66	2.39	0.5975
Subtotals	0.77	1.08	1.07	1.30	4.22	0.5275
Chickens—control	1.30	1.48	1.26	1.45	5.49	1.3725
Chickens—vitamin	1.40	1.52	1.40	1.52	5.84	1.4600
Subtotals	2.70	3.00	2.66	2.97	11.33	1.41625
Sheep—control	3.03	3.15	3.13	3.22	12.53	3.1325
Sheep—vitamin	3.10	3.18	3.17	3.25	12.70	3.1750
Subtotals	6.13	6.33	6.30	6.47	25.23	3.15375
Totals	9.60	10.41	10.03	10.74	40.78	
Means	1.60	1.735	1.672	1.790		1.69917

Before we start, a few general remarks about applying this transformation. Data with negative values cannot be transformed in this way. If there are zeros in the data, we are faced with the problem that the log of zero is minus infinity. To get around this, it is recommended that a 1 be added to each data point before transforming. Data containing a large number of zeros would probably be handled better by some other method. Logarithms to any base can be used, but common logarithms (to the base 10) are generally the easiest. Before transforming, it is legitimate to multiply all data points by a constant, since this has no effect on the subsequent analyses. This is a good idea if any of the data points are less than 1, for in this way we can avoid negative logarithms.

In the data we are working with, there are no zeros, but the lowest value is 0.18, so we will multiply all the data by 10 before taking the logs. This gives us a table of transformed values (Table 12.6).

The analysis of variance is:

Source of Variation	df	SS	MS	F
Blocks	3	0.12075	0.04025	13.77**
Treatments	5	28.60738	5.72148	1959.41**
Vitamins	1	0.04860	0.04860	16.62**
Species	2	28.54926	14.27463	4883.00**
S×V	2	0.00952	0.00476	1.63
Error	15	0.04385	0.00292	

This is certainly a more satisfying result than the analysis of the original data as far as positive results are concerned. We still do not get a significant interaction between species and vitamins, but we are now asking the question in a different way. Before, we were asking, "Does the amount of change in weight due to the addition of vitamins vary from species to species?" Now we are asking, "Does the *proportion* or *percent* change in weight due to vitamins vary from species to species?"

Did we get more positive results this time because we were simply "playing with figures" until we got a result we liked? Or was the transformation we used justified and is the new analysis valid? To be certain, we will check the assumptions of the analysis of variance with the new data.

As before, we construct a table of error terms by subtracting the mean, the treatment effects, and the block effects from each cell of the table (Table 12.7).

These error terms seem to be more randomly distributed and more nearly normally distributed than those of the original data.

TABLE 12.7.
Error components of transformed data

Treatment	Block I	II	III	IV
M-C	−0.10	−0.01	0.02	0.09
M-V	0.01	−0.03	0.05	−0.03
C-C	0.03	0.07	−0.08	−0.01
C-V	0.04	0.02	−0.03	−0.03
S-C	0.00	−0.02	0.02	0.00
S-V	0.02	−0.03	0.02	−0.02

To test the homogeneity of variance, we again carry out Bartlett's test from the data in Table 12.8

$$\chi^2 = 2.3026[(18 \times 0.9614) - (3 \times 5.11)] = 4.548$$

$$C = 1.13 \text{ as before}$$

$$\chi^2 \text{ adjusted} = \frac{4.548}{1.13} = 4.03$$

TABLE 12.8.
Bartlett's test applied to transformed data of vitamin experiment

Treatment	Mean	s_i^2	Coded s_i^2	Log coded s_i^2
M-C	0.4575	0.0243	24.3	1.39
M-V	0.5975	0.0040	4.0	0.60
C-C	1.3725	0.0118	11.8	1.07
C-V	1.4600	0.0048	4.8	0.68
S-C	3.1325	0.0062	6.2	0.79
S-V	3.1750	0.0038	3.8	0.58
Totals			54.9	5.11
Mean			9.15	
Log of mean			0.9614	

which, according to the χ^2 Table A.6, would be exceeded by chance more than 50% of the time. A glance at Table 12.8 shows that there is no indication of any relation between the means and the variances.

To carry out the test for additivity we calculate the block and treatment effects in Table 12.9 for the transformed data, just as we did in Table 12.5 with the raw data.

As before,

$$Q = \Sigma Y_{ij}(\bar{Y}_{i.} - \bar{Y}_{..})(\bar{Y}_{.j} - \bar{Y}_{..})$$

$$= 0.26(-1.24)(-0.10) + \ldots + 3.25(1.48)(0.09)$$

$$= -0.023768$$

$$\text{SS nonadditivity} = \frac{Q^2 \times \text{total experimental units}}{\text{SSTr} \times \text{SSB}}$$

$$= \frac{-0.023768^2 \times 24}{28.60738 \times 0.12075}$$

$$= 0.00392$$

TABLE 12.9.
Calculation of block and treatment effects for the transformed data

	Block				Mean	Treatment Effect
Treatment	I	II	III	IV	$(\bar{Y}_{i.})$	$\bar{Y}_{i.} - \bar{Y}_{..}$
M-C	0.26	0.48	0.45	0.64	0.46	−1.24
M-V	0.51	0.60	0.62	0.66	0.60	−1.10
C-C	1.30	1.48	1.26	1.45	1.37	−0.33
C-V	1.40	1.52	1.40	1.52	1.46	−0.24
S-C	3.03	3.15	3.13	3.22	3.13	1.43
S-V	3.10	3.18	3.17	3.25	3.18	1.48
Mean	1.60	1.74	1.67	1.79	1.70	
$\bar{Y}_{.j} - \bar{Y}_{..}$	−0.10	0.04	−0.03	0.09		

The sum of squares for error can now be partitioned as follows:

Source of Variation	df	SS	MS	F
Error	15	0.04385		
Nonadditivity	1	0.00392	0.00392	1.37
Residual	14	0.03993	0.00285	

The F value does not even approach the 10% level of significance for 1 and 14 degrees of freedom (required $F_{.10} = 3.10$).

We now feel confident that the new analysis is valid, since the transformed data satisfied *all* the assumptions of the analysis of variance. With the original data, none of the assumptions were true.

THE SQUARE ROOT TRANSFORMATION

Whenever we are dealing with counts of rare events, the data tend to follow a special distribution called a *Poisson distribution*. By a rare event, we mean one that has a very low probability of occurring in any individual. For example, suppose that in a lot of lettuce seed, 0.1% of the seed was carrying mosaic disease virus. The probability that any individual seed contains mosaic is then only 1/1000, so as far as a single seed is concerned, this is a rare event. If we take 100 samples of 1000 seeds each from such a lot, we will get approximately these results:

37 samples will contain 0 infected seeds
37 " " " 1 " "
18 " " " 2 " "
6 " " " 3 " "
2 " " " 4 " "

It is obvious that this looks very little like a normal distribution. This Poisson distribution has a very interesting characteristic—the variance is equal to the mean. In actual practice, the variance is generally somewhat larger than the mean because other factors, in addition to sampling variation, are affecting the occurrence of the events being counted. At any rate, the variance tends to be proportional to the mean, thus violating the assumption that the variances and means are not correlated.

Another example of data of this kind is found in insect counts, such as those made from a standard number of sweeps with a net. Here it is rather hard to define what we mean by an individual observation. We might consider it an individual site on which an insect could be found. In sweeping with a net, we are

TABLE 12.10.
Number of lygus per 50 sweeps

Treatment	Block I	II	III	IV	Total	Mean	s_i^2
A	7	5	4	1	17	4.25	6.25
B	6	1	2	1	10	2.50	5.67
C	6	2	1	0	9	2.25	6.92
D	0	1	2	0	3	0.75	0.92
E	1	0	1	2	4	1.00	0.67
F	5	14	9	15	43	10.75	21.58
G	8	6	3	6	23	5.75	4.25
H	3	0	5	9	17	4.25	14.25
I	4	10	13	5	32	8.00	18.00
J	6	11	5	2	24	6.00	14.00
K	8	11	2	6	27	6.75	14.25

sampling thousands of such sites and finding only a few insects. Thus the probability of finding an insect at a particular spot selected at random at one particular time is indeed a rare event.

Data of this kind can be made more nearly normal and at the same time the variances can be made relatively independent of the means by transforming them to square roots. Actually, it is better to use $\sqrt{Y+\frac{1}{2}}$, especially if there are counts under 10.

The data in Table 12.10 show the number of lygus bugs obtained in 50 sweeps in each plot of an experiment testing 10 insecticides and a check treatment, replicated four times in a randomized complete block design.

The analysis of variance is:

Source of Variation	df	SS	MS	F
Blocks	3	12.25	4.08	0.40
Treatments	10	380.00	38.00	3.70**
Error	30	308.00	10.27	

Transforming the data by taking $\sqrt{Y+\frac{1}{2}}$ gives Table 12.11.

TABLE 12.11.
Transformed lygus data

Treatment	Blocks I	II	III	IV	Total	Mean	s_i^2
A	2.74	2.35	2.12	1.22	8.43	2.11	0.41
B	2.55	1.22	1.58	1.22	6.57	1.65	0.39
C	2.55	1.58	1.22	0.71	6.06	1.52	0.60
D	0.71	1.22	1.58	0.71	4.22	1.06	0.18
E	1.22	0.71	1.22	1.58	4.73	1.18	0.13
F	2.35	3.81	3.08	3.94	13.18	3.29	0.54
G	2.92	2.55	1.87	2.55	9.89	2.45	0.19
H	1.87	0.71	2.35	3.08	8.01	2.00	0.99
I	2.12	3.24	3.67	2.35	11.38	2.84	0.53
J	2.55	3.39	2.35	1.58	9.87	2.47	0.55
K	2.92	3.39	1.58	2.55	10.44	2.61	0.59

The analysis of variance is:

Source of Variation	df	SS	MS	F
Blocks	3	0.532	0.177	0.36
Treatments	10	19.993	1.999	4.04**
Error	30	14.841	0.495	

The two analyses are not very different, since they both show a highly significant treatment effect. The F value is about 10% higher after transformation. Some important differences will occur in mean separation, as shown in Table 12.12.

You will note that in the transformed data, G and D, G and E, J and D, and J and E were declared significantly different, whereas they were not in the raw data.

The *weighted means* shown in Table 12.12 are obtained by "detransforming" the means of the transformed data back to the original units. This is done by squaring the transformed means and subtracting one-half. The means obtained in this way are smaller than those obtained directly from the raw data because more weight is given to the smaller variates. This is as it should be, since in a Poisson

TABLE 12.12.
Duncans' multiple-range test on raw and transformed data, (5% level)

Mean Separation of:	Treatments and Means										
	D	E	C	B	H	A	G	J	K	I	F
Raw data	0.75	1.00	2.25	2.50	4.25	4.25	5.75	6.00	6.75	8.00	10.75
Weighted means Transformed data	0.62	0.89	1.81	2.22	3.50	3.95	5.50	5.60	6.31	7.57	10.32

distribution the smaller variates are measured with less sampling error than the larger ones.

Actually in reporting the results of such an experiment, it is better to use these weighted means, making it clear in the report how they were obtained.

The general effect of the square root transformation is to increase the precision with which we can measure the differences between small means. This is highly desirable in insect control work, since we are generally not as interested in differences between two relatively ineffective treatments as we are in comparing treatments that give good control.

A glance at the variances in the two tables will show that before transformation there was a strong positive relation between means and variances. The coefficient of linear correlation between them was .89, significant at the 0.1% level. After transformation, the correlation was only .37, not even significant at the 10% level. Thus, one of the assumptions of the analysis of variance was violated in the original data, and this was remedied by the transformation.

As to the other assumptions in the analysis of variance, there do not appear to be any serious violations. An examination of the error components shows no striking deviation from a random and normal distribution. Carrying out Bartlett's test for homogeneity of variance on the raw data gives an adjusted chi-square value of 12.56, which has a 25% probability of being exceeded by chance alone. After transformation, this chi-square value was reduced to 4.81, which has a 90% probability of being exceeded by chance. Thus, transformation reduced the amount of heterogeneity over that in the raw data, but in neither case was it significant.

In carrying out Tukey's test for additivity, even with the raw data, the F value for nonadditivity was less than one.

In general, we can say that data requiring the square root transformation do not violate the assumptions of the analysis of variance nearly as drastically as data requiring a log transformation. Consequently, the changes in the analysis brought about by the transformation are not nearly so spectacular.

TABLE 12.13.
Number of lettuce seeds germinating in samples of 50

	Replicates					
Treatment	1	2	3	Mean	s_i^2	$\text{Log}(10 \times s_i^2)$
1	0	0	1	0.33	0.33	0.519
2	0	1	0	0.33	0.33	0.519
3	0	0	1	0.33	0.33	0.519
4	0	2	0	0.67	1.33	1.124
5	2	0	0	0.67	1.33	1.124
6	0	2	3	1.67	2.33	1.367
7	7	10	7	8.00	3.00	1.477
8	11	12	15	12.67	4.33	1.637
9	13	18	18	16.33	8.33	1.921
10	22	16	13	17.00	21.00	2.322
11	24	13	18	18.33	30.33	2.482
12	23	21	29	24.33	17.33	2.239
13	24	29	29	27.33	8.33	1.921
14	37	28	27	30.67	30.33	2.482
15	42	41	40	41.00	1.00	1.000
16	39	41	45	41.67	9.33	1.970
17	41	45	40	42.00	7.00	1.845
18	47	41	43	43.67	9.33	1.970
19	45	42	48	45.00	9.00	1.954
20	46	42	48	45.33	9.00	1.970
21	49	46	48	47.67	2.33	1.367
22	48	49	48	48.33	0.33	0.519
23	50	49	48	49.00	1.00	1.000
24	49	49	50	49.33	0.33	0.519
				Totals	178.00	35.767
				$10 \times$ Mean	74.167	
				$\text{Log}(10 \times \text{mean})$	1.8702	

THE ARCSINE OR ANGULAR TRANSFORMATION

Another kind of data that may require transformation is that based on counts expressed as percentages or proportions of the total sample. Such data generally have what is called a *binomial distribution* rather than a normal distribution. One of the characteristics of this distribution is that the variances are related to the means but in quite a different way than the types of data we have been considering. Up to now the cases we have discussed are those in which large means tend to have large variances and vice versa. In binomial data, variances tend to be small at the two ends of the range of values (close to zero and 100%), but larger in the middle (around 50%). This is actually a rather natural idea even to nonmathematicians. We are inclined to attach more importance to a difference between zero and 6%, or between 94% and 100%, than to a difference between 47% and 53%, even though these are all of the same magnitude.

The appropriate transformation for data of this kind is called the angular or arcsine transformation. It is obtained by finding the angle whose sine is the square root of the proportion (percentage/100). Written in mathematical shorthand, this is arcsine $\sqrt{Y}$ or sine $^{-1}\sqrt{Y}$. Table A.8 can be used to find the transforms directly from the percentages.

Data should be transformed if the range of percentages is greater then 40. Otherwise, it is scarcely necessary. The data in Table 12.13 are from a completely randomized experiment on lettuce seed with 24 treatments, each replicated three times. Treatments are arranged in order of magnitude of their means. Note that there is a strong tendency for the variances at the extremes to be smaller than those in the middle of the range. This is typical of binomial data. The logs of the variances (coded by multiplying by 10) are listed so that a Bartlett's test can be carried out.

$$\text{Unadjusted } \chi^2 = 2.3026\left[(\log \text{mean} \times \Sigma \text{df}) - (\text{df per sample} \times \Sigma \log \text{coded } s_i^2)\right]$$

$$= 2.3026\left[(1.8702 \times 48) - (2 \times 35.767)\right]$$

$$= 41.99$$

$$C = 1 + \frac{1}{3\,(\text{samples} - 1)} - \left(\frac{\text{Number of treatments}}{\text{df per treatment}} - \frac{1}{\Sigma \text{df}}\right)$$

$$= 1 + \frac{1}{3 \times 23}\left(\frac{24}{2} - \frac{1}{48}\right) = 1.1736$$

$$\text{Adjusted } \chi^2 = \frac{\chi^2}{C} = 35.78$$

This is just significant at the 5% level (required value 35.172), so we have fairly good evidence that the variances are not homogeneous.

Analyzing the raw data gives these results:

Source of Variation	df	SS	MS	F
Treatments	23	25266.0	1098.52	148.12**
Error	48	356.0	7.42	

The transformed data are shown in Table 12.14. Since the data in Table 12.13 were based on samples of 50, each variate had to be multiplied by 2 to convert it to a percentage. The pattern of variances observable in the raw data is no longer apparent in the transformed data.

Carrying out Bartlett's test:

$$\text{Unadjusted } \chi^2 = 2.3026(1.411 \times 48) - (31.39 \times 2)$$

$$= 11.3933$$

$$C = 1.1736 \text{ as before.}$$

$$\text{Adjusted } \chi^2 = \frac{\chi^2}{C} = 9.708$$

Referring to Table A.6, opposite 23 df we see that a value this large would be exceeded by chance more than 99% of the time.

An analysis of variance of the transformed data does not seem to lead us to a different conclusion than the analysis of the raw data:

Source of Variation	df	SS	MS	F
Treatments	23	59,487.8	2,586.43	100.29**
Error	48	1,237.9	25.79	

The important difference is not in the overall analysis, but in mean separation. A Duncan's multiple range test shows that:

1. Five differences were declared significant before transformation but not after: 7 − 8, 8 − 11, 10 − 12, 11 − 12, and 12 − 14.

2. Five differences were declared significant after transformation but not before: 18 − 22, 19 − 23, 19 − 24, 20 − 23, and 20 − 24.

TABLE 12.14.
The arcsine transformation of data in Table 12.13

Treatment	Replicates 1	2	3	Mean	s_i^2	$\text{Log}\, s_i^2$
1	0.0	0.0	8.1	2.70	21.870	1.34
2	0.0	8.1	0.0	2.70	21.870	1.34
3	0.0	0.0	8.1	2.70	21.870	1.34
4	0.0	11.5	0.0	3.83	44.083	1.64
5	11.5	0.0	0.0	3.83	44.083	1.64
6	0.0	11.5	14.2	8.57	56.863	1.75
7	22.0	26.6	22.0	23.53	7.053	0.85
8	28.0	29.3	33.2	30.17	7.323	0.86
9	30.7	36.9	36.9	34.83	12.813	1.11
10	41.6	34.4	30.7	35.57	30.723	1.49
11	43.9	30.7	36.9	37.17	43.613	1.64
12	42.7	40.4	49.6	44.23	22.923	1.36
13	43.9	49.6	49.6	47.70	10.830	1.03
14	59.3	48.4	47.3	51.67	44.003	1.64
15	66.4	64.9	63.4	64.90	2.250	0.35
16	62.0	64.9	71.6	66.17	24.243	1.38
17	64.9	71.6	63.4	66.63	19.063	1.28
18	75.8	64.9	68.0	69.57	31.543	1.50
19	71.6	66.4	78.5	72.17	36.843	1.57
20	73.6	66.4	78.5	72.83	37.043	1.57
21	81.9	73.6	78.5	78.00	17.410	1.24
22	78.5	81.9	78.5	79.63	3.853	0.59
23	90.0	81.9	78.5	83.47	34.903	1.54
24	81.9	81.9	90.0	84.60	21.870	1.34
				Totals	618.941	31.39
				Mean	25.789	
				Log mean	1.411	

Which set of conclusions should we accept? The answer is simple: we should accept the conclusions based on the more valid analysis, in this case, the analysis of the transformed data.

Remember, we do not transform data to give us results more to our liking. We transform data so that the analysis will be *valid* and the conclusions *correct*.

Another point to bear in mind when carrying out a transformation is that all tests of significance and mean separation should be carried out on the transformed data rather than on the raw data. Furthermore, it is better to calculate means of the transformed data before detransforming back to original units. In this way we obtain correctly weighted means.

PRETRANSFORMED SCALES

It often happens that we would like to express data in percentages but find it very difficult and time-consuming to make precise measurements. Consider, for example, the problem of evaluating the amount of scab on potato tubers. A convenient measure would be the percentage of tuber area covered with scab, but it is very difficult to measure this accurately. Another example would be the percentage of leaf area covered with disease lesions. Still another would be the percentage of weed control obtained by the application of various herbicides. In all of these cases we could, with a great deal of effort, measure these percentages fairly precisely, but the work involved would be so time-consuming that the number of plots we could measure would be severely limited. In order to make more measurements in a given amount of time, it is a common practice to make rough visual estimates of the percentages rather than precise measurements.

A scale is usually set up, such as the scale of zero to 10 commonly used in weed control work where zero represents no control and 10 represents 100% control. If the steps in this scale represent equal increments of percentages the data should be transformed by the angular transformation just as it should be for precise percentage measurements.

Why not pretransform our scale? In other words, we could select percentage steps such that, when they are transformed by the angular transformation, there will result a series of equally incremented steps that can be reduced to integers.

Suppose, for example, we wished to employ a scale from zero to five. The equal increments in terms of angles would be 90° divided by five, or 18°. We therefore need to find the percentages which, when transformed, give angles of 0, 18, 36, 54, 72, and 90°.

Referring to Table A.8, the closest entry in the table to 18 is 18.4, which is the angular transformation of 10%. The next step in the scale seems to pose a problem. Looking for 36° in the table, we see that there is an entry of 35.7 for the transform of 34%, and an entry of 36.3 for the transform of 35%. We might be tempted to specify this step in the scale as 34.5%, but this would give us a false sense of precision. After all, we are only planning to make rough visual estimates of percentages. In view of this, we are not justified in specifying fractional percentages in our scale except in the range below 5% or above 95%.

Table 12.15 gives the appropriate percentages for all of the commonly used scales.

These scales take advantage of the fact that it is generally easier to detect small differences in the vicinity of zero and 100% than around 50%. Actually, some scales have been used in the past which were deliberately or subconsciously

TABLE 12.15.
Pretransformed rating scales. Scale from zero to:

Rating	4	5	6	8	10	15	18	20	24
0	0	0	0	0	0	0	0	0	0
1	15	10	7	4	2.5	1	0.75	0.7	0.5
2	50	35	25	15	10	4	3	2.5	2
3	85	65	50	30	21	10	7	5	4
4	100	90	75	50	35	17	12	10	7
5		100	93	70	50	25	18	15	10
6			100	85	65	35	25	20	15
7				96	79	45	33	27	20
8				100	90	55	42	35	25
9					97.5	65	50	42	31
10					100	75	58	50	37
11						83	67	58	43
12						90	75	65	50
13						96	82	73	57
14						99	88	80	63
15						100	93	85	69
16							97	90	75
17							99.25	95	80
18							100	97.5	85
19								99.3	90
20								100	93
21									96
22									98
23									99.5
24									100

designed to conform to these percentage classes. In potatoes, a scale from zero to 10 has been used, which is based on photographic standards that roughly represent the percentages shown in Table 12.15. In apples, a starch rating has been employed that corresponds closely to the zero to 8 scale. In weed work, where a scale of zero to 10 is used, there is a tendency to use the rating of 1, for a small trace of control rather than 10%, and the rating of 9, for nearly complete control.

To determine which scale to use, we must decide how many steps we can distinguish with reasonable confidence. A scale with too many steps is unnecessarily complicated and implies greater accuracy than is justified. If we use a scale with too few steps, there is a tendency to record fractional ratings.

In analyzing data based on pretransformed rating scales, the data should *not* be transformed. Furthermore, means should be calculated from the ratings before transforming back to percentages. To make the back transformation we multiply the mean rating by the angular increment and find the corresponding percentage by reference to Table A.8. For example, if a treatment has a mean rating of 1.4 in a scale of zero to five, the angular increment is $90°/5=18°$, and $18°\times1.4=25.2°$. Referring to Table A.8, we see that 25.1° corresponds to 18%, and this would be the appropriate weighted mean to report. Reporting fractional percentages would hardly be justified except at extremely high or low values.

A word should be said about the ratings of *check* plots. It makes a difference whether these are included in the experiment as a zero level of some factor and are subject to the same variation as all other treatment levels or whether they are included as reference plots against which to compare the other plots. In the latter case, they are often arbitrarily given a rating of zero, and the other plots in a block are compared to them. If this is the case, data from the check plots should *not* be included in an analysis of variance. The check plots, arbitrarily assigned values of zero, have no variance. Their variance therefore differs from that of other treatments, so that the assumption of homogeneity of variance is automatically violated.

SUMMARY

1. The main assumptions basic to an analysis of variance are: random and normal distribution of error terms, homogeneity of variances, independence of variances and means, and additivity of main effects.

2. When these assumptions are seriously in error, an analysis of variance is not valid.

3. Transformations can often be made that will correct the failure of the data to meet the assumptions.

4. When *standard deviations* are linearly related to means, and main effects appear to be multiplicative, a log transformation will usually correct both situations.

5. Data based on counts of rare events, where *variances* are related to means, should be subjected to the square root transformation.

6. Data based on proportions or percentages should be given the arcsine or angular transformation.

7. Rating scales can be pretransformed by basing them on a variable scale of percentages.

8. When a transformation is used, all tests of significance and separation of means should be carried out with the transformed data.

9. If we wish to transform back to the original units, this should be done only after the means have been calculated from the transformed data.

13

LINEAR CORRELATION AND REGRESSION

THE IDEA

The terms *correlation* and *regression* may sound a bit formidable, but the basic ideas encompassed by the terms are so simple that we all use them in our everyday conversations. Consider, for example, the following familiar sayings:

"The bigger they are, the harder they fall."

"The more, the merrier."

"Easy come, easy go."

"The better the day, the better the deed."

"As the twig is bent, so is the tree inclined."

All these sayings have several ideas in common. Each implies two variable quantities, the magnitude of one depending on the magnitude of the other. Statisticians refer to these as the *independent* and *dependent variables*. Furthermore, in these particular sayings, there is the idea that as one variable increases, so does the other. In statistics this is called *direct* or *positive correlation*.

Consider another group of sayings:

"Much haste, little speed."

"Small pitchers have big ears."

"The best gifts come in small packages."

Here we have the same general idea of two variables, one dependent on the other, but there is a slight twist in the relationships. An *increase* in one variable is accompanied by a *decrease* in the other. This is called an *inverse* or *negative* correlation.

The idea of correlation is not confined to these simple clichés. Think of the questions that we encounter, time after time in agricultural work, that deal with the relations between two variables. How is the amount of applied fertilizer related to the yield of crop? What relationship is there between amount of feed consumed and weight gain in livestock? How is the price of a commodity affected by the supply? How is dosage of insecticide related to percentage of control, or to the amount of residue? What is the correlation between size of farm and income? The list of such questions could be extended indefinitely, but it should be clear by now that everyone is concerned with the subject of correlation, whether it is called by that name or not.

Another example of correlation that we encounter nearly every day is the common graph. Nearly every graph is essentially a picture of the correlation between two variables. The scale along the bottom, or abscissa, is usually the range of values of the independent variable. The values on the vertical scale, or ordinate, are those of the dependent variable. The graphing of data is often a very useful starting point in conducting a correlation analysis.

Now that we have looked at some common examples of correlation, we should be able to grasp an abstract definition of the term: The tendency of two variables to be related in a definite manner. Actually, the idea can be extended to more than two variables, such as in the law of supply and demand, where there are three variables involved: price, supply, and demand. To keep the discussion as simple as possible, we will limit it for the time being to correlations between two variables.

It is customary to consider one of the variables dependent on the other. The choice of which variable to call dependent and which one to call independent is usually obvious. For example, in studying the relation of yield to fertilizer, it would be logical to consider yield as dependent on fertilizer. With price and supply, we generally think of price as dependent on supply. On the other hand, there are situations in which supply is dependent on price. Often there is a time lapse between the measurement of one variable and the corresponding measurement of the other. In such cases, the first measured variable is called the independent one. It is sometimes useful to study the correlation between pairs of measurements on the same variable. For example, a study of the correlation between the prices of a commodity in successive years with the corresponding prices in the previous year may reveal a cyclic trend in the price pattern.

There are situations in which we really do not care which variable is designated as the dependent variable. We may simply want to describe the joint distribution of two variables where each one is distributed normally. Such a distribution is called a *bivariate normal distribution*. To describe this distribution we need an estimate of ρ (rho), which is one of the population parameters. The coefficient of correlation r, is the best estimate of ρ. Studying the correlation between the length of forearm and height would be an example of the situation where it would make no difference which variable was called dependent.

MEASURING CORRELATION

So far, we have talked about correlation as the general idea of two variables related in some definite manner. There has not been much mathematics or statistics involved. A simple observation that two variables seem to be related does not tell us much. We need answers to two important questions: how closely are the two variables related and is the relation real or could it have been an accident due to chance? To answer the first question we need a definite measure of the closeness of the relation between two variables. The measure is called the *coefficient of correlation*, designated by the symbol r. After defining a few more terms, we will be ready to show how this value is calculated and interpreted. The answer to the second question may be obtained by referring to the appropriate probability tables.

REGRESSION

The term *regression* has not been used in this discussion since the opening sentence. What does it mean? The dictionary is not of much help, for this is one of those unfortunate terms (like the term "error") that has undergone an evolution, so that its present meaning bears little resemblance to its original meaning. Briefly, regression is the *amount of change* in one variable associated with a *unit change* in the other variable. This definition may be open to criticism on the grounds that it is not sufficiently precise or general enough from a mathematical point of view, but for our purposes it should serve to point up the main distinction between correlation and regression. Note that correlation refers to the fact that two variables are related and to the *closeness* of this relationship. Regression, on the other hand, refers to the *nature* of the relationship.

Let us go back to some familiar sayings and see how the concept of regression crops up in our everyday thinking:

"A penny saved is a penny earned."

"A bird in the hand is worth two in the bush."

"A stitch in time saves nine."

"One picture is worth a thousand words."

Notice that all these sayings imply the correlation of two variables, but they go further and tell us in numerical terms *how* the two variables are related. Taking these sayings literally, we can set up a table such as Table 13.1.

We have followed the customary convention of calling the independent variable X and the dependent variable Y.

Column three of the table is headed *regression equation*. These are all equations of straight lines. The general equation for a straight line is $Y = a + bX$. The symbol a is called the *intercept*, since, when X has the value of zero, $Y = a$; hence the line crosses the Y-axis a units from the origin. When a is zero, the line passes through the origin, for when X equals zero, Y is also equal to zero. The symbol b is called the *slope* since it determines the steepness of the line. It is easy to see that b is the amont of change in Y, associated with a unit change in X. Now this is exactly the way we defined regression. Therefore, it is logical to call b the *regression coefficient*.

TABLE 13.1.
Sayings in mathematical terms

Independent Variable (X)	Dependent Variable (Y)	Regression Equation	Regression Coefficient
Pennies saved	Pennies earned	Y=X	1
Hand birds	Bush birds	Y=2X	2
Stitches in time	Stitches saved	Y=9X	9
Pictures	Words	Y=1000X	1000

CORRELATION VERSUS REGRESSION

For any given problem, which type of analysis should we employ? Some statisticians insist on drawing a sharp distinction between the two types of analysis. The distinction is based on whether the data conform to *model I*, in which the X values are fixed, or to *model II*, in which the values of X are random or subject to error.

Consider an experiment in which we deliberately apply several levels of some treatment, replicating each level several times. In this case we are primarily interested in the amount of change in Y associated with changes in the treatment level (X). This is regression. On the other hand, the coefficient of correlation (r), as an estimate of a population parameter (ρ), has no meaning. We are not dealing with a population that possesses such a parameter. However, the square of this coefficient (r^2), known as the *coefficient of determination*, has real meaning in such a problem. It represents the proportion of the total treatment sum of squares accounted for by regression.

Consider cases in which we are dealing with a *bivariate normal distribution* and neither variable can be designated as dependent on the other. Such cases definitely conform to model II, and we are primarily interested in the degree of association between the two variables, measured by the coefficient of correlation. A regression equation for estimating the value of one variable from the other is of little interest. Still we can calculate two such equations according to which variable we call independent. In reality, the coefficient of correlation is the geometric mean of the two regression coefficients so obtained.

From the above two cases we can see that the type of data will determine whether it is correlation or regression that is of primary interest, but we cannot completely separate the two types of analysis.

Between these two rather clear-cut cases there are many in which there is no question as to which variable to consider dependent, but there is some question as to whether the independent variable should be considered *random* or *fixed*. Even in experiments in which the treatments consist of specific amounts of some material, we cannot claim that each plot receives precisely the amount specified or that every replicate receives exactly the same amount. Nevertheless, these measurement errors are very small when compared with the sampling error in a random sample from a population with widely varying rates. Therefore the X values in such an experiment are considered fixed.

The situation is less clear when we are dealing, not with a planned experiment, but with pairs of measurements made on a series of individual units selected from a population. If the selection of individual units is made completely at random, then there is no question but that we are dealing with a model II regression problem where the X's are random. If, on the other hand, we select the individual units deliberately to provide us with a series of X values over a given range, then it is generally conceded that we can consider the X's as fixed.

We can see that the distinction between model I and model II regression problems is not a very sharp one. In the problems we shall consider, and in fact in most agricultural research, we are primarily interested in reasonably good fitting regression equations to describe the relation between variables. In addition, we are interested in determining how closely the regression equation fits the observed data, and for this purpose we calculate the coefficient of determination, or the square of the coefficient of correlation.

To illustrate the general methods of linear correlation and regression, we first use an example of a series of individual pairs of observations.

CALCULATING LINEAR CORRELATION

A familiar example of correlation is the relation of supply to price. Table 13.2 shows the supplies and prices of hogs from 1950 to 1959.

Is there a real relation between supply and price during this period? One of the first things we notice is that the highest price was accompanied by the lowest production and vice versa. This is encouraging evidence of the negative correlation we might expect. Next, let us get a better idea of the data by "drawing a picture." This we do easily by placing dots on a graph paper, letting the height above the X-axis represent the price, and the distance to the right of the Y-axis represent the number of hogs in the corresponding year (Fig. 13.1).

A graph of this type is called a *scatter diagram*. If we thought that the correlation between supply and price was very close, the rather haphazard scatter of these points might prove disappointing. Yet there does seem to be a general trend for the dots on the left to be higher than those on the right. The points seem to fall within a fairly long ellipse (Fig. 13.1), which is typical of diagrams representing a medium high correlation. Other types of scatter diagrams (Fig. 13.2) are guides to interpreting such graphs. The direction of the axis of the ellipse in our example indicates a negative correlation. Now we are ready to calculate just how close the relation is. First, we use a shortcut approximation.

TABLE 13.2.
Hog supplies and prices

Year	Hogs Marketed (millions) (X)	Price per cwt (dollars) (Y)
1950	73	18.0
1951	79	20.0
1952	80	17.8
1953	69	21.4
1954	66	21.6
1955	75	15.0
1956	78	14.4
1957	74	17.8
1958	74	19.6
1959	84	14.1

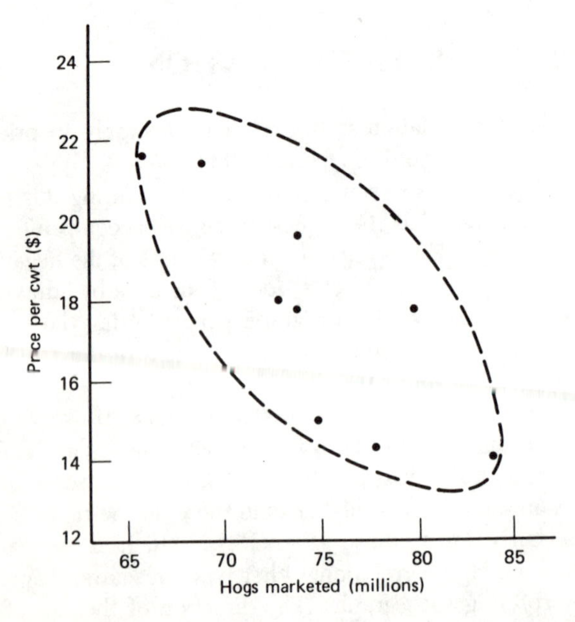

Figure 13.1. Scatter diagram showing relation between price of hogs and number of hogs marketed annually.

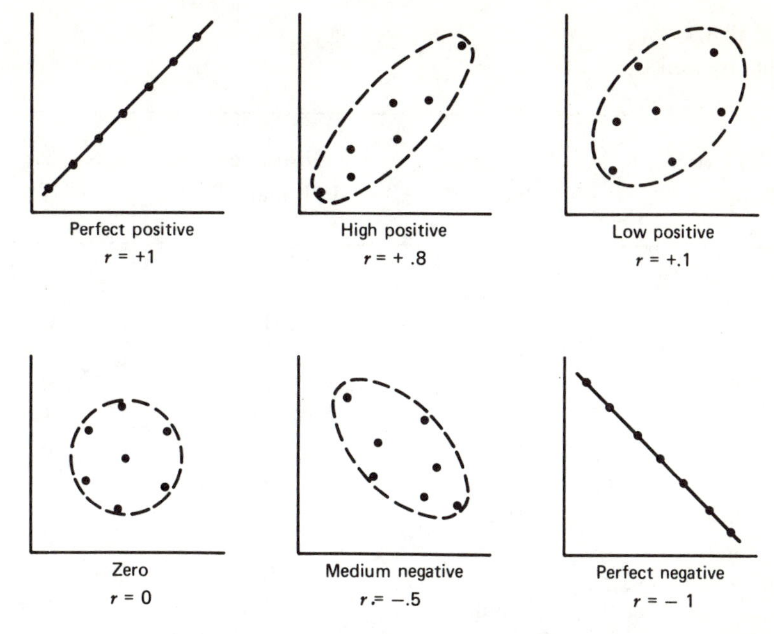

Figure 13.2. Various types of scatter diagrams with their associated coefficients of correlation.

Quick Shortcut Method

This is also known as the *rank difference method* and *Spearman's coefficient of rank correlation*. Neither of these names emphasizes the ease with which the method can be worked, even without the aid of a calculator. Although the method has serious drawbacks, it is very handy for obtaining a quick approximate estimate of the coefficient of correlation. Table 13.3 shows how the calculations are carried out with the hog price data.

First, rank the observations in each column from highest to lowest. In the case of ties, give each member of the tie the average rank. For example, in both columns above, ranks 6 and 7 are ties, so both are called 6.5.

Second, subtract the second from the first number in each row and enter the difference in the column headed d. The total of this column should always be zero, thus furnishing a check.

Third, square the figures in the d column and enter in the d^2 column. Actually, the second step can be omitted, since it is easy to square the numbers in one's head and write down the d^2 column directly.[1]

[1]Most of the d's will be small integers. If they end in a 0.5, squaring can be done mentally by using the following relation: $(X+0.5)^2=X(X+1)+0.25$. Thus, $4.5^2=4\times5+0.25=20.25$, $7.5^2=7\times8+0.25=56.25$, and so on.

TABLE 13.3.
Hog data by ranks

Rank of Supplies	Rank of Price	Difference in Ranks (d)	d^2
8	5	3	9
3	3	0	0
2	6.5	−4.5	20.25
9	2	7	49
10	1	9	81
5	8	−3	9
4	9	−5	25
6.5	6.5	0	0
6.5	4	2.5	6.25
1	10	−9	81
	Totals	0.0	280.5

Fourth, obtain the total of the d^2 column. This total is written Σd^2.
Fifth, calculate the coefficient of correlation, r, by means of the formula:

$$r = 1 - \left[\frac{6\Sigma d^2}{n(n-1)(n+1)} \right]$$

where n is the number of pairs of observations.
In our example,

$$r = 1 - \frac{6 \times 280.5}{10 \times 9 \times 11}$$
$$= 1 - 1.70$$
$$= -.70$$

The answer will always be between +1 and −1. Plus or minus one represents perfect correlation, while zero indicates no correlation at all. Thus, in our example, there appears to be a fairly high negative correlation, so we will calculate the coefficient more accurately using the standard method.

Standard Method

This is known more precisely as the *product-moment method for the coefficient of linear correlation.*

In Chapter 2 we indicated that the deviation of an individual Y from the mean of Y's $(Y-\bar{Y})$ can be represented by an italicized lower case y. Likewise we can use the symbol x for $(X-\bar{X})$. Adopting these shorter symbols greatly simplifies many of the expressions we will encounter, and they will be used frequently in this and succeeding chapters.

The formula for the coefficient of correlation can be written in several forms. It is convenient to write these in terms of r^2 first, then find r by taking the square root of the final answer.

$$r^2 = \frac{\left[\Sigma(X-\bar{X})(Y-\bar{Y})\right]^2}{\Sigma(X-\bar{X})^2\Sigma(Y-\bar{Y})^2} \tag{1}$$

Since $x = X-\bar{X}$ and $y = Y-\bar{Y}$, we can write (1) in abbreviated form:

$$r^2 = \frac{(\Sigma xy)^2}{\Sigma x^2 \Sigma y^2} \tag{2}$$

While these forms are simple, they usually are not easy to calculate directly because they involve the squaring of cumbersome decimals. To avoid this, we take advantage of the relation

$$\Sigma x^2 = \Sigma(X-\bar{X})^2 = \Sigma X^2 - \frac{(\Sigma X)^2}{n}$$

By substituting y for x where necessary, we can rewrite (2) in this form:

$$r^2 = \left[\Sigma XY - \frac{\Sigma X \Sigma Y}{n}\right]^2 \Big/ \left[\left(\Sigma X^2 - \frac{(\Sigma X)^2}{n}\right)\left(\Sigma Y^2 - \frac{(\Sigma Y)^2}{n}\right)\right] \tag{3}$$

This is called the "computational form."

Particular attention should be paid to the expression in brackets in the numerator of equation (3). This is called the *sum of cross-products*. Unlike the familiar sums of squares found in the denominator, which must always be positive, sums of cross-products can be either positive or negative.

Using formula (3), we can now compute the coefficient of correlation for the data in our example, using the standard method. We will need ΣX, ΣY, ΣX^2, ΣY^2 and ΣXY. From the data we find $\Sigma X = 752$, $\Sigma Y = 179.7$, $\Sigma X^2 = 56{,}804.0$, $\Sigma Y^2 = 3{,}297.53$ and $\Sigma XY = 13{,}420.40$. Therefore

$$r^2 = \left[13{,}420.40 - \frac{752 \times 179.7}{10}\right]^2 \Big/ \left[\left(56{,}804.0 - \frac{752^2}{10}\right)\left(3{,}297.53 - \frac{179.7^2}{10}\right)\right]$$

$$= [13{,}420.40 - 13{,}513.44]^2 / [(56{,}804.0 - 56{,}550.4)(3{,}297.53 - 3{,}229.21)]$$

$$= (-93.04)^2 / (253.6 \times 68.32)$$

$$= 0.4996$$

$$r = \sqrt{r^2} = \sqrt{0.4996} = -0.707$$

Note that the sign of r, must be the same as the sign of Σxy,—in this case, negative. The answer by the shortcut method was -0.70, very close to the answer by the standard method, -0.707. Do not be too enthusiastic about this coincidence. The answers by the two methods will not usually be this close. In Chapter 14 we illustrate a case in which the shortcut method gives perfect correlation and is extremely misleading. Other cases could be found in which the shortcut method would give an answer which was much too low.

One would use the shortcut method for a quick check without the use of a calculator or when only an approximate answer was deemed sufficient. For a more efficient estimate of the coefficient of correlation, and a test of significance, one should use the standard method.

STATISTICAL SIGNIFICANCE

In the last paragraph we mentioned *significance*. The general idea is the same as it was in the analysis of variance. We assume the hypothesis that there is no correlation between the two variables and that any apparent relationship is simply due to chance. This is, as usual, called the *null hypothesis*. Then we ask the question, "If this null hypothesis were true, what is the probability that a value of r would be obtained as large or larger than we observed?" If this probability is 5%, we call the correlation *significant*. If we claim that the correlation is real, we run a 5% risk of being wrong. If the probability is 1% or less, we call the correlation *highly significant* and reject the null hypothesis with only a 1% risk of being wrong.

Fortunately, the difficult computations required to find the required probabilities have been made and summarized (Table A.7). Looking at the table on the line opposite 8 degrees of freedom, we find that a coefficient of correlation of .7 would occur by chance somewhere between 1% and 5% of the time. We can say, therefore, that the correlation is significant. We should be very careful in interpreting data of this type. Even if the correlation is significant, we need to be cautious about claiming that a fluctuation in supply *causes* a fluctuation in price. Price and supply may both be related to time, a third variable that has not been considered in the calculations. At the end of this chapter we discuss some of the pitfalls encountered in working with correlation, and an example will be given to show how risky it is to interpret the correlation between two variables that are both related to time.

Why 8 degrees of freedom? We have been accustomed to using *one* less than the number of items as the degrees of freedom, but, now with 10 pairs of observations, we use *two* less, or 8, as the number of degrees of freedom. For the first time, it becomes obvious why care was used in saying that degrees of freedom were *usually* one less than the number of items. Here is the first exception we have encountered. The reason commonly given for subtracting two is that one degree is lost in calculating the mean and the other is lost for regression.

To make matters simpler, let us look at it another way. Suppose we have two pairs of observations—any two pairs providing they are not identical. They can be represented on a graph as two points, and a line can be drawn through them. We call this line the regression line, and the two points fit it perfectly. Since this would be true for any two pairs of observations, no matter how unrelated, it would be ridiculous to attach any meaning to a coefficient of correlation based on only two pairs. Just as one observation cannot tell us anything about variability, two pairs of observations tell us nothing about correlation.

To use a simple illustration of these points, suppose that this morning's paper reports that the Dodgers made 8 runs last night, and a certain stock closed at 51. The day before, the Dodgers made 4 runs and the same stock closed at 49. From these data we can conclude that both Dodger runs and the price of this stock are subject to variation. We can even estimate the amount of variation in both cases, but the estimate will be very rough, since in each case it is based on only 1 degree of freedom $(n-1)$. What about the relation between the two variables? It is easy to verify that

$$r^2 = \frac{(\Sigma xy)^2}{\Sigma x^2 \Sigma y^2} = \frac{4^2}{(8)\,(2)} = 1$$

Hence, $r = 1$

Wouldn't it be absurd to maintain that there was perfect correlation between the number of Dodger runs and the price of a certain stock on the same day? Yet that is what the coefficient of correlation apparently says. We get around such absurdity if we say that this correlation was based on $(n-2)$ or zero degrees of freedom and is, therefore, meaningless.

How often have you heard people draw sweeping conclusions regarding correlations based on a very few observations? Imagine a person flying from San Francisco to Denver for the first time and generalizing, "the farther east one goes, the colder it gets." (Or, to sound profound, the person might say, "I have observed a positive correlation between temperature and longitude.") This illustration is not so farfetched, for it is not unusual to find people making broad generalizations from scanty observations. It is a fault we must try to avoid, and the science of statistics is designed to help us avoid this pitfall.

THE REGRESSION LINE

So far, in our example dealing with supply and price, we have determined only the closeness of the relation and the probability that it was due to chance. We have not learned anything about *how* the two variables are related.

If we assume the relation is linear, that is, best described by a straight line, the question is reduced to that of finding the particular straight line that fits the data the closest. What do we mean by the *closest fit*? It is obvious from looking at the graph of the data that no straight line can be constructed passing through all the points. No matter what line we construct, several points will deviate from that line. We measured variation among a single set of observations by taking the sum of squares of deviations from the mean. It seems logical, then, to measure the variation from a line by taking the sum of squares of deviations from the line. Using this measure as the criterion for closeness of fit, we try to find the straight line that will make the sum of squares of deviations as small as possible. Such a procedure is called a *least-squares method.* Those familiar with calculus will immediately recognize this problem as a typical one involving finding the minimum value of a function.

The solution to the problem turns out to be very simple. In terms of deviations from the means of X and Y, the equation of the best fitting line is:

$$\hat{y} = \left(\frac{\Sigma xy}{\Sigma x^2}\right)x$$

($\hat{y}$ is read: "the estimated value of y").

The expression $\Sigma xy/\Sigma x^2$ is the *regression coefficient,* since it tells us the estimated change in y, with each unit change in x. This fits our definition of regression, and we have already called the regression coefficient b, so we can now say: $b = \Sigma xy/\Sigma x^2$. More precisely, we should call this "the regression coefficient of Y on X," and use the symbol b_{yx}. Generally, if b is used with no subscript, this is the coefficient understood.

The equation given above can be rewritten in terms of the observations themselves, instead of in terms of deviations from means. We can write: $(\hat{Y} - \bar{Y}) = b(X - \bar{X})$ which can be rewritten: $\hat{Y} = (\bar{Y} - b\bar{X}) + bX$.

If we let $\bar{Y} - b\bar{X} = a$, the equation can be written $\hat{Y} = a + bX$, which is the slope-intercept form of a straight-line equation mentioned at the beginning of our discussion on regression.

Now, let us see how to apply this equation to our data. We already have all of the sums we need from the calculation of r, the coefficient of correlation. There we found that

$$\Sigma X = 752; \text{ so } \bar{X} = \frac{752}{10} = 75.2$$

$$\Sigma Y = 179.7; \text{ so } \bar{Y} = 17.97$$

$$\Sigma xy = -93.04$$

$$\Sigma x^2 = 253.6; \text{ so } b = -\frac{93.04}{253.6} = -.367$$

Therefore, substituting in the equation

$$\hat{Y} = (\bar{Y} - b\bar{X}) + bX$$

TABLE 13.4.
Observed and estimated hog prices

X	Y	$\hat{Y}=45.57-.367X$	$d=Y-\hat{Y}$	d^2
73	18.0	18.8	−0.8	0.64
79	20.0	16.6	3.4	11.56
80	17.8	16.2	1.6	2.56
69	21.4	20.2	1.2	1.44
66	21.6	21.4	0.2	0.04
75	15.0	18.1	−3.1	9.61
78	14.4	16.9	−2.5	6.25
74	17.8	18.4	−0.6	0.36
74	19.6	18.4	1.2	1.44
84	14.1	14.7	−0.6	0.36
		Totals	0.0	34.26

we get

$$\hat{Y}=\left[17.97-(-.367)75.2\right]+(-.367)X$$

$$\hat{Y}=45.57-.367X$$

This equation can be put into these words: "Starting with a base price of $45.57 per cwt, every unit (million) increase in annual hog marketings is associated with an average reduction in price of 0.367 dollars per cwt."

Table 13.4 compares the observed values of Y with the estimated values ($\hat{Y}$'s), based on the regression equation.

The fact that the sum of deviations is zero serves as a check on the calculations. This will always be true (except for rounding errors). The sum of squares of deviations can be calculated in a much simpler way from the following formula:

$$\Sigma d^2=(1-r^2)\Sigma y^2$$

In our example

$$\Sigma d^2=(1-.4996)68.32=34.19$$

which is an answer very close to 34.26 shown in Table 13.4. The small difference is due to rounding.

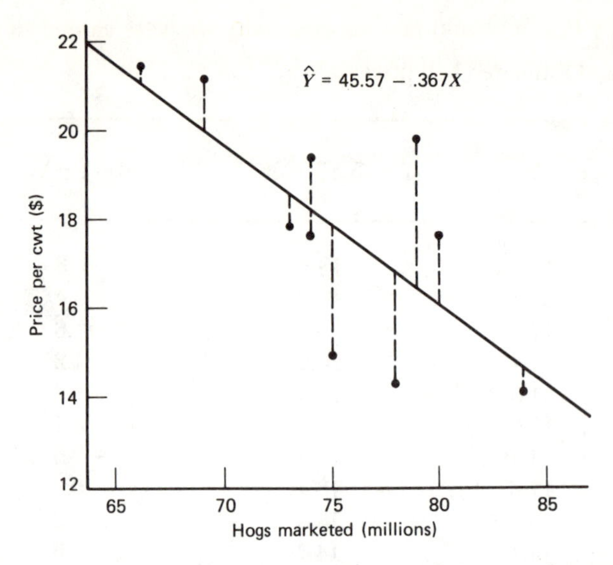

Figure 13.3. Regression line for hog data showing deviations from regression.

This sum of squares, Σd^2, is called the *sum of squares due to deviation from regression* and the square root of the quantity $\Sigma d^2/(n-2)$ is called the *standard error of estimate*. This is just another kind of standard error, similar to the ones we have encountered before. It is a measure of the amount of variation from the regression line.

It is not usually necessary to go to all the trouble of constructing a table like Table 13.4 to check the correctness of the regression line. Constructing the line on the scatter diagram will usually reveal any gross errors. The construction of the line is very simple, since only two points are necessary to determine any line. One point can be on the Y-axis, *a* units (in this case, 45.57) from the origin. Another can be the point representing $\bar{X}$ (the mean of X) and $\bar{Y}$ (the mean of Y). The line passing through these two points will be the required regression line. Figure 13.3 shows the line in our example drawn through the observed points. The dotted lines drawn from the observed points to the regression line represent the deviations. Note that the scales at the bottom and side of the graph do not begin at zero. They are designed to include just slightly more than the range of the observations.

You will notice that the deviations are represented as *vertical* lines. It is the sum of squares of these deviations that we have minimized to come up with the *closest fitting line*. Suppose we decide to construct a line such that the sum of squares of the *horizontal* deviations from the points to the line is a minimum. Will this give the same line? The answer is no, unless there is perfect correlation. This new line will have the equation

$$\hat{x} = \left(\frac{\Sigma xy}{\Sigma y^2}\right) y$$

The expression $\Sigma xy/\Sigma y^2$ is called the *regression coefficient of X on Y* and is designated by b_{XY}. It should now be clear why we were careful to point out that the symbol b is understood to mean b_{YX}, the regression of Y on X, unless otherwise specified.

There is a reason for mentioning that there are two *best-fitting lines* according to which way the deviations are taken. Note that

$$b_{YX} \cdot b_{XY} = \frac{\Sigma xy}{\Sigma x^2} \cdot \frac{\Sigma xy}{\Sigma y^2} = r^2$$

This brings out the relation between the regression coefficients and the coefficient of correlation.

We can now answer the questions raised about data in our example.

1. How close was the relation between supply and price?
   ANSWER: Fairly close. The coefficient of correlation was $-.7$, and ± 1 would be perfect.

2. What is the probability that such a correlation could be due to chance?
   ANSWER: A correlation of this size from 10 pairs of observations would occur between 5 and 1% of the time by chance alone.

3. What equation would best describe the relation between price (Y) and supply (X) from these data?
   ANSWER: $\hat{Y} = 45.57 - .367X$

4. How well does this line fit the data?
   ANSWER: The sum of squares of deviations of the observed points from the line was 34.19 or about one-half the total price variation. Thus, only half the price variation was in some way associated with variation in supply. A simple analysis of variance table shows this (Table 13.5).

TABLE 13.5.
Regression analysis arranged in an analysis of variance form

Source of Variation	Degrees of freedom	Sum of squares	Mean Square	F
Total	9	$\Sigma y^2 = 68.32$		
Regression	1	$r^2 \Sigma y^2 = 34.13$	34.13	7.99*
Deviation from regression	8	$(1 - r^2) \Sigma y^2 = 34.19$	4.27	

*See page 127 for note.

Notice from this table that r^2 is the proportion of the total sum of squares accounted for by regression, and $(1-r^2)$, sometimes called the *coefficient of alienation*, is the proportion not accounted for.

The fact that the F value of 7.99 lies between the required F value at the 5% point (5.32) and the 1% point (11.26) for 1 and 8 degrees of freedom, verifies our previous finding in answer to question 2.

In fact, it does not matter whether we look up the F value in Table A.3 or the r value in Table A.7. The two tests are identical, as can be easily shown. From the analysis of variance in Table 13.5 we can see that, in symbolic terms,

$$F=\frac{r^2\Sigma y^2}{(1-r^2)\Sigma y^2/(n-2)}=r^2(n-2)/(1-r^2)$$

Solving this equation for r^2 gives

$$r^2=\frac{F}{F+n-2}$$

We can substitute a required F value in this equation and take the square root to find the required r value. For example, the required 1% F value is 11.26 for 1 and 8 degrees of freedom. Substituting this in the equation above, we get

$$r^2=\frac{11.26}{11.26+10-2}=.5846$$

$$r=.7646$$

This is the value in Table A.7 for 8 degrees of freedom at the 1% level.

CONFIDENCE LIMITS

The *deviation mean square* (DMS) provides the basic quantity for the calculation of several confidence limits. The variance of the regression coefficient is

$$s_b{}^2=\frac{DMS}{SSX}$$

and the confidence limits are

$$b\pm t(s_b)$$

In our example of hog supplies and prices,

$$s_b{}^2=\frac{4.27}{253.6}=0.0168$$

$$s_b=0.1298$$

The tabular t value for 8 degrees of freedom at the 5% level is 2.306, so the 5% confidence limits are

$$-0.367 \pm 2.306(0.1298)$$

$$= -0.367 \pm 0.299 = -0.666 \text{ and } -0.068$$

The t value at the 1% level is 3.355, so the 1% confidence limits are:

$$-0.367 \pm 3.355(0.1298)$$

$$= -0.367 \pm 0.435 = -0.802 \text{ and } +0.068$$

Notice that the 5% confidence limits do not bracket zero, but the 1% limits do. This agrees with the previous conclusions that the regression is significant at the 5% level but not at the 1% level.

The estimates of Y designated as $\hat{Y}$ are subject to two kinds of error: the variance of the mean and the variance of the regression coefficient. The variance of $\hat{Y}$ is

$$s_{\hat{Y}}^2 = \text{DMS}\left(\frac{1}{n} + \frac{x^2}{\text{SSX}}\right)$$

Notice that the size of this variance depends on the value of x (the deviation of X from the mean $\bar{X}$). The confidence limits for $\hat{Y}$ are

$$\hat{Y} \pm t(s_{\hat{Y}})$$

In our example,

$$s_{\hat{Y}}^2 = 4.27\left(\frac{1}{10} + \frac{x^2}{253.6}\right)$$

$$= .427 + 0.0168x^2$$

Confidence limits associated with several values of x are given in Table 13.6. Plotting these values gives a "confidence belt" around the regression line bounded by two curves, shown as the inner belt in Figure 13.4.

The confidence limits we have just calculated apply to the means of populations of Y values associated with specific values of X. It is often of more interest to set confidence limits on the predictions of single values of Y, given specific values

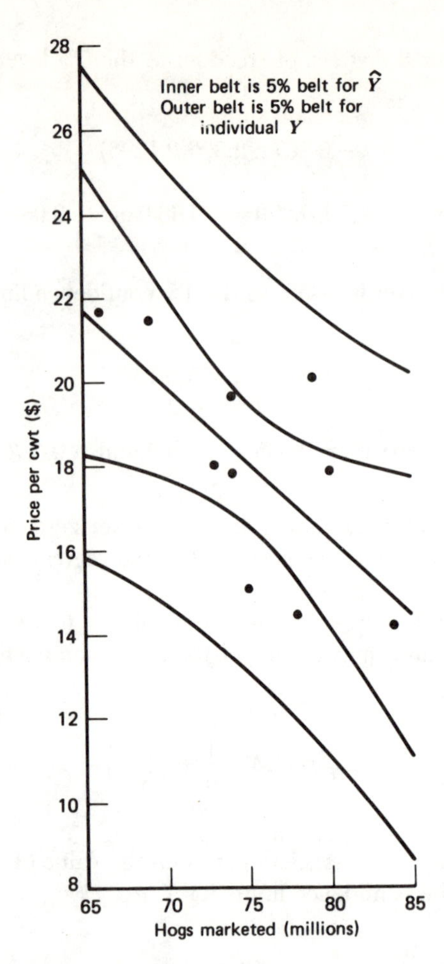

Figure 13.4. Confidence belts around regression line of hog data.

of X. Here we must take into account an additional source of error. In addition to the error of the regression coefficient and the error of the mean, we have the variation of individuals around the estimated mean.

The total variance of Y for a specific value of $X-\bar{X}$ or x is

$$s_Y^2 = \text{DMS}\left(1 + \frac{1}{n} + \frac{x^2}{\text{SSX}}\right)$$

and the confidence limits are

$$\hat{Y} \pm t(s_Y)$$

Limits for selected values of x are given in Table 13.6 and plotted in Figure 13.4 as the outer confidence belt.

TABLE 13.6.
Variances, standard errors and 5% confidence limits of $\hat{Y}$ and Y associated with selected values of X in hog price and supply example[a]

X	x	$\hat{Y}$	$s_{\hat{Y}}^2$	$s_{\hat{Y}}$	$t(s_{\hat{Y}})$	Lower Limit	Upper Limit	s_Y^2	s_Y	$t(s_Y)$	Lower Limit	Upper Limit
65.2	−10	21.64	2.11	1.45	3.35	18.29	24.99	6.28	2.53	5.82	15.82	27.46
67.2	−8	20.91	1.50	1.23	2.83	18.08	23.74	5.77	2.40	5.54	15.37	26.45
69.2	−6	20.17	1.03	1.02	2.34	17.83	22.51	5.30	2.30	5.31	14.86	25.48
71.2	−4	19.44	0.70	0.83	1.92	17.52	21.36	4.97	2.23	5.14	14.30	24.58
73.2	−2	18.71	0.49	0.70	1.61	17.10	20.32	4.76	2.18	5.03	13.68	23.74
75.2	0	17.97	0.43	0.65	1.51	16.46	19.48	4.70	2.17	5.00	12.97	22.97
77.2	2	17.23	0.49	0.70	1.61	15.62	18.84	4.76	2.18	5.03	12.20	22.26
79.2	4	16.50	0.70	0.83	1.92	14.58	18.42	4.97	2.23	5.14	11.36	21.64
81.2	6	15.77	1.03	1.02	2.34	13.43	18.11	5.30	2.30	5.31	10.46	21.08
83.2	8	15.04	1.50	1.23	2.83	12.21	17.87	5.77	2.40	5.54	9.50	20.58
85.2	10	14.30	2.11	1.45	3.35	10.95	17.65	6.28	2.53	5.82	8.48	20.12

[a]Note: The t value used in these calculations was 2.306, the tabular t value at the 5% level for 8 degrees of freedom.

REGRESSION IN REPLICATED EXPERIMENTS

We have shown in Chapters 6, 9, and 10 how we can use an orthogonal set of coefficients to find the sum of squares due to linear regression. This method is applicable only to certain sets of treatment levels, but the general methods of this chapter can be used for any series of treatment levels.

The data from Chapter 10 will be used to illustrate the general methods, and later in Chapter 15 the same data will be analyzed with the shortcut method.

In table 13.7, we designate as Y the totals of the five harvest dates. Working with totals rather than means reduces the amount of rounding errors. We will fit a straight regression line to these values and test its significance.

$$\Sigma x^2 = \Sigma X^2 - \frac{(\Sigma X)^2}{5} = 55 - \frac{15^2}{5} = 10$$

$$\Sigma y^2 = \Sigma Y^2 - \frac{(\Sigma Y)^2}{5} = 571{,}372.24 - \frac{1600^2}{5} = 59{,}372.24$$

TABLE 13.7.
Totals (Y) from time of harvest treatments (X) in the sugar beet experiment in Chapter 10

X	Y	X^2	XY	Y^2	$\hat{Y}$	$Y-\hat{Y}$	$(Y-\hat{Y})^2$
1	140.0	1	140.0	19,600.00	169.8	−29.8	888.04
2	267.2	4	534.4	71,395.84	244.9	22.3	497.29
3	335.2	9	1,005.6	112,359.04	320.0	15.2	231.04
4	417.0	16	1,668.0	173,889.00	395.1	21.9	479.61
5	440.6	25	2,203.0	194,128.36	470.2	−29.6	876.16
Total 15	1,600.0	55	5,551.0	571,372.24	1600.0	0.0	2,972.14

$$\Sigma xy = \Sigma XY - \frac{\Sigma X \Sigma Y}{5} = 5{,}551.0 - \frac{15(1600)}{5} = 751.0$$

$$b = \frac{\Sigma xy}{\Sigma x^2} = \frac{751}{10} = 75.1$$

$$a = \overline{Y} - b\overline{X} = 320 - 75.1(3) = 94.7$$

$$r^2 = \frac{(\Sigma xy)^2}{(\Sigma x^2)(\Sigma y^2)} = \frac{751^2}{(10)(59{,}372.24)} = .94994$$

$$\text{SSY (on a per-plot basis)} = \frac{\Sigma y^2}{16} = \frac{59{,}372.24}{16} = 3710.765$$

$$\text{SS Regression} = r^2\ (\text{SSY}) = .94994(3710.765) = 3525.004$$

$$\text{SS Deviation} = (1 - r^2)(\text{SSY}) = .05006(3710.765) = 185.761$$

The sum of squares of deviations shown in the last column of Table 13.7 can be reduced to a per-plot basis by dividing by 16: 2,972.14/16 = 185.759, which agrees with the value given above except for rounding errors.

The regression equation was calculated from the treatment totals, and if we wish to have an equation for estimating the means, we divide a and b by 16, giving us the equation

$$\hat{Y} = 5.91875 + 4.69375X$$

The sums of squares can be summarized in an analysis of variance table, Table 13.8.

TABLE 13.8.
ANOVA table for the regression of sugar beet yield on harvest date (Table 13.7)

Source of Variation	df	SS	MS	F	Required F 5%	Required F 1%
Harvest dates	4	3710.765	927.691	111.92	3.26	5.41
Regression	1	3525.004	3525.004	425.26	4.75	9.33
Deviation	3	185.761	61.920	7.47	3.49	5.95
Error	12	99.467	8.289			

Notice that in the hog data, where we were dealing with individual pairs of observations, we used the *deviation mean square* for testing the *regression mean square*. In a replicated experiment, however, we have an error term that we can use for testing both the regression mean square and the deviation mean square.

In our example, the highly significant regression mean square tells us that there is a highly significant trend for the yield of sugar beets to increase as the harvest date is advanced (within the range of dates employed in this experiment). The F value for deviation from regression, while not nearly as large as for regression, is still highly significant. This tells us that there is some highly significant source of variation in addition to the positive linear trend that is affecting the yields. We will examine some of the possible sources in the next chapter.

PITFALLS

Probably no part of statistics is subject to more abuse and misinterpretation than correlation and regression. The statement that "one can prove anything with statistics" is true only if one ignores some of the basic principles involved. The two principles most often ignored in correlation are:

1. The full name of the coefficient of correlation is the coefficient of *linear* correlation, and

2. Nothing in the definition of correlation indicates or implies that the relation between two variables is one of cause and effect. The following are examples of how easy it is to get into trouble.

A LOW CORRELATION DOESN'T ALWAYS MEAN LACK OF RELATION. Look at the following pairs of figures:

X	Y
0	0
1	144
2	256
3	336
4	384
5	400
6	384
7	336
8	256
9	144
10	0

If we calculate the coefficient of correlation between X and Y, we find that it is zero. (Try it and see.) However, if we conclude that there is no relation between X and Y, we would be completely wrong. X is the elapsed time in seconds after shooting an arrow vertically at 160 ft/sec. Y is the elevation of the arrow in feet. Of course, it is utterly ridiculous to contend that there is no relation between the height of an arrow and its time in flight. What is wrong with this paradox? The important word *linear*, implied when we speak of the coefficient of correlation, was ignored. It is true that no straight line will come close to fitting these data, but the equation $Y = 160X - 16X^2$ will give a perfect fit. This is the equation of a parabola.

The moral of this example is that one should be on the lookout for *curvilinear* relations that might fit the data better than a simple linear relation. Ways to handle data of this kind will be presented later.

A HIGH CORRELATION DOES NOT NECESSARILY MEAN A CAUSE AND EFFECT RELATIONSHIP. Consider Table 13.9 from which we can calculate the coefficient of correlation and the regression equation.

The high value of the coefficient of correlation, .937, indicates a close relation between X and Y. One might be tempted to say that each unit change in X *causes* a change of .643 in Y. Now let us see what X and Y represent. The X's are the number of cigarettes used annually in the United States (in billions) from 1944 to 1958. The Y's are the index numbers of production per man-hour for hay and forage crops during the same period. It would require a big stretch of the imagination to think of any direct cause and effect relation between cigarette consumption and efficiency in the hay business. It just happened that both of these variables showed a steady increase with time during the period being considered.

The moral of this example is that the coefficient of correlation will measure the closeness of *relation* between two variables, but it tells us nothing about

TABLE 13.9.
Fifteen pairs of highly correlated data

X	Y		
		$\Sigma X=$	5669
295	73	$\bar{X}=$	377.9
339	78	$\Sigma X^2=$	2,163,935
343	85	$(\Sigma X)^2/15=$	2,142,504
344	91	$\Sigma x^2=$	21,431
357	100	$\Sigma Y=$	1768
359	109	$\bar{Y}=$	117.9
368	119	$\Sigma Y^2=$	218,482
395	125	$(\Sigma Y)^2/15=$	208,388
414	129	$\Sigma y^2=$	10,094
406	135	$\Sigma XY=$	681,962
385	142	$(\Sigma X \Sigma Y)/15=$	668,186
394	139	$\Sigma xy=$	13,776
404	140		
420	147		
446	156		

$$r^2=(\Sigma xy)^2/\Sigma x^2 \Sigma y^2$$
$$=(13{,}776)^2/(21{,}431\times 10{,}094)$$
$$=189{,}778{,}176/216{,}324{,}514=.8773$$
$$r=\sqrt{.8773}=.937 \text{ (coefficient of correlation)}$$
$$b=\Sigma xy/\Sigma x^2=13{,}776/21{,}431=.643 \text{ (regression coefficient)}$$
$$a=\bar{Y}-b\bar{X}=117.9-243.0=125.1 \text{ (intercept)}$$
$$\hat{Y}=-125.1+.643X \text{ (regression equation)}$$

whether this relation is one of *cause* and *effect*. That decision is up to the investigator and must be based on a great deal of knowledge of the variables under study.

WATCH FOR PART–WHOLE CORRELATIONS. Several years ago, a paper presented at a meteorological meeting dealt with studies on length of growing seasons between killing frosts. It was reported that there was little or no correlation between the last frost in the spring and the first frost in the fall over a long period of time. The next conclusion reported was that there was a rather high correlation between dates of last frost in the spring and length of seasons.

If we examine this second conclusion, we note that the length of season is completely determined by two parts, the beginning (last spring frost), and the end (first fall frost). It can be easily proven that, if a variable is made up of two or more independent parts, there is automatically a correlation between any one of the parts and the whole. The relation is simple: r=(standard deviation of part)/(standard deviation of whole). In the case of the frost data, if spring frost dates and fall frost dates are about equally variable, then we expect the correlation between spring frost dates and length of season to be about $\sqrt{.5}$ or .707. The conclusion about the correlation between spring frost and length of season, while correct, was trivial.

EXTRAPOLATION IS TEMPTING BUT DANGEROUS. Often a series of observations fall within a rather restricted range of values for the two variables under study. If they show a high coefficient of correlation, there is a great temptation to extend the regression line beyond the range of observations and try to predict what would happen to the values of Y if X were to take on values above or below those actually observed. This is called *extrapolation*. It is a dangerous practice, because many variables that are related in a curvilinear fashion will give a high linear correlation if only a short section of the curve is sampled.

Table 13.10 gives the measurements of 10 onion bulbs with diameters between 50 and 70 mm. with their corresponding weights in grams.

TABLE 13.10.
Measurements of ten onion bulbs

Diameter (X)	Weight (Y)
51.0	63.4
66.2	115.3
69.2	140.6
69.5	132.6
56.9	80.7
67.1	125.6
58.1	80.0
53.9	78.7
63.0	112.8
60.0	96.2

The calculation of r, the coefficient of correlation and of the regression equation, is as follows:

$\Sigma X = 614.9$	$\Sigma Y = 1031.9$	
$\bar{X} = 61.49$	$\bar{Y} = 103.19$	
$\Sigma X^2 = 38{,}192.17$	$\Sigma Y^2 = 113{,}247.79$	$\Sigma XY = 65{,}014.60$
$(\Sigma X)^2/n = 37{,}810.20$	$(\Sigma Y)^2/n = 106{,}481.76$	$\Sigma X \Sigma Y/n = 63.451.53$
$\Sigma x^2 = 381.97$	$\Sigma y^2 = 6{,}766.03$	$\Sigma xy = 1{,}563.07$

$$r^2 = (1{,}563.07)^2/(381.97 \times 6{,}766.03) = .9454$$

$$r = \sqrt{.9454} = .97 \text{ (coefficient of correlation)}$$

$$b = 1{,}563.07/381.97 = 4.092 \text{ (regression coefficient)}$$

$$a = 103.19 - (4.092)(61.49) = -148.43 \text{ (intercept)}$$

$$\hat{Y} = 4.092X - 418.43 \text{ (regression equation)}$$

The correlation of .97 between diameter and weight is very high. (This is not surprising.) Within the range of 50 to 70 mm, a straight line equation describes the relation between the two variables very well.

Now let us extrapolate and see what happens. A bulb measuring 92.4 mm was found to weigh 300.2 g, but our estimate of weight from the regression equation is

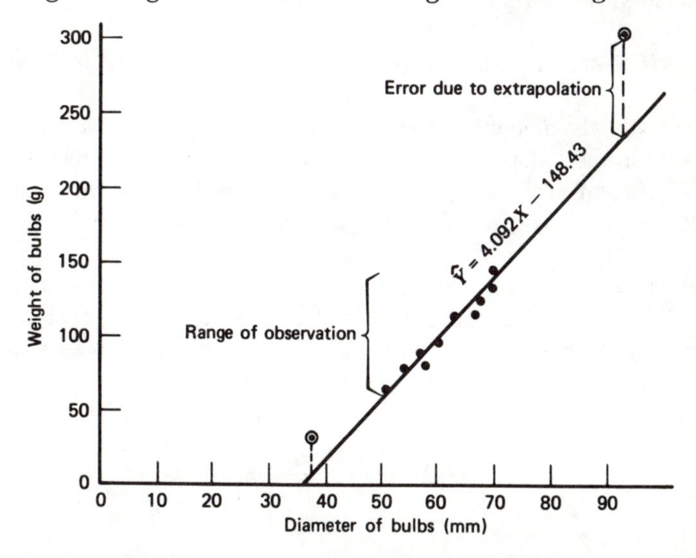

Figure 13.5. Regression line drawn through onion data over a limited range, showing the danger of extrapolation from limited observations.

229.7. Extrapolation caused us to err by 70.5 g in our estimate. Going in the other direction, a bulb measuring 37.8 mm weighed 27.8 g, but extrapolation gave an estimate of 6.2 g. Extrapolating for still smaller values of X soon gives us completely absurd estimates of Y. For example, a 36.27 mm bulb would be estimated to weigh nothing, and all bulbs smaller than this, less than nothing. Figure 13.5 shows the line fitted to the data and the effects of extrapolating.

It is easy to see why extrapolation leads us so far astray in this case. The linear regression equation implies that a given amount added to the diameter of a bulb will add a certain fixed amount to the weight. It should be obvious, however, that this cannot be so. One centimeter added to a 9 cm bulb will certainly result in a greater increase in weight than 1 cm added to a 2 cm bulb.

If one wishes to find out how two variables are related outside the range of his observations, the safest procedure is to make more observations in the region of interest.

SUMMARY

Correlation is the tendency of two variables to be related in a definite manner.

The two variables are called *independent* and *dependent*, according to which one is viewed as depending on the other. The independent variable is called X and the dependent variable Y.

The *coefficient of correlation* measures the *closeness* of the relationship.

Regression is the amount of change in the dependent variable associated with a unit change in the independent variable. A linear regression equation is written $\hat{Y} = a + bX$, where $\hat{Y}$ is the *estimated value of* Y, a is the *intercept* or point where the line crosses the Y axis, and b is the *slope* or *regression coefficient*.

Graphing a set of data made up of pairs of variates produces a *scatter diagram*. This is usually a convenient first step in regression analysis. A *quick shortcut method* known as the *rank difference method* gives an easy-to-calculate approximation to the coefficient of correlation. The formula is

$$r = 1 - \frac{6\Sigma d^2}{n(n-1)(n+1)}$$

where r is the coefficient of correlation, d is the difference in rank in each pair of observations, and n is the number of pairs.

The *standard method* or *product-moment method* can be expressed by several formulas:

$$r^2 = \frac{\left[\Sigma(X-\bar{X})(Y-\bar{Y})\right]^2}{\Sigma(X-\bar{X})^2\Sigma(Y-\bar{Y})^2} \text{ (direct observational form)}$$

$$r^2 = \frac{(\Sigma xy)^2}{\Sigma x^2 \Sigma y^2} \text{ (deviation from mean form)}$$

$$r^2 = \left[\Sigma XY - \frac{\Sigma X \Sigma Y}{n}\right]^2 \Big/ \left[\left(\Sigma X^2 - \frac{(\Sigma X)^2}{n}\right)\left(\Sigma Y^2 - \frac{(\Sigma Y)^2}{n}\right)\right] \text{ (computational form)}$$

$$r^2 = b_{YX} b_{XY} \text{ (regression form)}$$

$r = \pm\sqrt{r^2}$ (The sign will correspond to the sign of the number inside the brackets of the numerators in equations 1 to 3. It will correspond to the sign of b_{XY} in equation 4.)

The significance of the coefficient of correlation can be determined by reference to a special r table, using $n-2$ degrees of freedom; where n is the number of pairs of observations.

Correlations based on only two pairs of observations will always be plus or minus one, but they are meaningless.

The *regression coefficient* is: $b = \Sigma xy / \Sigma x^2$.

The intercept is: $a = \bar{Y} - b\bar{X}$.

When a and b are determined, we can write the regression equation, $\hat{Y} = a + bX$.

Lack of agreement between observed and estimated values of Y is measured by the *sum of squares due to deviation from regression*, obtained from the relation: $\Sigma d^2 = (1 - r^2)\Sigma y^2$.

The sum of squares due to deviations divided by its degrees of freedom (n−2) gives the *deviation mean square* (DMS). The square root of the deviation mean square is called the *standard error of estimate*.

The sum of squares due to regression can be obtained directly:

$$SSR = r^2 \Sigma y^2$$

or by subtraction:

$$SSR = \Sigma y^2 - \Sigma d^2$$

The *mean square due to regression* is the same as the sum of squares, since it has only 1 degree of freedom. A test of significance based on F=(regression mean square)/(deviation mean square) can be checked in an F table under 1 and (n−2) degrees of freedom. This will give the same test as reference to an r table.

The variance of the regression coefficient is ${s_b}^2 = DMS/SSX$. The confidence limits for b are: $b \pm t(s_b)$. The variance of an estimated $\hat{Y}$ is ${s_{\hat{Y}}}^2 = DMS(1/n + x^2/SSX)$. The confidence limits for $\hat{Y}$ are: $\hat{Y} \pm t(s_{\hat{Y}})$. The variance of an individual prediction of Y is: ${s_Y}^2 = DMS(1 + 1/n + x^2/SSX)$. The confidence limits for a single estimated Y are: $\hat{Y} \pm t(s_Y)$.

In a replicated experiment, the regression mean square and the deviation mean square can be tested with the same error term used to test the total treatment mean square.

It should always be remembered that the ordinary coefficient of correlation assumes a *linear* relation between the two variables. Also, it cannot help us decide whether the relation is one of *cause* and *effect*.

A low coefficient of correlation doesn't always mean a lack of relation. There may be a very close *curvilinear relation*.

A high coefficient of correlation does not imply a direct cause and effect relationship. The two variables may simply both be related to a third variable, such as time.

Avoid correlation of a variable with one of its component parts. The conclusions reached are trivial.

Avoid extrapolation of a regression line beyond the range of observations.

14

CURVILINEAR RELATIONS

In the previous chapter we warned repeatedly to keep in mind that the usual coefficients of correlation and regression are based on a *linear* relationship between two variables. A linear relation is the simplest type of relation found between variables. Even if there are pronounced deviations from linearity for extreme values of X and Y, it often happens that, within the useful or practical range of values of the variables, a straight line is sufficient to characterize the relationship. For example, in fertilizer tests we often notice that there is a steady increase in yield with increased application of some nutrient up to a point. Above that point the increase in yield may be less pronounced, and finally the yield will actually decrease as we use excessive amounts of fertilizer. If we are interested only in low to medium fertilizer applications, a straight line may be satisfactory for describing the relation between yield and fertilizer. If we wish to describe this relation through the whole range of applications from zero to extremely high, we will probably have to use a curve that reaches a maximum and then decreases.

DECIDING WHAT CURVE TO USE

Since there are so many different kinds of curves we might use to express the relation between two variables, we first have to decide what kind of curve we are going to try to fit to the data. It would be desirable to find one that expresses some *natural* relation between the two variables, but this is not always possible. Sometimes a thorough knowledge and experience with the variables we are studying enables us to select one type of curve that is more logical than others. We will cite some examples of this as we go along. Sometimes the converse is true. Finding a curve that fits the data closely may give us an important clue as to a natural relation that exists between two variables. Many of our natural laws were discovered in this way; for example, Boyle's law, Charles' law, and the law of falling bodies.

With biological data, the relation between two variables may be so complex that no simple equation can suffice to describe the relationship. We often must be content to find an equation that fits the data reasonably well without making any claims that the equation expresses any natural relation. It is always possible to find a curve that will fit the data perfectly, but such a curve may be strictly artificial and completely devoid of physical or biological meaning.

From a multitude of types of curves, we have selected five for consideration. These were chosen first, because they are the most common ones encountered in biological and economic data and second, because only elementary mathematical ideas need be utilized in discussing them.

The Power Curve

This is a curve where Y is a function of some power of X. The general form of the equation for a curve of this type is

$$Y = aX^b$$

If we take the logarithm of both sides of this equation, we get

$$\log Y = \log a + b \log X$$[1]

If we let the logs of X and Y be the variables, calling them X′ and Y′, and the constant $\log a$ is called a', we can rewrite the equation:

$$Y' = a' + bX'$$

This is easily recognizable as the general equation for a straight line discussed in the preceding chapter. Therefore, all we have to do to analyze data of this type is to transform the observations to logarithms, then proceed exactly as we did with linear correlation and regression.

The value of b can be positive or negative and a whole number or a fraction. Figure 14.1 shows some of the wide variety of curve shapes that result from different values of b. After transformation of X and Y to logarithms, all of these curves become straight lines with slope b, as shown on the right-hand side of the figure.

The effect of a on the original curves is to compress or expand the scale on one of the axes, while its effect on the log transformed line is simply to move it up or down without changing its slope.

Since only positive numbers have logarithms, the log form of the equations has no meaning for negative values of X. Thus, we should apply the log transformation only to data where all the observations of X and Y are positive. This is not really a very serious restriction, since many physical measurements, such as weight, length, area, and so forth, take only positive values.

How do we know whether it is plausible to use the log transformation? Here again, the use of a graph gives a good start. Graphing can be done in two ways. The observed values of X and Y can be converted to logarithms and plotted on ordinary graph paper. An even simpler method is to plot the original values on a graph paper called log paper. With either method, a scatter diagram will result. If

[1]For those who do not recall the rules of logarithms and exponents, a review will be helpful in this discussion. Any elementary algebra text can be consulted.

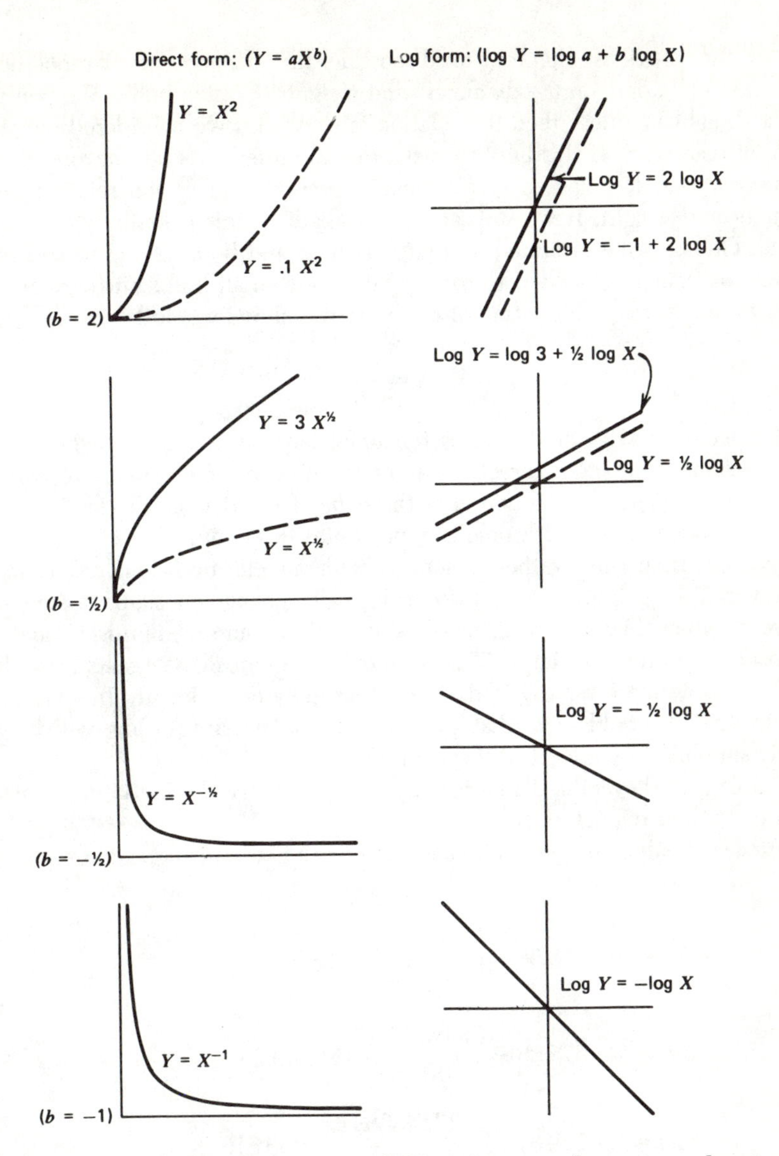

Figure 14.1. Various shaped curves with their log transformations showing how logs can convert curves to straight lines.

this scatter diagram has the appearance of a long narrow ellipse, typical of linearly correlated data, we can proceed to analyze the logarithms of X and Y.

From a logical point of view, we would expect data based on measurements involving two different numbers of dimensions to fit curves of the form $Y = aX^b$. For example, height is one-dimensional, while weight, being related to volume, is three-dimensional. Therefore, in correlating height with weight, it would be logical to try the log transformation. The same would be true with measurements of width and area, length and volume, surface and diameter, and the like.

In the previous chapter, in discussing the dangers of extrapolation, we presented some data on the diameters and weights of onion bulbs. We pointed out that a straight line described the relation fairly well if we considered only a short range of diameters. If this line was extended in either direction beyond the range of observations, it failed to give a good representation of the relation between diameter and weight. If we stop and think about it, this is really what we would expect. One centimeter added to a large bulb would be expected to add more to the weight than 1 cm added to the diameter of a small bulb. Furthermore, if the bulbs were spheres, the relation of diameter to volume would be

$$V = \frac{\pi d^3}{6}$$

If the specific gravity of the bulbs remained fairly constant throughout all bulb sizes, weight would be a direct linear function of volume. Therefore, we would expect weight (Y) to be a function of the *cube* of the diameter (X).

The true situation with onions is not quite this simple, since they are rarely spherical in shape but rather spheroids with an elliptical longitudinal section. Moreover, as the bulbs grow, they continually change in shape, being prolate spheroids when they are small, nearly spherical at some medium size, and oblate spheroids when they are large. This constant change in shape results from the fact that they grow more rapidly in diameter than they do in length. In spite of these complexities, it would seem that the type of data we are dealing with might be greatly simplified by a log transformation.

Table 14.1 shows the diameters and weights observed with 30 bulbs, arranged in order of their diameters.

First, we calculate the coefficients of correlation and regression equation for the original data.

$$\Sigma x^2 = 118{,}958.58 - \frac{(1817.2)^2}{30} = 8{,}884.72$$

$$\Sigma y^2 = 542{,}675.26 - \frac{(3383.6)^2}{30} = 161{,}050.29$$

$$\Sigma xy = 241{,}772.67 - \frac{1817.2(3383.6)}{30} = 36{,}816.74$$

$$r^2 = \frac{(\Sigma xy)^2}{\Sigma x^2 \Sigma y^2} = \frac{(36{,}816.74)^2}{8{,}884.72(161{,}050.29)} = .9473$$

$$r = \sqrt{.9473} = .973 \quad b = \frac{\Sigma xy}{\Sigma x^2} = \frac{36{,}816.74}{8{,}884.72} = 4.144$$

$$a = \bar{Y} - b\bar{X} = \frac{3383.6}{30} - \left[4.144\left(\frac{1817.2}{30}\right)\right] = -138.20$$

$$\hat{Y} = -138.20 + 4.144X$$

TABLE 14.1.
Diameters and weights of onion bulbs

	Diameter (X)	Weight (Y)	Log X(X′)	Log Y(Y′)
	35.1	24.3	1.54531	1.38561
	35.3	24.1	1.54777	1.38202
	35.5	24.4	1.55023	1.38739
	37.8	27.8	1.57749	1.44404
	37.8	28.7	1.57749	1.45788
	41.4	42.0	1.61700	1.62325
	41.7	34.5	1.62014	1.53782
	44.8	56.1	1.65128	1.74896
	44.9	49.0	1.65225	1.69020
	47.9	58.4	1.68034	1.76641
	51.0	63.4	1.70757	1.80209
	53.9	78.7	1.73159	1.89597
	56.9	80.7	1.75511	1.90687
	58.1	80.0	1.76418	1.90309
	60.0	96.2	1.77815	1.98318
	63.0	112.8	1.79934	2.05231
	66.2	115.3	1.82086	2.06183
	67.1	125.6	1.82672	2.09899
	69.2	146.6	1.84011	2.16613
	69.5	132.6	1.84198	2.12254
	70.7	142.8	1.84942	2.15473
	73.1	137.1	1.86392	2.13704
	73.1	163.2	1.86392	2.21272
	77.4	180.0	1.88874	2.25527
	81.7	198.0	1.91222	2.29667
	81.7	207.8	1.91222	2.31765
	82.3	190.8	1.91540	2.28058
	83.1	225.5	1.91960	2.35315
	84.6	237.0	1.92737	2.37475
	92.4	300.2	1.96567	2.47741
Totals	1817.2	3383.6	52.90339	58.27655
Sums of squares	118,958.58	542,675.26	93.80268806	116.4541216
Sums of cross-products	241,772.67		104.0495715	

At first glance it looks as though a straight line has given us an excellent fit to the data. The coefficient of correlation, .973, is very high. However, if we look at the graph of the data with the superimposed regression line (Fig. 14.2), we notice a disturbing thing. All the deviations from the line at the ends of the range are positive, while those in the middle of the range are negative. If the deviations were more or less random, we would be satisfied, but this systematic grouping of deviations leads us to expect that a curve would describe the observations still better. There is another even more compelling reason to try to fit a curve. The straight line we have fitted to the data simply does not make sense for diameters less than about 34 mm, for it indicates that bulbs smaller than this would have negative weights.

Now we fit a straight line to the logs of X and Y and see whether these difficulties are overcome. The calculations are exactly the same, except that we replace X with $X' = \log X$, and Y with $Y' = \log Y$.

$$\Sigma x'^2 = 93.80268806 - \frac{(52.90339)^2}{30} = .51039894$$

$$\Sigma y'^2 = 116.4541216 - \frac{(58.27655)^2}{30} = 3.2489123$$

$$\Sigma x'y' = 104.0495715 - \frac{52.90339(58.27655)}{30} = 1.2820031$$

$$r^2 = \frac{1.2820031^2}{.51039894(3.2489123)} = .991129$$

$$r = \sqrt{.9911} = .996$$

$$b = \frac{1.2820031}{.51039894} = 2.5118$$

$$a' = \frac{58.27655}{30} - 2.5118\left(\frac{52.90339}{30}\right) = -2.4869$$

$$\hat{Y}' = -2.4869 + 2.5118X'$$

The coefficient of correlation, .996, indicates an extremely close fit, even higher than that obtained from the untransformed data. The improvement in the correlation is not, however, the main reason for preferring the use of the transformed data in this case. It can be seen from Figure 14.3, that the deviations of the points from the regression line are more or less randomly distributed as to direction. Moreover, the relation between X and Y expressed in the new equation implies that as the diameter approaches zero the weight also approaches zero.

The regression equation in the log form can be transformed back to the

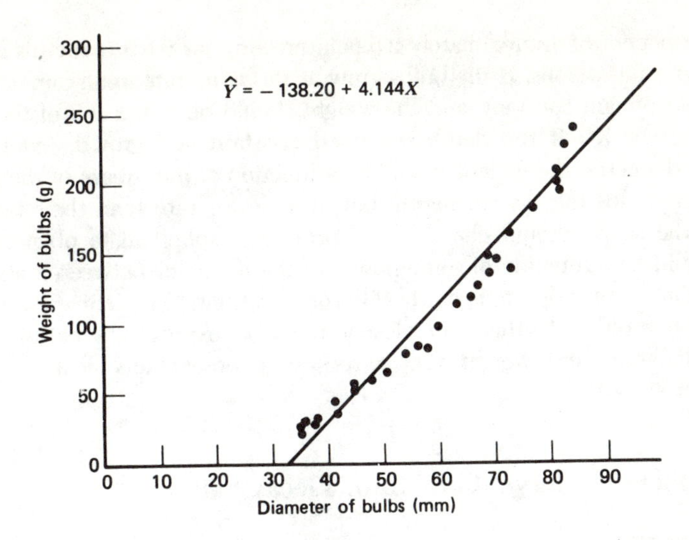

Figure 14.2. Onion data from a wider range of observations than Figure 13.5 showing nonrandom deviations from the regression line.

original measurements by taking the antilog of a' to find a, and substituting:

$$\text{equation: } Y = aX^b$$

$$\text{log form: } Y' = -2.4869 + 2.5118X'$$

$$\text{original form: } Y = .00326(X^{2.5118})$$

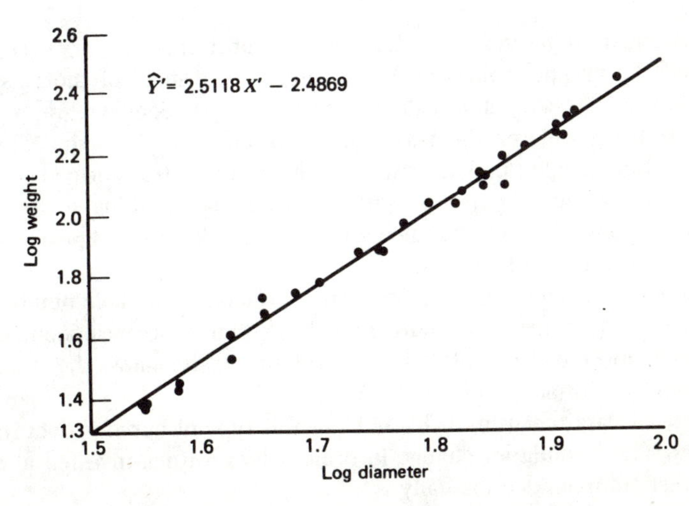

Figure 14.3. The same onion data as Figure 14.2 converted to logs, showing the improved fit to a straight line.

The exponent of approximately 2.5 is interesting for what it reveals about the growth pattern of onions. If the bulbs grew at the same rate in all dimensions, the shape would remain constant, and the weight should be a function of the cube of the diameter or X^3. If the depth remained constant and growth involved only increase in diameter, the weight should be a function of the square of the diameter or X^2. If the bulbs increase in depth, but at a slower rate than their increase in diameter, the shape should change from prolate to spherical to oblate, and the weight should be a function of some power of the diameter between 2 and 3. The last situation is exactly in accord with observations. The equation we have developed not only fits the data closely but also expresses a natural relation between diameter and weight that agrees with other facts dealing with the geometry of growth.

The Exponential Curve (Growth or Decay Curve)

In this curve, X appears as an exponent, and the coefficient b describes the rate of growth or decay. The general equation for this type of curve is

$$Y = ab^x$$

If we take the logarithm of both sides of the equation, we get:

$$\log Y = \log a + (\log b)X$$

Letting $\log Y = Y', \log a = a'$, and $\log b = b'$, then

$$Y' = a' + b'X$$

Again, transformation has yielded a straight line, but in this type of curve, it is the log of Y and the original values of X that are used, instead of the logs of both variables. For this reason, it is called a *semilog* type. Semilog graph paper is available with a log scale on the Y-axis and an ordinary scale on the X-axis. Data can be plotted on semilog paper, or the Y values can be transformed to logs and plotted on ordinary graph paper. In either case, if the resulting scatter diagram looks like linear data, it is worth calculating the coefficients of linear correlation and regression of the log of Y on X.

The values of X can be positive or negative, fractions or whole numbers, but b can be only a positive number. Figure 14.4 shows two typical exponential curves, one with $b = 2$, and the other with $b = 1/2$. The figure also shows the straight lines resulting from transformation of Y to $\log Y$.

The type of data that is most likely to fit this type of curve is data related to *interest rates*. The formula for change in principal with time, invested at constant rate of interest compounded annually is

$$A = P(1 + r)^t$$

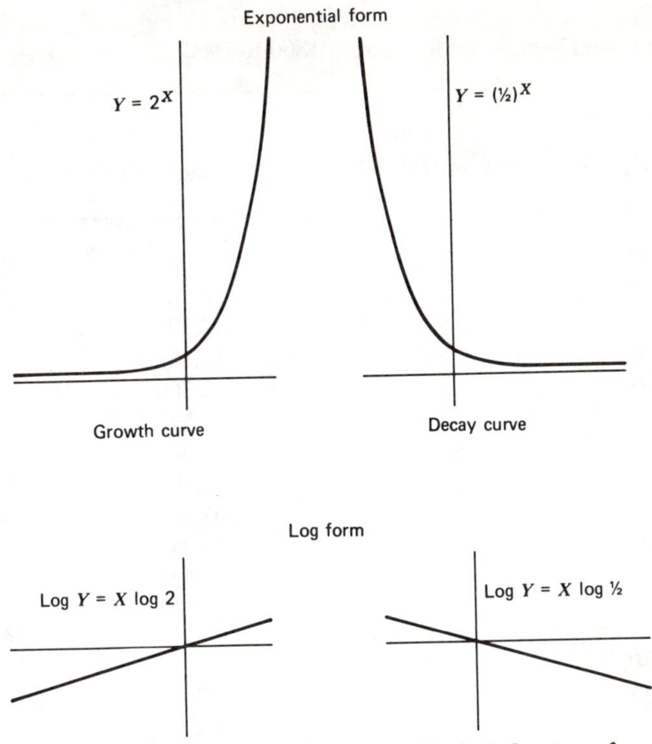

Figure 14.4. Typical exponential curves with their log transformation.

where A is the amount at the end of time t, P is the original principal, r is the annual rate of interest, and t is the time in years.

Where do we find anything like this in agriculture? Many organisms have a fairly constant growth, at least during the early stages of growth, and therefore follow the compound interest law. If we are studying the relation between time and size of an organism or a population, it is often profitable to see whether the data fit this type of curve.

Another situation in which this type of curve will be useful is in dealing with physical laws that are *exponential* in character. Consider for example, Van Hoff's law, which states that the rate of reaction approximately doubles with each 10°C rise in temperature. Many plant responses are known to follow this law fairly well, at least through a limited temperature range. Thus, temperature and rate of spoilage in fruits and vegetables often can be studied easily by assuming that they are related *exponentially*.

The rate of cooling of produce placed in a refrigerated room follows this kind of curve. In this case, we are not dealing with increase or growth but with decrease or *decay*. A decay curve has a b value of less than one, while in a growth curve b is greater than one. Other examples of decay curves are the curve of degradation of certain insecticides in the soil and the decay of radioactive isotopes.

TABLE 14.2.
Population of San Diego, California, from 1860 to 1960

Year of Census	Decades from 1860 (X)	Population (Y)	Log Y
1860	0	731	2.864
1870	1	2,300	3.362
1880	2	2,636	3.421
1890	3	16,159	4.208
1900	4	17,700	4.248
1910	5	39,578	4.597
1920	6	74,361	4.871
1930	7	147,995	5.170
1940	8	203,341	5.308
1950	9	334,387	5.524
1960	10	573,224	5.758
Totals	55		49.331
Sums of squares	385		230.393503
Sum of X log Y			277.981

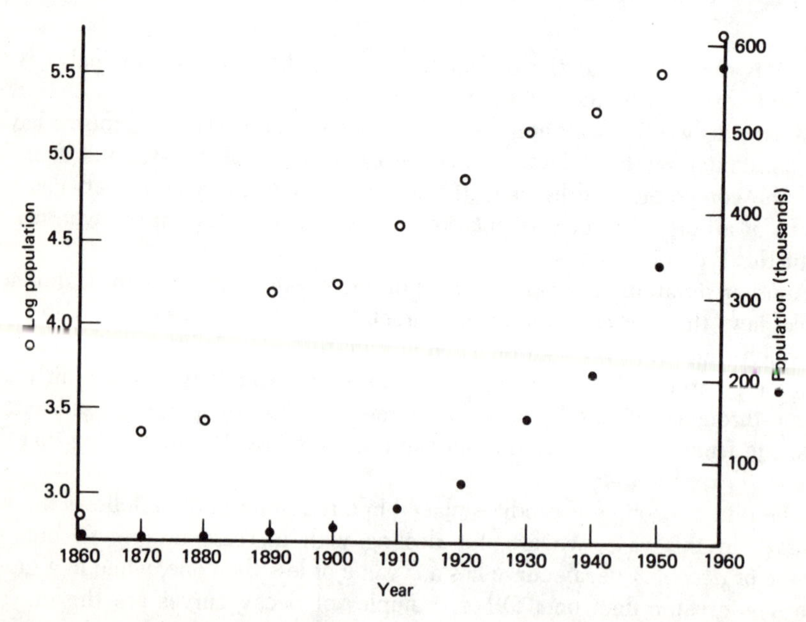

Figure 14.5. Populations of San Diego for 11 decades plotted directly and on a semilog scale.

As an example of data that can be analyzed by transforming Y to log Y, we will take the relation of population (Y) to time (X) for the city of San Diego, California, through 11 censuses (Table 14.2).

A graph of the populations against time (Fig. 14.5) shows at once that it is useless to calculate a linear regression equation for these data. This is a striking example of a case where the shortcut method would give extremely misleading results. Since the rank of the populations is exactly the same as the rank of the years, the shortcut method would give us a coefficient of correlation of +1. It would fail to reveal the fact that the data are decidely curvilinear. However, when the log of population is plotted against time, we see that a straight line appears reasonable for representing the relation.

The calculations are straightforward if we replace Y with $Y' = \log Y$ as one of the variables.

$$\Sigma x^2 = 385 - \frac{(55)^2}{11} = 110 = \Sigma X^2 - \frac{(\Sigma X)^2}{n}$$

$$\Sigma y'^2 = 230.393503 - \frac{(49.331)^2}{11} = 9.161907 = \Sigma Y'^2 - \frac{(\Sigma Y')^2}{n}$$

$$\Sigma xy' = 277.982 - \frac{55(49.331)}{11} = 31.326 = \Sigma XY' - \frac{(\Sigma X \Sigma Y')}{n}$$

$$r^2 = \frac{(\Sigma xy')^2}{\Sigma x^2 \Sigma y'^2} = \frac{31.326^2}{110(9.161907)} = 0.9737$$

$$r = \sqrt{0.9737} = 0.987$$

$$b' = \frac{\Sigma xy'}{\Sigma x^2} = \frac{31.326}{110} = 0.2848$$

$$a' = \bar{Y}' - b'\bar{X} = \frac{49.331}{11} - 0.2848\left(\frac{55}{11}\right) = 3.0606$$

Regression equation: $\hat{Y}' = 3.0606 + 0.2848X$
Taking the antilog of both sides, gives the exponential equation: $\hat{Y} = 1{,}150(1.927)^x$
This equation tells us that, on the average, the population increased by 92.7% every 10 years.

There is no question that the exponential curve fits the data much better than any straight line that could be used. However, even when we use the logs of the population against time and fit a straight line, the fit is not ideal, and there is a slight but definite tendency for the points to form a curve. Deviations in the middle of the line are positive, while those at the ends are negative. It appears from the graph that the rate of growth has not been constant but has had a tendency to slow down.

If the curve were extrapolated to 1970, the estimated population would be 1,561,000. Later, we will show how a still better equation can be devised to express the relation of population to time.

Asymptotic Curves

These are special cases of the exponential curve discussed in the last section. If the coefficient b in the equation $Y = ab^x$ is less than one, Y approaches zero as X increases without limit. A line approached by a curve in this way is called an *asymptote*. In the above case, the asymptote is the X-axis. There are cases where the asymptote is some value of Y other than zero. For example, the temperature of a crate of produce placed in a refrigerator will approach the temperature of the air in the refrigerator. The uptake of a cation in plants will show a very marked increase associated with small increases of the cation in the nutrient medium at low levels. Once the level in the medium reaches a level adequate for normal plant growth, the increase of uptake associated with additional increases in the medium is very small. The uptake approaches an upper limit which can be considered an asymptote.

If Y decreases as X increases and approaches an asymptote from above, an equation of the form $Y = c + ab^X$ may give a good fit. If Y increases as X increases and approaches an asymptote from below, the equation would be $Y \doteq c - ab^X$. The asymptote in either of these cases is $Y = c$. There is no simple, straightforward method for fitting data to these equations. The difficulty lies in finding the value of c. In some cases this value is fairly obvious, as in the case of a cooling curve where we expect the value of the asymptote to be the temperature of the cooling medium. In other cases, all we can do is make a reasonable estimate.

In the case of the descending curve, we can rewrite the equation as $(Y - c) = ab^X$. Taking the logs of both sides gives us the linear equation: $\log(Y - c) = \log a + X \log b$. For any chosen value of c, we can fit a straight line of this form to the data. We can try various values of c and compare the values of r^2 to try to maximize the closeness of fit.

It should be noted that c must be less than the smallest value of Y, since $Y - c$ must be positive in order to have a logarithm.

The case of the ascending curve is similar. Here the equation can be written $(c - Y) = ab^X$, and the log form is: $\log(c - Y) = \log a + X \log b$. In this case, c must be greater than the largest observed value of Y. Computer programs can easily be written to try successive values of c until one is found that gives the smallest sum of squares of deviations from the calculated line.

This fairly simple approach is open to criticism on the grounds that it is not a least squares solution in the sense that the sum of $(Y - \hat{Y})^2$ is a minimum. It is the sum of squares of the differences between the observed and calculated values of $\log(c - Y)$ or $\log(Y - c)$ that is being minimized.

It may be that these logs display more homogeneity of variance over the range of X values than do the Y variates themselves. This can be tested only when there are several values of Y for each value of X as in a replicated experiment (see

Chapter 12). If the variances of the logs are more nearly homogeneous than the original Y variates, then it is valid to fit a straight line to $\log(c-Y)$ or $\log(Y-c)$ instead of computing a least squares curve based on the untransformed Y variates.

If it is desired to find an equation that makes the sum of $(Y-\hat{Y})^2$ a minimum, a detailed method is presented in *Statistical Methods*, 6th edition, by Snedecor and Cochran (pp. 467–471). Actually, the results obtained by fitting a straight line to $\log(c-Y)$ or $\log(Y-c)$ generally give equations very close to those obtained by the more involved "true" least squares method.

The Polynomial Type

This type of curve has the general equation $Y = a + bX + cX^2 + dX^3 + \ldots$. The row of dots means we can have as many terms as we like. If the equation has only the first two terms on the right-hand side, we can recognize it as the equation of a straight line. If it ends with the third term (cX^2), it is a *second-degree* or *quadratic* equation. The curve represented by a quadratic equation has a special name, a *parabola*. An equation ending in dX^3 is called a *third-degree* or *cubic* equation. The highest power of X appearing in the equation determines the degree, and special names are given to the more common degrees. Corresponding to the first five degrees are the terms linear, quadratic, cubic, quartic, and quintic, respectively.

The polynomial is by far the most widely used expression for describing the relation between two variables. Sometimes it may not be a particularly "natural" expression, that is, one that expresses a cause and effect relation between the variables. However, it is so flexible and so easily handled mathematically that it is very useful.

Figure 14.6 shows a few of the many shapes of curves that can be represented by a polynomial equation. A striking property of this type of equation is that no matter how many pairs of observations we have, it is possible to calculate a polynomial curve that will exactly fit every point, providing there is only one value of Y for each value of X. The degree of the polynomial required to do this is, at most, one less than the number of pairs of observations. In actual practice, one seldom calculates more than a third- or fourth-degree equation. The calculations beyond this are formidable, and the results are usually a meaningless, meandering curve.

We noted that a straight line was simply a special case of the general polynomial equation—a first-degree or linear polynomial. To find an expression for the curvilinear relation of two variables, we try to do the same as we did in fitting a straight line. That is, we seek the curve of a given degree that will make the sum of squares of deviations a minimum.

The problem is to find the coefficients a, b, c, d, and so forth that will give a polynomial meeting the requirement that the sum of squares of deviations be a minimum. To do this, we make use of what are known as *normal equations*. We need as many equations as there are coefficients, or one more than the degree of the equation we wish to fit.

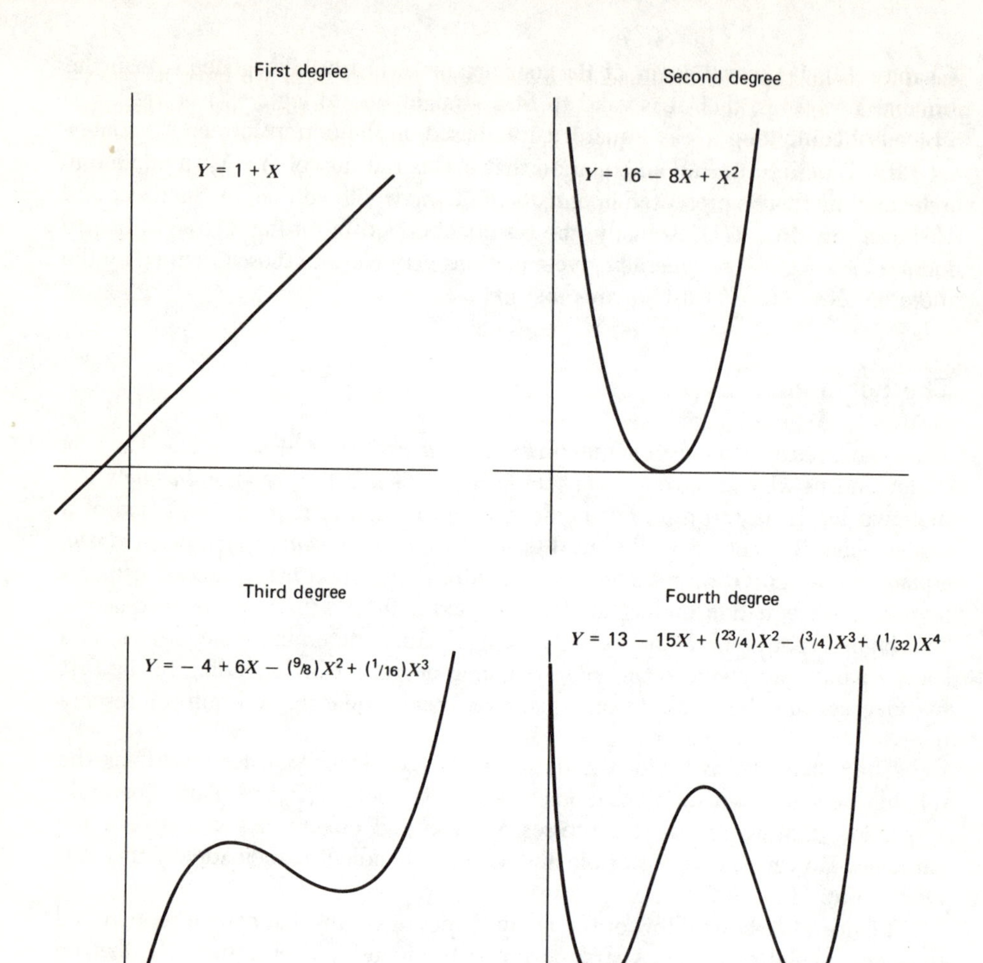

Figure 14.6. Typical shapes of polynomial curves of the first four degrees.

The normal equations are as follows:

$$an + b\Sigma X + c\Sigma X^2 + d\Sigma X^3 + \ldots = \Sigma Y$$
$$a\Sigma X + b\Sigma X^2 + c\Sigma X^3 + d\Sigma X^4 + \ldots = \Sigma XY$$
$$a\Sigma X^2 + b\Sigma X^3 + c\Sigma X^4 + d\Sigma X^5 + \ldots = \Sigma X^2Y$$
$$a\Sigma X^3 + b\Sigma X^4 + c\Sigma X^5 + d\Sigma X^6 + \ldots = \Sigma X^3Y$$

. . . .

The dots mean that we continue with the same pattern until we have as many terms to the left of the equal sign and as many equations as there are coefficients to be calculated. Thus, for a straight line we need only the first two terms of the first two equations. For a quadratic or second-degree curve, we need the first three terms of the first three equations, and so on.

From the data, we need to calculate the sums of powers of X and sums of products called for in the equation. For an nth power equation, we need the sums of all the powers of X up to X^{2n}, and the sums of products up to X^nY. The mathematics is simple, but the arithmetic is overpowering if we try to fit polynomials of high degree.

As an example, we will use some data on the yield of green lima beans at different ages of the field at picking time (Table 14.3). The date of the earliest pick is used as the base date and given an X value of zero. The values of X for subsequent pickings are the number of days from the base date. Yield in pounds is the dependent variable, designated by Y. The data are expected to be curvilinear, since at the first there should be an increase in yield with age of the field, but as the beans increase in maturity, they turn from green to pale and white. Therefore, the yield of greens will decrease after reaching a maximum.

We now have all the sums we need for the normal equations up to the third degree. We will first fit a straight line to the data, using the normal equations:

$$an + b\Sigma X = \Sigma Y$$

$$a\Sigma X + b\Sigma X^2 = \Sigma XY$$

Filling in the known values in these equations, we have

$$6a + 52b = 229.7 \qquad (1)$$

$$52a + 658b = 1{,}978.1 \qquad (2)$$

TABLE 14.3.
Yield in pounds of green lima beans (Y) on six dates (X)

X	Y	X^2	X^3	X^4	X^5	X^6	XY	X^2Y	X^3Y
0	27.4	0	0	0	0	0	0	0	0
4	39.3	16	64	256	1,024	4,096	157.2	628.8	2,515.2
7	46.2	49	343	2,401	16,807	117,649	323.4	2,263.8	15,846.6
10	47.8	100	1,000	10,000	100,000	1,000,000	478.0	4,780.0	47,800.0
13	44.5	169	2,197	28,561	371,293	4,826,809	578.5	7,520.5	97,766.5
18	24.5	324	5,832	104,976	1,889,568	34,012,224	441.0	7,938.0	142,884.0
Totals 52	229.7	658	9,436	146,194	2,378,692	39,960,778	1,978.1	23,131.1	306,812.3

Multiplying equation (1) by 52 and equation (2) by 6 we get

$$
\begin{array}{rll}
312a+2{,}704b= & 11{,}944.4 & (3) \\
312a+3{,}948b= & 11{,}868.6 & (4) \\
\hline
1{,}244b= & -75.8 & \\
b= & -75.8/1{,}244=-.06093 &
\end{array}
$$

and subtracting (3) from (4),

Substituting this value of b in equation (1), we get

$$6a=229.7+52(.0609)=232.868$$

$$a=38.8114$$

The regression equation is therefore

$$\hat{Y}=38.81-.0609X$$

We could have arrived at the same equation by using the standard formulas:

$$b=\frac{\Sigma xy}{\Sigma x^2} \qquad \text{and} \qquad a=\bar{Y}-b\bar{X}$$

The purpose of going through the normal equation procedure was to gain some practice in the process we will follow for curves of higher degree.

We can see by the graph of this line (Fig. 14.7) that it gives a poor fit. For the

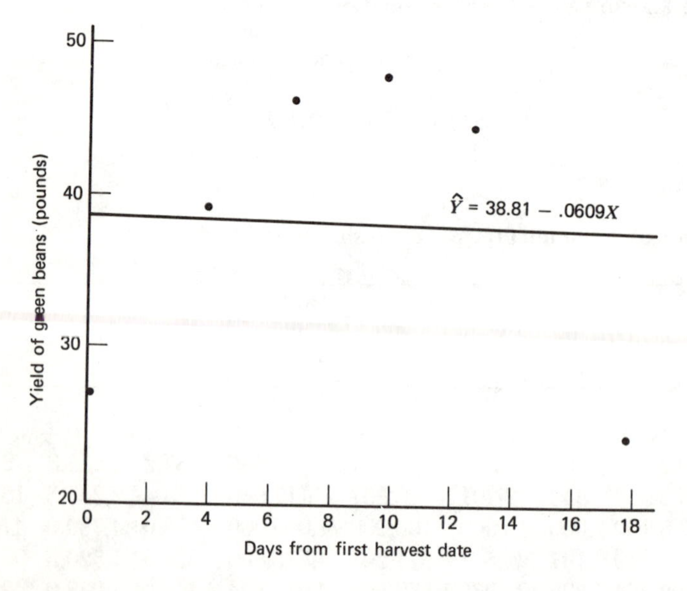

Figure 14.7. Graph of lima bean data showing the complete failure of linear regression to express the relation between yield and age of crop.

coefficient of correlation, we need ΣY^2, which is 9,295.03. Then,

$$r^2 = \frac{(\Sigma xy)^2}{\Sigma x^2 \Sigma y^2} = \left[1978.1 - \left(\frac{52(229.7)}{6}\right)\right]^2 \Bigg/ \left(658 - \frac{(52)^2}{6}\right)\left(9295.03 - \frac{(229.7)^2}{6}\right)$$

$$= (-12.6)^2/(207.33)(501.35) = .00153$$

$$r = \sqrt{.00153} = -.039$$

The coefficient is close to zero and obviously not significant. We have a good example of one of the pitfalls described in Chapter 13, "A low coefficient of correlation does not necessarily mean a lack of relation." Although the coefficient in the present example is almost zero, it would be ridiculous to conclude that there was no relation between yield of green limas and the age of the crop at picking.

We will now fit a second-degree or quadratic curve to the data. We need three normal equations:

$$an + b\Sigma X + c\Sigma X^2 = \Sigma Y$$

$$a\Sigma X + b\Sigma X^2 + c\Sigma X^3 = \Sigma XY$$

$$a\Sigma X^2 + b\Sigma X^3 + c\Sigma X^4 = \Sigma X^2 Y$$

Filling in the observed values from the table, we get

$$6a + 52b + 658c = 229.7 \qquad (1)$$

$$52a + 658b + 9436c = 1,978.1 \qquad (2)$$

$$658a + 9436b + 146,194c = 23,131.1 \qquad (3)$$

Multiply (1) by 52 and (2) by 6 and subtract:

$$312a + 2704b + 34,216c = 11,944.4$$

$$\underline{312a + 3948b + 56,616c = 11,868.6}$$

$$1244b + 22,400c = -75.8 \qquad (4)$$

Now multiply (1) by 658 and (3) by 6 and subtract:

$$3948a + 34,216b + 432,964c = 151,142.6$$

$$\underline{3948a + 56,616b + 877,164c = 138,786.6}$$

$$22,400b + 444,200c = -12,356.0 \qquad (5)$$

The two preceding steps eliminated a and gave us two equations in two unknowns. Now, multiply (4) by 22,400; (5) by 1244; and subtract:

$$\begin{array}{r} 27{,}865{,}600b + 501{,}760{,}000c = \ -1{,}697{,}920 \\ \underline{27{,}865{,}600b + 552{,}584{,}800c = -15{,}370{,}864} \\ 50{,}824{,}800c = -13{,}672{,}944 \\ c = -.2690 \end{array}$$

Substituting c back in (4): $1244b - 6025.6 = -75.8$

$$1244b = 5949.8$$
$$b = 4.7828$$

Substituting b and c in (1): $6a + 248.7056 - 177.0020 = 229.7$

$$6a = 157.9964$$
$$a = 26.3327$$

We can now write the second-degree equation:

$$\hat{Y} = 26.3327 + 4.7828X - .2690X^2$$

Let us see how much of an improvement this is over the linear equation. We call the linear estimate $\hat{Y}_L$ and the quadratic estimate $\hat{Y}_Q$. Table 14.4 shows these two estimates compared with the original values.

The results can be summarized in an analysis of variance table as follows:

Source of variation	SS	df
Total	501.35	5
Linear	0.83	1
Deviations from linear	500.52	4
Quadratic component	492.76	1
Deviations from quadratic	7.76	3

Thus we see that fitting a straight line accounted for only about 0.2% of the variability in Y (0.83/501.35), and the quadratic curve accounted for (492.76 + 0.83)/501.35 or 98.5%.

The proportion of the variability of Y accounted for by the linear plus quadratic components (0.985) is designated as R^2 and called the "multiple coefficient of determination." This will be discussed in more detail in Chapter 16.

When a quadratic equation seems to fit the data very well as in the lima bean example, it is often useful to find the value of X that will give the maximum (or minimum) value of Y. This is a simple problem in calculus which leads to the

TABLE 14.4.
Observed and calculated lima bean yields

X	Y	$\hat{Y}_L$	$d_L = Y - \hat{Y}_L$	d_L^2	$\hat{Y}_Q$	$d_Q = Y - \hat{Y}_Q$	d_Q^2
0	27.4	38.81	−11.41	130.19	26.33	1.07	1.14
4	39.3	38.57	0.73	0.53	41.16	−1.86	3.46
7	46.2	38.38	7.82	61.15	46.63	−0.43	0.18
10	47.8	38.20	9.60	92.16	47.26	0.54	0.29
13	44.5	38.02	6.48	41.99	43.05	1.45	2.10
18	24.5	37.71	−13.21	174.50	25.27	−0.77	0.59
Totals			0.01	500.52		0.00	7.76

solution:

$$X_{max} = \frac{-b}{2c}$$

In our example, $X_{max} = -4.7828/2(-0.2690) = 8.9$, or approximately 9 days after the base date. Substituting this value of X in the quadratic equation gives 47.59 as the estimated maximum value of Y.

Since only 1.5% of the variability in Y remains unaccounted for after fitting the quadratic equation, in practice we would generally conclude the regression analysis at this point. However, to illustrate the method, we will fit a third-degree curve. The normal equations are

$$an + b\Sigma X + c\Sigma X^2 + d\Sigma X^3 = \Sigma Y$$

$$a\Sigma X + b\Sigma X^2 + c\Sigma X^3 + d\Sigma X^4 = \Sigma XY$$

$$a\Sigma X^2 + b\Sigma X^2 + c\Sigma X^4 + d\Sigma X^5 = \Sigma X^2 Y$$

$$a\Sigma X^3 + b\Sigma X^4 + c\Sigma X^5 + d\Sigma X^6 = \Sigma X^3 Y$$

Substituting the observed values, we have the following equations, which we want to solve for a, b, c, and d:

$$6a + 52b + 658c + 9{,}436d = 229.7 \qquad (1)$$

$$52a + 658b + 9{,}436c + 146{,}194d = 1{,}978.1 \qquad (2)$$

$$658a + 9{,}436b + 146{,}194c + 2{,}378{,}692d = 23{,}131.1 \qquad (3)$$

$$9{,}436a + 146{,}194b + 2{,}378{,}692c + 39{,}960{,}778d = 306{,}812.3 \qquad (4)$$

We first eliminate a as follows: Equation (2) times 6 minus equation (1) times 52 gives

$$1{,}244b + 22{,}400c + 386{,}492d = -75.8 \qquad (5)$$

Equation (3) times 6 minus equation (1) times 658 gives

$$22{,}400b + 444{,}200c + 8{,}063{,}264d = -12{,}356.0 \qquad (6)$$

Equation (4) times 6 minus equation (1) times 9,436 gives

$$386{,}492b + 8{,}063{,}264c + 150{,}726{,}572d = -326{,}575.4 \qquad (7)$$

Now we eliminate b by the following steps: Equation (6) times 1,244 minus equation (5) times 22,400 gives

$$50{,}824{,}800c + 1{,}373{,}279{,}616d = -13{,}672{,}944 \qquad (8)$$

Equation (7) times 1,244 minus equation (5) times 386,492 gives

$$1{,}373{,}279{,}616c + 38{,}127{,}789{,}500d = -276{,}963{,}704 \qquad (9)$$

To eliminate c we take equation (8) times 1,373,279,616 minus equation (9) times 50,824,800 and divide both sides by 10,000,000 and round off to reduce the large numbers to 10 digit figures. This gives

$$5{,}194{,}037{,}206d = -38{,}232{,}948$$

$$d = -.00736$$

Substituting d in equation (8) and solving for c gives

$$c = -.07015$$

Substituting d and c in equation (5) gives

$$b = 3.48886$$

Finally, substituting d, c, and b in equation (1) gives

$$a = 27.31449$$

And the third degree or cubic equation is

$$\hat{Y}_c = 27.31449 + 3.48886X - .07015X^2 - .00736X^3$$

Calculating the estimated values $\hat{Y}_c$, we find a substantial improvement over the fit of the quadratic curve.

X	Y	$\hat{Y}_c$	$d = Y - \hat{Y}_c$	d^2
0	27.4	27.31	.09	.01
4	39.3	39.68	−.38	.14
7	46.2	45.78	.42	.18
10	47.8	47.83	−.03	.00
13	44.5	44.64	−.14	.02
18	24.5	24.46	.04	.00
Totals			.00	.35

The sum of squares for deviation from quadratic can now be partitioned as follows:

Source of variation	SS	df	MS	F	Required F	
					5%	1%
Deviation from quadratic	7.76	3				
Cubic component	7.41	1	7.41	42.3	18.51	98.49
Deviation from cubic	0.35	2	0.175			

The improved fitting achieved by calculating a cubic equation, while appreciable, was significant only at the 5% point. With so few degrees of freedom, this is not surprising, since an F value of 98.49 is required for significance at the 1% level.

Figure 14.8 shows the quadratic and cubic curves, drawn over a much wider range than the observations, to bring out their difference in shape. Throughout the range of observations, the two curves are not very different, but the superior fit of the cubic is evident.

You probably noticed how increasingly cumbersome the calculations became as we went from linear to quadratic to cubic curves. Various methods have been devised for systematizing these calculations; the most common are the Doolittle and the abbreviated Doolittle methods. A treatment of these is beyond the scope of this discussion but can be found in some advanced statistics texts. Programs are also available for calculating coefficients to almost any desired degree on an electronic computer.

In cases where the values of X are equally spaced, there are extremely simple shortcut methods that will be presented in the next chapter.

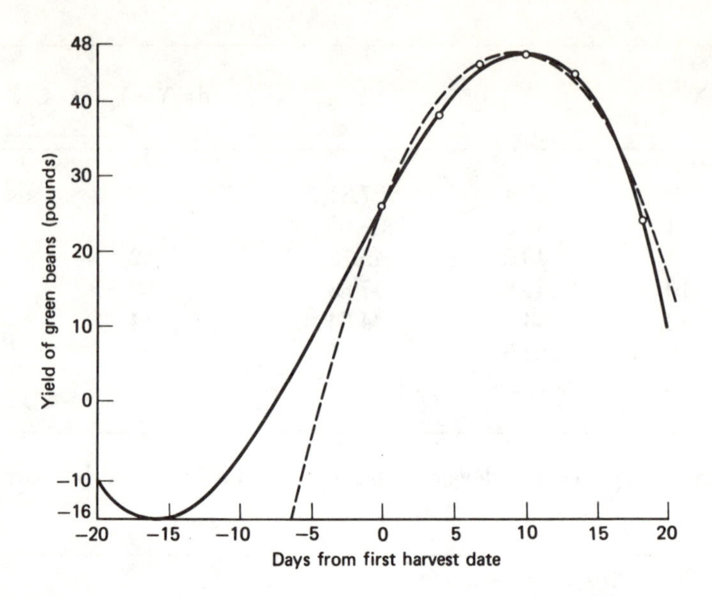

Figure 14.8. The same lima bean data as Figure 14.7, showing the good fit of a quadratic curve (dotted line), and the even closer fit of the cubic curve (solid line).

POLYNOMIALS IN REPLICATED EXPERIMENTS. When our data consists only of single values of Y for each value of X, the only way to test the significance of a regression component is to test its mean square against the residual mean square. In replicated experiments, on the other hand, we have an error mean square, which can be used for testing not only each regression component but also the residual mean square.

In the last chapter we fit a straight line to the yields of sugar beets at five harvest dates from Table 10.1. We found that while the mean square for linear regression was highly significant, there was also a significant amount of deviation from linearity.

We will now fit a quadratic equation to these data to see whether a second-degree curve will account for a large portion of the deviation from a straight line. Some of the sums we need for the normal equations have already been calculated in Table 13.7. The others will be found in Table 14.5.

We now have all the sums needed for the normal equations.

$$5a + 15b + 55c = 1600.0 \quad (1)$$

$$15a + 55b + 225c = 5551.0 \quad (2)$$

$$55a + 225b + 979c = 21912.6 \quad (3)$$

TABLE 14.5
Fitting a quadratic equation to sugar beet time of harvest data

X	Y	X^3	X^4	X^2Y	$\hat{Y}$	$(Y-\hat{Y})$	$(Y-\hat{Y})^2$
1	140.0	1	1	140.0	142.1714	−2.1714	4.7150
2	267.2	8	16	1068.8	258.7142	8.4858	72.0088
3	335.2	27	81	3016.8	347.6284	−12.4284	154.4651
4	417.0	64	256	6672.0	408.9140	8.0860	65.3834
5	440.6	125	625	11015.0	442.5710	−1.9710	3.8848
15	1600.0	225	979	21912.6	1599.9990	0.0010	300.4571

Equation (1) multiplied by 3 and subtracted from equation (2) and equation (1) multiplied by 11 and subtracted from equation (3) give us two equations in two unknowns:

$$10b + 60c = 751.0 \tag{4}$$

$$60b + 374c = 4312.6 \tag{5}$$

Equation (4) multiplied by 6 and subtracted from equation (5) gives

$$14c = -193.4$$

$$c = -13.8143$$

Substituting the value of c in equation (4) gives

$$b = 157.9857$$

and substitution of b and c in equation (1) gives

$$a = -2.0000$$

The quadratic equation is therefore

$$\hat{Y} = -2 + 157.9857X - 13.8143X^2$$

In Table 14.5, we have entered the values of $\hat{Y}$, the differences between these and the observed values, and the squares of the differences. The sum of the deviations is essentially zero, as it should be, and the sum of squares of deviations is 300.4571, which must be reduced to a per-plot basis since we were working with totals. Since there were 16 plots entered into each harvest date total, 300.4571/16 equals 18.7786 as the sum of squares for deviations from the quadratic curve. Since the sum of squares for deviation from linear regression was 185.7587, the sum of squares for quadratic regression is

$$185.7587 - 18.7786 = 166.9802$$

All of this can be summarized in an analysis of variance table:

Source of Variation	df	SS	MS	F
Harvest dates	4	3710.7650	927.691	111.92
Linear	1	3525.0062	3525.006	425.26
Quadratic	1	166.9802	166.980	20.14
Residual	2	18.7786	9.389	1.13
Error	12	99.4670	8.289	

We can see that the quadratic regression accounted for a very large portion of the significant deviation from linear. The residual sum of squares is not significant and, in fact, would not be significant if all of it were associated with a single degree of freedom, so there is no need to continue further with the regression analysis.

We have used a rather long and laborious process to find the quadratic equation and the sums of squares due to quadratic regression and deviation from regression. In the next chapter we will learn a shortcut method for finding the quadratic equation. We have already had some experience in finding the sum of squares for regression by use of the coefficients in Table A.11. Under the portion of the table for n=5, we see that the quadratic coefficients are: 2, −1, −2, −1, and 2.

$$SS = \frac{(\Sigma c_i T_i)^2}{r(\Sigma c_i^2)} = \frac{[(2)140.0 - 267.2 - (2)335.2 - 417.0 + (2)440.6]^2}{16(14)}$$

$$= \frac{(-193.4)^2}{224} = 166.9802$$

which is the same as we obtained indirectly.

Combining Curve Types

We have discussed four general types of curves and shown how to fit observed data to them. Sometimes it is worthwhile to use a combination of two types. For

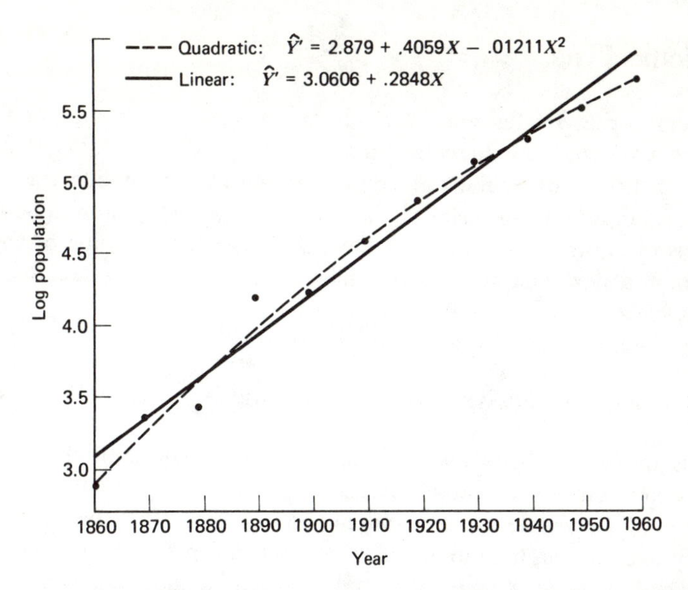

Figure 14.9. San Diego population data with a straight line fitted to the logs of the population (solid line) and the improvement obtained by fitting a quadratic equation (dotted line).

example in the data on the population of San Diego, we found that plotting the *logs* of the population against years gave a much closer approximation to a straight line than when we plotted just population against years. However, a glance at Figure 14.5 shows that even the transformed data do not quite form a straight line, but rather have a definite tendency to curve. The rate of increase seems to be slowing down with time.

We can easily fit a second-degree curve to the data again using $Y' = \log Y$ as the dependent variable instead of Y. The calculations are left to the interested reader as a good exercise in fitting a second-degree curve. The equation obtained is

$$Y' = 2.87906 + .40590X - .01211X^2$$

Figure 14.9 shows the comparison between the straight line and the second-degree curve in relation to the logs of the population. We have already pointed out that extrapolation of the straight line would give a prediction of 1,561,000 for 1970. Extrapolation of the second-degree curve gives a prediction of 756,800.[2] In view of the closer agreement of the second-degree curve with past trends, the lower prediction is probably more reasonable.

[2]The 1970 census figures are now available and give the population of San Diego as 697,000, which is 8% below the predicted figure.

The Periodic Type

This is a curve that relates some variable to time and is repeated at fixed time intervals. It is known in mathematical texts as a *Fourier curve* and is useful for any kind of data that tends to fluctuate up and down at regular intervals. Very few statistics texts discuss fitting data of this kind, but we have found it so useful for many kinds of agricultural data, that we will give a brief outline of the general method. In the next chapter we will take up a shortcut method for handling special cases.

The general equation for a periodic curve is

$$Y = a_0 + a_1 \cos CX + b_1 \sin CX + a_2 \cos 2CX + b_2 \sin 2CX + a_3 \cos 3CX + b_3 \sin 3CX \ldots$$

where X is an observed time expressed as units from some arbitrary starting time, and C is a constant equal to 360° divided by the number of units in a cycle.

Suppose, for example, we are studying hourly fluctuations of some variable in 24-hour cycles, and we take midnight as the starting point. An observation made at 9 A.M. would have an X value of 9, and C would be 360°/24, or 15°. The value of CX would therefore be $9 \times 15°$ or, 135°.

The row of dots at the right of the general equation means that we can continue adding pairs of terms as long as the total number of terms does not exceed the number of time periods for which we have observations.

This curve has many features similar to the polynomial curve. It has the same remarkable property that if there is a single value of Y for each value of X, an equation can be found that will exactly pass through every point.

You will recall that a first-degree polynomial is a straight line with the equation $Y = a + bX$. This line is completely described with two numbers, the intercept a, and the slope b. A first-degree Fourier curve is a simple wave curve with the equation $Y = a_0 + a_1 \cos CX + b_1 \sin CX$. To describe this curve we need *three* numbers. The term a_0 gives the central value around which the wave fluctuates. It can be looked on as a weighted mean. A second value A $= \sqrt{a_1^2 + b_1^2}$, is called the *semiamplitude* and tells us how far the curve fluctuates above and below the central point. The total range from the highest to the lowest point on the wave is 2A and is called the *amplitude*. The third value needed to describe the wave is the *phase angle*. This tells us the point in the cycle where the wave reaches its maximum value. To find this we first find θ'(theta) = arc tan (b_1/a_1), read "the angle whose tangent is b_1/a_1." We then find the phase angle by applying the following rules:

If b_1 is positive and a_1 is positive $\theta = \theta'$
If b_1 is positive and a_1 is negative $\theta = 180° - \theta'$
If b_1 is negative and a_1 is negative $\theta = 180° + \theta'$
If b_1 is negative and a_1 is positive $\theta = 360° - \theta'$

In the polynomial we obtained more complicated curves by adding terms with successive powers of X, such as cX^2 dX^3, and so forth. With the Fourier curve we obtain more complicated wave forms by adding *pairs* of terms such as $a_2 \cos 2CX + b_2 \sin 2CX$, $a_3 \cos 3CX + b_3 \sin 3CX$, and so forth. The effect of the second-degree pair is to superimpose on the first wave a second wave with two complete oscillations per cycle. The third-degree pair superimposes another curve with three complete oscillations per cycle, and so on.

The method of fitting a Fourier curve is also very similar to the method for fitting a polynomial. We use a set of normal equations in which we substitute sums calculated from the observed data and solve these for the required coefficients.

To simplify the normal equations, it is convenient to adopt two symbols, U and V:

$$U_i = \cos i(CX)$$

$$V_i = \sin i(CX) \qquad \text{Thus } \Sigma U_2 V_1 \text{ means } \Sigma \cos 2(CX) \sin(CX).$$

The normal equations are as follows:

$$a_0 n + a_1 \Sigma U_1 + b_1 \Sigma V_1 + a_2 \Sigma U_2 + b_2 \Sigma V_2 + \ldots = \Sigma Y$$

$$a_0 \Sigma U_1 + a_1 \Sigma U_1^2 + b_1 \Sigma U_1 V_1 + a_2 \Sigma U_1 U_2 + b_2 \Sigma U_1 V_2 + \ldots = \Sigma U_1 Y$$

$$a_0 \Sigma V_1 + a_1 \Sigma U_1 V_1 + b_1 \Sigma V_1^2 + a_2 \Sigma U_2 V_1 + b_2 \Sigma V_1 V_2 + \ldots = \Sigma V_1 Y$$

$$a_0 \Sigma U_2 + a_1 \Sigma U_1 U_2 + b_1 \Sigma U_2 V_1 + a_2 \Sigma U_2^2 + b_2 \Sigma U_2 V_2 + \ldots = \Sigma U_2 Y$$

$$a_0 \Sigma V_2 + a_1 \Sigma U_1 V_2 + b_1 \Sigma V_1 V_2 + a_2 \Sigma U_2 V_2 + b_2 \Sigma V_2^2 + \ldots = \Sigma V_2 Y$$

As with the polynomial, we need as many terms on the left-hand side of these equations and as many equations as we have coefficients to calculate. For a polynomial of the nth degree we needed $n+1$ equations each with $n+1$ terms on the left-hand side. For the Fourier curves, we need $2n+1$ equations, each with $2n+1$ terms.

To illustrate the procedure, we will fit a first-degree Fourier curve to the mean temperatures observed in nine months at Stockton, California. Table 14.6 shows the observed data and the necessary columns for filling in the terms of the normal equations.

We can now write the three normal equations required to find a_0, a_1, and b_1.

$$9a_0 + 2.366a_1 - 1.366b_1 = 518.9 \qquad (1)$$

$$2.366a_0 + 4a_1 + 0.866b_1 = 77.894 \qquad (2)$$

$$-1.366a_0 + 0.866a_1 + 5b_1 = -103.202 \qquad (3)$$

Multiplying equation (1) by .866 and equation (2) by 1.366 and adding gives

$$11.026a_0 + 7.513a_1 = 555.771 \qquad (4)$$

TABLE 14.6.
Mean monthly temperatures for nine months at Stockton, California

(Cycle = 12 months, C = 360°/12 = 30°)

Y (Temp)	Month	X	CX	$U_1 = \cos$ (CX)	$V_1 = \sin$ (CX)	U_1^2	V_1^2	U_1V_1	YU_1	YV_1
44.7	Jan.	0	0	1.000	0.000	1.00	0.00	0.000	44.700	0.000
49.0	Feb.	1	30°	0.866	0.500	0.75	0.25	0.433	42.434	24.500
53.7	Mar.	2	60°	0.500	0.866	0.25	0.75	0.433	26.850	46.504
59.7	Apr.	3	90°	0.000	1.000	0.00	1.00	0.000	0.000	59.700
76.2	Aug.	7	210°	−0.866	−0.500	0.75	0.25	0.433	−65.989	−38.100
72.7	Sep.	8	240°	−0.500	−0.866	0.25	0.75	0.433	−36.350	−62.958
64.0	Oct.	9	270°	0.000	−1.000	0.00	1.00	0.000	0.000	−64.000
53.0	Nov.	10	300°	0.500	−0.866	0.25	0.75	−0.433	26.500	−45.898
45.9	Dec.	11	330°	0.866	−0.500	0.75	0.25	−0.433	39.749	−22.950
Totals 518.9				2.366	−1.366	4.00	5.00	0.866	77.894	−103.202

Multiplying equation (1) by 5 and equation (3) by 1.366 and adding gives

$$43.134a_0 + 13.013a_1 = 2453.526 \tag{5}$$

Multiplying equation (4) by 13.013 and equation (5) by 7.513 and subtracting gives

$$-180.584a_0 = -11{,}201.093 \text{ and } a_0 = 62.027$$

Substituting this value of a_0 in equation (4) gives

$$(11.026 \times 62.027) + 7.513a_1 = 555.771$$

$$7{,}513a_1 = -128.139$$

$$a_1 = -17.056$$

Substituting a_0 and a_1 in equation (3) gives

$$(-1.366 \times 62.027) + (0.866 \times -17.057) + 5b_1 = -103.202$$

$$-84.729 - 14.770 + 5b_1 = -103.202$$

$$5b_1 = 84.729 + 14.770 - 103.202 = -3.703$$

$$b_1 = -0.741$$

We can now write our equation:

$$Y = 62.027 - 17.056\cos(CX) - 0.741\sin(CX)$$

Substituting the values of cos(CX) and sin(CX) for each month gives us predicted values which we can compare with the observed values.

The figures in parentheses in Table 14.7 represent the data for months which we assumed were not available when we computed the curve and therefore did not enter into the calculations. It will be noted that the curve we calculated from the available data overestimated the actual means for the missing months.

The fit of the curve to the observed data is very close. The total sum of squares of the observed temperatures is 1032.942, and we can partition this in an analysis of variance as follows:

Source of variation	df	SS	MS	F
Total	8	1032.942		
Due to regression	2	1016.187	508.094	181.85***
Deviation from regression	6	16.755	2.794	

TABLE 14.7.
Observed and predicted temperatures at Stockton, California in nine months

Month	Y (Observed)	$\hat{Y}$ (Predicted)	$(Y-\hat{Y})$	$(Y-\hat{Y})^2$
January	44.7	44.97	−0.27	0.0729
February	49.0	46.89	2.11	4.4521
March	53.7	52.86	0.84	0.7056
April	59.7	61.29	−1.59	2.5281
(May)	(66.2)	(69.91)	(−3.71)	
(June)	(72.8)	(76.43)	(−3.63)	
(July)	(78.2)	(79.08)	(−0.88)	
August	76.2	77.17	−0.97	0.9409
September	72.7	71.20	1.50	2.2500
October	64.0	62.77	1.23	1.5129
November	53.0	54.14	−1.14	1.2996
December	45.9	47.63	−1.73	2.9929
Totals	518.9	518.92	−0.02	16.7550

Regression has 2 degrees of freedom, since we calculated two parameters, a_1 and b_1 in addition to the mean. The sum of squares for regression is obtained by subtracting the sum of squares of deviations from the total. The proportion of the total sum of squares associated with regression is 1016.187/1032.942 = 0.9838 and is designated as R^2.

The value of 62.027 for a_0 is interesting. We referred to this earlier as a weighted mean. It is an estimate of what the mean would be if we had data for the whole year. It is indeed very close to the true annual mean of 61.34 based on complete records. Obviously the mean of the observed data, 518.9/9 = 57.656 would be a very poor estimate of the annual mean, since the missing data were all from warm months. However, the value of a_0 obtained by fitting a Fourier curve, enables us to arrive at a close estimate in spite of the missing data.

The values of a_1 and b_1 can be used to find the semiamplitude and phase angle.

$$\text{Semiamplitude} = A = \sqrt{a_1^2 + b_1^2} = \sqrt{(-17.056)^2 + (-0.741)^2} = 17.1$$

$$\theta' = \tan^{-1} b_1/a_1 = \text{angle whose tangent is } -0.741/-17.056 = 2.5^\circ$$

by the rules of signs $\theta = 180^\circ + \theta' = 182.5^\circ$.

Since 1 month = 30°, 182.5° is equivalent to 6.1 months. This says that the maximum point in the curve occurs about 6.1 months after the starting date. We

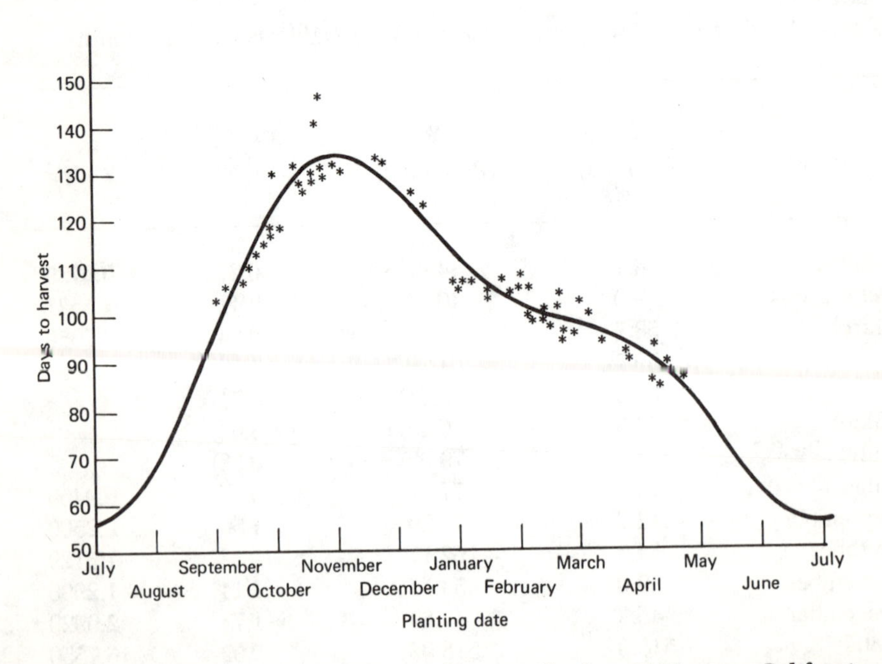

Figure 14.10. Planting date and days to harvest of celery in Ventura, California.

used the *mean* for January as our starting date, so we call this date January 15. Therefore our calculated maximum is 6.1 months after January 15, or about July 18.

We have gone through the steps in fitting data to a simple one degree Fourier curve. If it becomes necessary to fit data in this way to a curve of 2 or more degrees, the calculations become quite formidable, since two additional equations must be added for each degree. Such problems can be handled very easily on a computer. Figure 14.10 shows a curve, relating planting date to length of time to harvest in celery, which was calculated and plotted on a computer. Ten years of data were used in calculating this curve.

Fortunately, if we have data taken at equal intervals throughout a complete cycle, the calculations become greatly simplified, and in the next chapter we describe the shortcut methods for handling data of this kind.

SUMMARY

If the scatter diagram of two variables shows a tendency for the points to be scattered around a curve rather than around a straight line, it is advisable to analyze the curvilinear relation between the variables. Failure to do so can be very misleading.

If the logs of the two variables form a scatter diagram that appears to fit a straight line, the curve describing the relation is of the form: $Y = aX^b$ and is called a power curve. Variables involving different numbers of dimensions are most likely to fit this type of curve.

To analyze such data, transform the original variables X and Y to new variables $X' = \log X$ and $Y' = \log Y$. Then proceed exactly as with linear correlation and regression, finding the regression equation for the straight line: $Y' = a' + bX'$.

If the log of Y plotted against X forms a straight line scatter diagram, the appropriate curve is of the form: $Y = ab^x$, called an exponential curve. Data in which the variable Y tends to have a fairly constant *rate* of increase or decrease can be expected to fit this type of curve.

To analyze, transform Y only to $Y' = \log Y$ and proceed as with linear regression, fitting to the equation:

$$Y' = a' + b'X$$

A special type of exponential curve in which Y approaches some value other than zero is called an asymptotic curve. It has the equation: $Y = c \pm ab^X$, where c is the asymptote. This equation can be transformed to a straight line by transforming Y to $Y' = \log(Y - c)$ or $Y' = \log(c - Y)$, but the best value of c must be found by trial and error.

Curvilinear data that do not approach linear data under either a log or semilog transformation can be fitted to a polynomial of the form:

$$Y = a + bX + cX^2 + dX^3 + \ldots$$

using as many terms as necessary to obtain a satisfactory fit.

To find the unknown coefficients a, b, c, d, etc., solve the set of simultaneous equations, known as *normal* equations.

$$an + b\Sigma X + c\Sigma X^2 + d\Sigma X^3 + \ldots = \Sigma Y$$

$$a\Sigma X + b\Sigma X^2 + c\Sigma X^3 + d\Sigma X^4 + \ldots = \Sigma XY$$

$$a\Sigma X^2 + b\Sigma X^3 + c\Sigma X^4 + d\Sigma X^5 + \ldots = \Sigma X^2Y$$

$$a\Sigma X^3 + b\Sigma X^4 + c\Sigma X^5 + d\Sigma X^6 + \ldots = \Sigma X^3Y$$

The number of equations and the number of terms to the left of the equal sign must each be equal to the number of coefficients needed, or one more than the degree of the regression equation.

Equations of the first few degrees have special names, as do some of the curves:

Degree	Name of Equation	Name of Curve
First	Linear	Straight line
Second	Quadratic	Parabola
Third	Cubic	Cubic parabola
Fourth	Quartic	Quartic parabola
Fifth	Quintic	Quintic parabola

If the deviations of the observations from a computed curve appear to be more or less random, fitting a higher degree curve usually is not worthwhile. If the deviations are systematic or in definite groups as to sign, it is generally advantageous to calculate the equation of next higher degree.

In replicated experiments, the mean square for deviations from regression can be tested by the error mean square.

Calculations of coefficients for equations higher than cubic should be attempted only by mastering special methods (such as the Doolittle method) or with an electronic computer. When values of X are equally spaced, much time will be saved by using the shortcut methods described in Chapter 15. Combining log and polynomial methods will sometimes result in a much better fit to the data than either method alone.

Data that fluctuate up and down with time in a rather regular pattern can be fitted to a periodic (Fourier) curve of the form:

$$Y = a_0 + a_1 \cos CX + b_1 \sin CX + a_2 \cos 2CX + b_2 \sin 2CX + \ldots$$

The *normal* equations for finding the unknown coefficients are

$$a_0 n + a_1 \Sigma U_1 + b_1 \Sigma V_1 + a_2 \Sigma U_2 + b_2 \Sigma V_2 + \ldots = \Sigma Y$$

$$a_0 \Sigma U_1 + a_1 \Sigma U_1^2 + b_1 \Sigma U_1 V_1 + a_2 \Sigma U_1 U_2 + b_2 \Sigma U_1 V_2 + \ldots = \Sigma U_1 Y$$

$$a_0 \Sigma V_1 + a_1 \Sigma U_1 V_1 + b_1 \Sigma V_1^2 + a_2 \Sigma U_2 V_1 + b_2 \Sigma V_1 V_2 + \ldots = \Sigma V_1 Y$$

$$a_0 \Sigma U_2 + a_1 \Sigma U_1 U_2 + b_1 \Sigma U_2 V_1 + a_2 \Sigma U_2^2 + b_2 \Sigma U_2 V_2 + \ldots = \Sigma U_2 Y$$

$$a_0 \Sigma V_2 + a_1 \Sigma U_1 V_2 + b_1 \Sigma V_1 V_2 + a_2 \Sigma U_2 V_2 + b_2 \Sigma V_2^2 + \ldots = \Sigma V_2 Y$$

where $U_i = \cos i(CX)$ and $V_i = \sin i(CX)$.

When data are obtained from equally spaced time intervals throughout a complete cycle, shortcut methods, described in Chapter 15, can be used.

15

SHORTCUT REGRESSION METHODS

It frequently happens that we make observations on a dependent variable Y associated with equally spaced values of an independent variable, X. For example, if the independent variable is time, and we make readings of Y at daily, weekly, monthly, or yearly intervals, the X's or times are equally spaced. Another case in which we frequently have equally spaced intervals of X is in experiments involving rates of fungicides, insecticides, fertilizers, and the like. An experiment in which the treatment rates are equally spaced has real advantages from the standpoint of ease of analysis.

There are other advantages besides ease of computation in the use of equally spaced rates. If we wish to learn something about the trend of response to treatment levels, it is best to have the information provided by the experiment evenly distributed through the range of treatment levels. There is very little justification, for example, in a 0, 1, 2, 4 series of treatment levels, although this series is very commonly used in experimental work. The series is neither arithmetic nor geometric. The information obtained in the lower portion of the range is more complete than in the upper part. Suppose we find an increase in yield with increasing levels of X from 0 to 2 but a marked reduction in yield with treatment level 4. It would be useful to know where, in the range between 2 and 4, this reversal in trend occurs. A treatment level of 3 would be most helpful.

The shortcut method we are about to describe was discussed in the section on trend comparisons in Chapter 6. The method is so useful that it seems worthwhile to extend that discussion and to relate it to the previous chapter of this section dealing with curvilinear regression. Statisticians usually refer to this as the *method of orthogonal polynomials*. Those of you who suffer a mental block when confronted by such an imposing title can think of it as the "shortcut method for measuring trends." You will find it easy to use and a tremendous timesaver.

POLYNOMIAL CURVE FITTING

The heart of the method for fitting polynomials is Table A.11,[1] the use of which eliminates many of the laborious computations ordinarily required in curvilinear

[1]This table, calculated by the authors, is used rather than one of the many similar tables found in other publications. To the best of our knowledge the K values do not appear in any other published tables.

regression. The table can be used to (1) find the linear, quadratic, cubic, and quartic regression equations for any number of equally spaced observations up to 25, and (2) partition the treatment sum of squares in an analysis of variance into linear, quadratic, cubic, quartic and residual components for up to 25 equally spaced treatments or observations.

At the top of the table are values of n, the number of observations or treatments. For any given problem we need use only the portion of the table under the appropriate value of n. The first column of coefficients, headed c_1, in addition to being used for various computations, consists of *coded values of* X. The coding is done in such a manner as to result in the smallest possible whole numbers. Regardless of the values of equally spaced X's, if n is odd, we can take: $X' = (X - \bar{X})/L$, where L is the interval between successive values of X. If n is even, we take: $X' = (X - \bar{X})2/L$. These transformations will give the values in the c_1 column.

It is not necessary to know how the other coefficients in the table are obtained in order to use them. However, the curious student will find the following relations of interest:

The coefficients in the c_2 column can be found from the following relation: $c_{2i} = (c_{1i}^2 n - \Sigma c_{1i}^2)/GCD$. After the numerators are calculated for all values of i from 1 to n, the greatest common denominator (GCD) must be determined so that the coefficients can be reduced to the lowest possible set of integers. The coefficients of the c_3 column are found from the following: $c_{3i} = (c_{1i}^3 \Sigma c_{1i}^2 - c_{1i} \Sigma c_{1i}^4)/GCD$, and those of the c_4 column from:

$$c_{4i} = \left(c_{1i}^4 n \Sigma c_{1i}^2 c_{2i} - c_{1i}^2 n \Sigma c_{1i}^4 c_{2i} - \Sigma c_{1i}^4 \Sigma c_{1i}^2 c_{2i} + \Sigma c_{1i}^2 \Sigma c_{1i}^4 c_{2i}\right)/GCD$$

It can be seen that the calculations become very cumbersome, especially for larger values of n, so being provided with a table is a great timesaver.

The calculation of the K values is most easily handled by utilizing some of the concepts in theory of numbers which are beyond the scope of this book.

The steps in finding the linear, quadratic, cubic and quartic regression equations are as follows:

1. Arrange the values of Y in a column according to the ascending values of the associated X's, starting with the Y corresponding to the lowest value of X.

2. Multiply the values of Y by the coefficients for c_1, c_2, c_3 and c_4 shown in the table, giving four columns.

3. Find the sum of each column, observing the plus and minus signs. These sums are called ΣY, P_1, P_2, P_3, and P_4.

4. Using the values of P obtained and the values of K from the table the linear, quadratic, cubic, and quartic equations can be written from these relations:

   Linear equation: $\hat{Y}_L = \bar{Y} + (K_2P_1)X'$
   Quadratic: $\hat{Y}_Q = (\bar{Y} - K_1P_2) + (K_2P_1)X' + (K_4P_2)X'^2$
   Cubic: $\hat{Y}_C = (\bar{Y} - K_1P_2) + (K_2P_1 - K_3P_3)X' + (K_4P_2)X'^2 + (K_5P_3)X'^3$
   Quartic: $\hat{Y}_4 = (\bar{Y} - K_1P_2 + K_3P_4) + (K_2P_1 - K_3P_3)X' + (K_4P_2 - K_7P_4)X'^2 + (K_5P_3)X'^3 + (K_6P_4)X'^4$

Note that these equations are in terms of *coded values of* X.

5. If the values of Y in step 1 were totals of several observations or replicates at each level of X, and we want the equations to be in terms of means, we must divide each term in the equations by the number of replicates. (This must be the same for all levels of X.)

Table 15.1 shows the daily total milk production of 37 cows, in pounds, recorded once a month for the 10 months from freshening to the end of lactation. We will apply the five preceding steps to these data.

The coefficients c_1, c_2, c_3 and c_4 were taken from Table A.11 and multiplied by the corresponding values of Y (milk production). The totals of these columns gave the values fo ΣY, P_1, P_2, P_3, and P_4. We are now ready to apply step 4 and write the equations.

$$\hat{Y}_L = 1{,}959.48 + (1/330)(-22{,}266.6)X' = 1.959.48 - 67.475X'$$

$$\hat{Y}_Q = [1{,}959.48 - (1/32)(-1{,}048.8)] - 67.475X' + (1/1{,}056)(-1{,}048.8)X'^2$$

$$= 1{,}992.26 - 67.475X' - 0.9932X'^2$$

$$\hat{Y}_C = 1{,}992.26 + [-67.475 - (293/205{,}920)(4{,}798.2)]X'$$

$$-0.9932X'^2 + (1/46{,}184)(4{,}798.2)X'^3$$

$$= 1{,}992.26 - 74.302X' - 0.9932X'^2 + 0.11651X'^3$$

$$\hat{Y}_4 = [1{,}992.26 + (9/1{,}280)(-5{,}384.6)] - 74.302X'$$

$$+[-0.9932\text{-}(41/54912)(-5384.6)]X'^2$$

$$+0.11651X'^3 + (1/109{,}824)(-5{,}384.6)X'^4$$

$$= 1{,}954.40 - 74.302X' + 3.0272X'^2 + 0.11651X'^3 - 0.049029X'^4$$

TABLE 15.1.
Milk production records of 37 cows for 10 months

Milk Production (Y)	Month (X)	X′ (c_1)	c_1Y	c_2	c_2Y	c_3	c_3Y	c_4	c_4Y
2,442.3	1	−9	−21,980.7	6	14,653.8	−42	−102,576.6	18	43,961.4
2,517.6	2	−7	−17,623.2	2	5,035.2	14	35,246.4	−22	−55,387.2
2,334.4	3	−5	−11,672.0	−1	−2,334.4	35	81,704.0	−17	−39,684.5
2,166.1	4	−3	−6,498.3	−3	−6,498.3	31	67,149.1	3	6,498.3
2,030.0	5	−1	−2,030.0	−4	−8,120.0	12	24,360.0	18	36,540.0
1,903.9	6	1	1,903.9	−4	−7,615.6	−12	−22,846.8	18	34,270.2
1,779.5	7	3	5,338.5	−3	−5,338.5	−31	−55,164.5	3	5,338.5
1,630.6	8	5	8,153.0	−1	−1,630.6	−35	−57,071.0	−17	−27,720.2
1,485.7	9	7	10,399.9	2	2,971.4	−14	−20,799.8	−22	−32,685.4
1,304.7	10	9	11,742.3	6	7,828.2	42	54,797.4	18	23,484.6
Totals 19,594.8			$P_1 = -22,266.6$		$P_2 = -1,048.8$		$P_3 = 4,798.2$		$P_4 = -5,384.6$

These equations are based on the total milk production of 37 cows. If we want them on a per-cow basis, we simply divide each term by 37 and obtain:

$$\hat{Y}_L = 52.959 - 1.8236X'$$

$$\hat{Y}_Q = 53,845 - 1.8236X' - 0.02684X'^2$$

$$\hat{Y}_C = 53.845 - 2.0082X' - 0.02684X'^2 + 0.003149X'^3$$

$$\hat{Y}_4 = 52.822 - 2.0082X' + 0.08182X'^2 + 0.003149X'^3 - 0.0013251X'^4$$

In actual practice it is not necessary to construct a table like Table 15.1, since the required P values can be found by accumulating the products on a calculating machine without writing down each individual product. Close attention must be paid to the signs of the coefficients. Where a coefficient is negative, its product with the corresponding Y value must be subtracted from the accumulated sum.

It is very important to keep in mind that the equations we have calculated are in terms of X′, the *coded values* of X. These are identical to the c_1 coefficients. Suppose in our example we wish to calculate the predicted milk production per cow from the quadratic equation for the third month. Referring to Table 15.1, we see that X′ for the third month is -5, so we substitute -5 for X′ in the quadratic equation:

$$\hat{Y}_Q = 53.845 - 1.8236(-5) - 0.02684(-5)^2$$
$$= 53.845 + 9.118 - 0.671 = 62.292$$

A common mistake made by students is to substitute the c_1 coefficients in the linear equation, the c_2 coefficients in the quadratic equation, and so on. It is the c_1 coefficients that are the coded values of X in every equation, regardless of the degree.

It is generally easiest to work with the equations in this form, but if the results are to be published in a scientific paper, they should appear in terms of the original values of X. To do this, it is necessary to substitute $(X - \overline{X})/L$ or $(X - \overline{X})2/L$ for X′ in the equations, depending on whether n is odd or even. To show how this is done, we will write our quadratic equation $\hat{Y}_Q = 53.845 - 1.8236X' - 0.02684X'^2$ in terms of X.

In this case $n = 10$ was even, so we substitute $(X - \overline{X})2/L$ for X′. The interval between successive values of X was 1, so $L = 1$. The value of $\overline{X}$ was 5.5, so we have $X' = (X - 5.5)2/1$ or $2X - 11$. Substituting this in our equation gives

$$\hat{Y}_Q = 53.845 - 1.8236(2X - 11) - 0.02684(2X - 11)^2$$
$$= 53.845 - 1.8236(2X - 11) - 0.02684(4X^2 - 44X + 121)$$
$$= 53.845 - 3.6472X + 20.0596 - 0.10736X^2 + 1.18096X - 3.23764$$

TABLE 15.2.
Observed and calculated monthly milk production of 37 cows

Observed Y	$\hat{Y}_L$	$Y-\hat{Y}_L$	$\hat{Y}_Q$	$Y-\hat{Y}_Q$	$\hat{Y}_C$	$Y-\hat{Y}_C$	$\hat{Y}_4$	$Y-\hat{Y}_4$
2,442.3	2,566.8	−124.5	2,519.1	−76.8	2,495.6	−53.3	2,461.7	−19.4
2,517.6	2,431.8	85.8	2,415.9	101.7	2,423.7	93.9	2,465.2	52.4
2,334.4	2,296.9	37.5	2.304.8	29.6	2.324.4	10.0	2,356.4	−22.0
2,166.1	2,161.9	4.2	2,185.7	−19.6	2,203.1	−37.0	2,197.4	−31.3
2,030.0	2.027.0	3.0	2.058.7	−28.7	2,065.5	−35.5	2,031.6	−1.6
1,903.9	1.892.0	11.9	1.923.8	−19.9	1,917.1	−13.2	1,883.2	20.7
1,779.5	1,757.1	22.4	1,780.9	−1.4	1,763.6	15.9	1,757.9	21.6
1,630.6	1,622.1	8.5	1,630.1	0.5	1,610.5	20.1	1,642.5	−11.9
1,485.7	1,487.2	−1.5	1,471.3	14.4	1,463.4	22.3	1,504.9	−19.2
1,304.7	1,352.2	−47.5	1,304.5	0.2	1,328.0	−23.3	1,294.1	10.6
Σ dev		−0.2		0.0		−0.1		−0.1
$\Sigma(\text{dev})^2$		27,268.90		18,930.76		16,258.59		6,105.23
$\Sigma(\text{dev})^2/37$		737.00		511.64		439.42		165.01

Collecting terms gives

$$\hat{Y}_Q = 70.65696 - 2.46624X - 0.10736X^2$$

Let us use this equation to again calculate Y_Q for the third month. Substituting 3 for X in this new equation gives

$$\hat{Y}_Q = 70.65696 - 2.46624(3) - 0.10736(3)^2 = 62.292, \text{ the same as before}$$

Let us see how much work we have saved. Using the methods of Chapter 14 (which we must use if the X's are not equally spaced), to find the four regression equations we would need to find ΣX, ΣX^2, ΣX^3, ΣX^4, ΣX^5, ΣX^6, ΣX^7, ΣX^8, ΣY, ΣXY, ΣX^2Y, ΣX^3Y and ΣX^4Y. These values would have to be substituted in the normal equations and we would have to solve sets of simultaneous equations, two for the linear coefficients on up to five for the quartic. If you worked through the examples in Chapter 14, you can appreciate what a laborious task this would be. Contrast all of these calculations with the shortcut method. Using this, we need only ΣY, P_1, P_2, P_3, and P_4. Substituting these values in the standard equations of step 4 gives us directly the four required regression equations. We have only five sums to calculate instead of 13, and there are no simultaneous equations to solve.

Now that we have the four equations, we can see how the values calculated from them compare with the observed milk production for each month. It is better to work with the totals rather than the means, since fewer rounding errors are introduced. Table 15.2 shows the values calculated from each equation and the deviations of these from the observed values.

There are several things to notice about this table. The sum of the deviations for all of the curves should add up to zero except for small rounding errors. This furnishes a check on the calculations. The sum of squares of deviations from a curve furnishes a measure of the closeness of fit; the smaller this sum of squares, the closer the fit of the curve to the data. Each added degree results in a reduction in this sum of squares. This must always be true; if it is not, look for an error in the computations. (The question is whether the improvement of fit is significant; we will show how to test this shortly.) For now, simply note that there is a moderate reduction in sum of squares as we go from the linear to the quadratic curve, a very small reduction as we go from quadratic to cubic, and a large reduction as we go from cubic to quartic. Finally, note that the signs of the deviations seem to fall in rather definite patterns in the first three degrees, while those from the quartic are more or less at random. Also, we can see that the quartic curve is the only one that shows an increase in milk production from the first to the second month. This is known to be characteristic of most milk production curves in cattle.

Partitioning the Sum of Squares

Finding all the calculated values and their deviations from the observed values and then finding the sums of squares of these deviations was a laborious procedure. The second feature of the shortcut method of analyzing equally spaced data is the ease with which these sums of squares can be calculated. Looking at Table A.11 under any value of n, you might recognize that the c values are really orthogonal sets of coefficients. Each column of coefficients adds up to zero, and the products of the corresponding coefficients of any two columns also add to zero. We learned in Chapter 6 that the sum of squares associated with a single degree of freedom can be found from a set of coefficients by applying the general formula

$$SS = \frac{(\Sigma c_i T_i)^2}{r \Sigma c_i^2}$$

As calculated previously, P_1 is the same as $\Sigma c_i T_i$ when the c's are the linear coefficients. Likewise $P_2 = \Sigma c_i T_i$ when we use the quadratic coefficients, and so on. The divisors shown in Table A.11 are the sums of squares of the coefficients. Therefore, the sum of squares due to linear regression is simply P_1^2/(divisor times number of replicates). Likewise the sum of squares for quadratic regression is P_2^2/(divisor times number of replicates), and so on up to the quartic component. After calculating the sums of squares for each component, we can find the residual sum of squares by subtracting the component sums of squares from the total sum of squares. This residual sum of squares is the same as the sum of squares of deviations of the observed data from the curve.

Let us apply this method of partitioning to the milk production data. The value of P_1 that we found was $-22{,}266.6$, so the linear SS is

$$\frac{-22{,}266.6^2}{330 \times 37} = 40{,}606.18$$

The total sum of squares of Y was 41,343.01 so the residual sum of squares in $41{,}343.01 - 40{,}606.18 = 736.83$. This is the same (except for a small difference due to rounding) as the sum of squares of deviations from linear found by a much more difficult method in Table 15.2.

Since P_2 was found to be $-1{,}048.8$, the sum of squares for quadratic is

$$\frac{-1{,}048.8^2}{132 \times 37} = 225.22$$

Subtracting this from 736.83 leaves a residual of 511.62. The value calculated in Table 15.2 was 511.64.

P_3 was 4,798.2 so the sum of squares for cubic is

$$\frac{4{,}798.2^2}{8{,}580 \times 37} = 72.52$$

leaving a residual of 439.09 (compared to 439.42 in Table 15.2).

Finally P_4 was $-5{,}384.6$ so the sum of squares for quartic is

$$\frac{-5{,}384.6^2}{2{,}860 \times 37} = 273.99$$

leaving a residual of 165.10.

All these results can be summarized in an analysis of variance table (Table 15.3) in which the sums of squares for cows, and error, were obtained from the individual cow records.

There was a highly significant difference among cows and among months. Neither of these results is surprising, but we want to know more about the pattern of change in milk production from month to month. The very high F value for the linear component tells us there is a highly significant downward trend. The significant deviation from linear indicates that a straight line does not fully account for the month-to-month variation. The significant quadratic component shows that a simple curve is an improvement over a straight line, but there is still a significant amount of residual variation. Fitting a cubic curve did not result in a significant improvement, and the residual left is not significant. At this point, many workers are inclined to stop. Often, as in this case, this is a mistake. The quartic component accounted for such a high proportion of the remaining sum of squares that it was highly significant. The deviation from quartic is not significant. The likelihood of finding another significant component is very small, for even if a single component accounted for 80% of the remaining variability, it would not be significant. We are therefore justified in terminating the analysis at this point.

TABLE 15.3.
Analysis of variance of milk production records

Source of variation	df	SS	MS	F
Total	369	76,167.74		
Cows	36	23,464.56	651.79	18.59**
Months	9	41,343.01	4,593.67	131.02**
Linear	1	40,606.18	40,606.18	1,158.19**
Deviation from Linear	8	736.83	92.10	2.63*
Quadratic	1	225.22	225.22	6.42*
Deviation from Quadratic	7	511.61	73.09	2.08*
Cubic	1	72.52	72.52	2.07ns
Deviation from Cubic	6	439.09	73.18	2.09ns
Quartic	1	273.99	273.99	7.81**
Deviation from Quartic	5	165.10	33.02	.94ns
Error	324	11,360.17	35.06	

Comparison of Shortcut and Regular Methods

In Chapter 14, we fitted a quadratic equation to the yield of sugar beets at five harvest dates. To do this, we first had to find seven sums of powers and products. Then from these sums we obtained three simultaneous equations which we had to solve for three unknowns. We now contrast this with the shortcut method.

We first find, using the coefficients from Table A.11 under $n=5$,

$$P_1=(-2)140.0+(-1)267.2+(1)417.0+(2)440.6=751$$

$$P_2=(2)140.0+(-1)267.2+(-2)335.2+(-1)417.0+(2)440.6=-193.4$$

Using these values and the K values from Table A.11, we can immediately write the quadratic equation

$$\hat{Y}_Q=320-(1/7)(-193.4)+(1/10)751X'+(1/14)(-193.4)X'^2$$

$$=347.6286+75.1X'-13.8143X'^2$$

To convert to original X units, we substitute $(X-3)$ for X' and $(X-3)^2=X^2-6X+9$ for X'^2. This gives

$$\hat{Y}=-2.0+157.9857X-13.8143X^2$$

exactly the same as obtained by the longer method.

Unequally Spaced Treatments

We have pointed out the advantages of equally spaced treatments, but if we have an experiment with unequally spaced treatments, it is still possible to find a set of orthogonal coefficients for calculating regression sums of squares. The formulas for finding these coefficients are much more complicated than in the case of equally spaced treatments. Also there is no simple way of writing the equations directly by the use of K values.

In Table A.11a we have given sets of orthogonal coefficients and divisors for some of the more commonly encountered treatment levels. These will at least make the determination of the regression sums of squares easier in such cases.

PERIODIC CURVE FITTING

Table A.12 gives sets of orthogonal coefficients for fitting periodic data when the observations are equally spaced throughout a complete cycle. The table is constructed for selected values of n most commonly encountered in dealing with daily, weekly, or yearly cycles.

Unlike the sets of coefficients we have been dealing with, these cannot be reduced to small integers. For this reason the calculation of P values is somewhat more difficult, but in other respects the calculation of equations, and partitioning of sums of squares are even easier than with polynomials, since no special divisors or K values are needed.

The reason that dealing with equally spaced intervals is so much simpler than dealing with irregular data is that most of the terms in the normal equations given in Chapter 14 drop out. Thus $\Sigma U_i = \Sigma V_i = 0$ where i is any subscript. Also $\Sigma U_i^2 = \Sigma V_i^2 = n/2$. Therefore the first normal equation, which is

$$na_0 + a_1\Sigma U_1 + b_1\Sigma V_1 + a_2\Sigma U_2 + b_2\Sigma V_2 + \ldots = \Sigma Y$$

reduces to $na_0 = \Sigma Y$, or $a_0 = \Sigma Y/n = \bar{Y}$. Likewise the other normal equations reduce to

$$a_1\left(\frac{n}{2}\right) = \Sigma U_1 Y \quad \text{or} \quad a_1 = \frac{2\Sigma U_1 Y}{n}$$

$$b_1 = \frac{2\Sigma V_1 Y}{n}$$

$$a_2 = \frac{2\Sigma U_2 Y}{n}$$

$$b_2 = \frac{2\Sigma V_2 Y}{n}$$

TABLE 15.4.
Monthly mean temperatures at Stockton, California with the calculations for fitting a second degree periodic curve ($C = 1/12 \times 360° = 30°$)

Month (X)	Temp (Y)	cos CX (U_1)	U_1Y	sin CX (V_1)	V_1Y	cos 2CX (U_2)	U_2Y	sin 2CX (V_2)	V_2Y
0	44.7	1.0	44.7000	0.0	0.0000	1.0	44.7000	0.0	0.0000
1	49.0	0.866	42.4340	0.5	24.5000	0.5	24.5000	0.866	42.4340
2	53.7	0.5	26.8500	0.866	46.5042	−0.5	−26.8500	0.866	46.5042
3	59.7	0.0	0.0000	1.0	59.7000	−1.0	−59.7000	0.0	0.0000
4	66.2	−0.5	−33.1000	0.866	57.3292	−0.5	−33.1000	−0.866	−57.3292
5	72.8	−0.866	−63.0448	0.5	36.4000	0.5	36.4000	−0.866	−63.0448
6	78.2	−1.000	−78.2000	0.0	0.0000	1.0	78.2000	0.0	0.0000
7	76.2	−0.866	−65.9892	−0.5	−38.1000	0.5	38.1000	0.866	65.9892
8	72.7	−0.5	−36.3500	−0.866	−62.9582	−0.5	−36.3500	0.866	62.9582
9	64.0	0.0	0.0000	−1.0	−64.0000	−1.0	−64.0000	0.0	0.0000
10	53.0	0.5	26.5000	−0.866	−45.8980	−0.5	−26.5000	−0.866	−45.8980
11	45.9	0.866	39.7494	−0.5	−22.9500	0.5	22.9500	−0.866	−39.7494
Totals	736.1	$PU_1 = -96.4506$		$PV_1 = -9.4728$		$PU_2 = -1.6500$		$PV_2 = 11.8642$	
$a_0 =$	61.34	$a_1 = -16.0751$		$b_1 = -1.5788$		$a_2 = -0.2750$		$b_2 = 1.9774$	

$$\hat{Y} = 61.34 - 16.0751 \cos CX - 1.5788 \sin CX - 0.275 \cos 2CX + 1.9774 \sin 2CX$$

TABLE 15.5.
Observed and calculated mean monthly temperatures at Stockton, California

Month	Y Observed	$\hat{Y}_1$ 1st Degree	$(Y-\hat{Y}_1)$	$\hat{Y}_2$ 2nd Degree	$(Y-\hat{Y}_2)$
January	44.7	45.26	−0.56	44.99	−0.29
February	49.0	46.63	2.37	48.20	0.80
March	53.7	51.94	1.76	53.79	−0.09
April	59.7	59.76	0.06	60.04	−0.34
May	66.2	68.01	−1.81	66.44	−0.24
June	72.8	74.47	−1.67	72.62	0.18
July	78.2	77.42	0.78	77.14	1.06
August	76.2	76.05	0.15	77.63	−1.43
September	72.7	70.74	1.96	72.59	0.11
October	64.0	62.92	1.08	63.19	0.81
November	53.0	54.67	−1.67	53.09	−0.09
December	45.9	48.21	−2.31	46.36	−0.46
Totals			0.02		0.02
Σd^2			28.86		4.99

and so on, following the same pattern except in the case where n is even, in which case the last coefficient that can be calculated is

$$a_{(n/2)} = \frac{\Sigma U_{(n/2)} Y}{n}$$

(We would seldom carry an analysis this far, since there would then be no residual sum of squares. In other words, an equation carried this far would exactly fit all of the data points, which is analogous to fitting a straight line to two points.)

We will adopt a symbol similar to one used in fitting polynomials, designating $\Sigma U_i Y$ as PU_i, and $\Sigma V_i Y$ as PV_i. Notice that in the case of the polynomial we had a single P value for each degree, but in fitting a periodic curve we need two P values called PU and PV for each degree[2] of fit.

The general terms in the equation are

$$a_0 = \bar{Y}, \qquad a_i = \frac{2PU_i}{n}, \qquad b_i = \frac{2PV_i}{n}$$

[2]We have designated each pair of terms added to the general periodic regression equation as a *degree* to maintain the analogy with the general polynomial. Technically, these are referred to as *harmonics*.

Let us apply this method for fitting a periodic curve to the complete data on monthly mean temperatures at Stockton, California, shown in Table 15.4.

The equation we have calculated is a general one in which we can substitute any value of X and look up the appropriate sines and cosines in a trigonometric table. However, if we are interested only in calculating values corresponding to the observed data points, we can simply substitute U_1 for cos CX, V_1 for sin CX, U_2 for cos 2CX, and V_2 for sin 2CX in the equation. For example, to find $\hat{Y}$ for March (month number 2, since January was called month 0), we calculate

$$\hat{Y}_2 = 61.34 - 16.0751(0.5) - 1.5788(0.866) - .275(-0.5) + 1.9774(0.866) = 53.79$$

If we want the calculated value for only the first-degree curve, we simply use the first three terms of the above equation:

$$\hat{Y}_1 = 61.34 - 16.0751(0.5) - 1.5788(0.866) = 51.94$$

The calculated values for the first- and second-degree equations are shown in Table 15.5 along with the deviation of the observed values from these two curves.

Partitioning the Sum of Squares

As with the polynomial, there is a very easy way to partition the total sum of squares without constructing a table like Table 15.5. The sum of squares for first-degree regression is $2(PU_1^2 + PV_1^2)/n$, and for second-degree, it is $2(PU_2^2 + PV_2^2)/n$, and so on. Unlike the polynomial, we do not need a different divisor for each degree. Sums of squares for deviations from observed data can be obtained by subtraction. From Table 15.4, we found that PU_1 was -96.4506 and PV_1 was -9.4728. Therefore the first-degree sum of squares is

$$\frac{2\left[(-96.4506)^2 + (-9.4728)^2\right]}{12} = 1565.41$$

The total sum of squares for Y was 1594.33, so that the sum of squares for deviation is

$$1{,}594.33 - 1{,}565.41 = 28.92$$

a result that differs from the value 28.86 found in Table 15.5 because of rounding.

Likewise, the sum of squares due to second-degree regression is

$$\frac{2\left[(-1.65)^2 + (11.8642)^2\right]}{12} = 23.91$$

The residual or deviation from second-degree sum of squares is $28.92 - 23.91 = 5.01$ (compared to 4.99 in Table 15.5). These results are summarized in Table 15.6.

TABLE 15.6.
Analysis of variance of temperature data

Source of Variation	df	SS	MS	F value
Months	11	1594.33		
1st degree	2	1565.41	782.705	243.61**
Deviation	9	28.92	3.213	
2nd degree	2	23.91	11.955	16.70**
Deviation	7	5.01	0.716	

Notice that each degree has 2 degrees of freedom. This is because two coefficients, a and b, had to be calculated for each degree. The mean square for each degree is tested against its residual component to make an F test. In this case, both the first and second degrees were highly significant.

We have fitted a curve to the mean monthly temperatures and partitioned the sum of squares for months into several components. If we wish to take into consideration the individual yearly records from which these means were computed, the analysis of variance is considerably more complicated. The student is referred to Bulletin 615 of the Connecticut Agricultural Experiment Station, 1958 entitled *Periodic Regression in Biology and Climatology*, by C. I. Bliss, for a detailed discussion of this subject.

The second-degree curve we have calculated is really made up ot two simple sine curves, one added to the other. The first has a semiamplitude

$$A=\sqrt{a_1^{\,2}+b_1^{\,2}}\,,\qquad \text{so}\qquad A=\sqrt{(-16.0751)^2+(-1.5788)^2}=16.13$$

The phase angle is $\tan^{-1}(b_1/a_1)+180°$ = angle whose tangent is $0.0982+180°=185°36'$ which converted to time is about six months and five days after the beginning of the cycle. Since our cycle begins with the January mean, we can call it January 15, so the maximum of our curve will fall on July 20, and the minimum 6 months earlier on January 20.

Referring to Figure 15.1, looking at the solid curve in the bottom half of the figure, we see that the observed temperatures tend to lie above the curve in the first and third quarters and below the curve in the second and fourth quarters. The second-degree curve largely adjusts for these discrepancies. It has a semiamplitude

$$A=\sqrt{a_2^{\,2}+b_2^{\,2}}=\sqrt{(-.275)^2+(1.9774)^2}=2.00$$

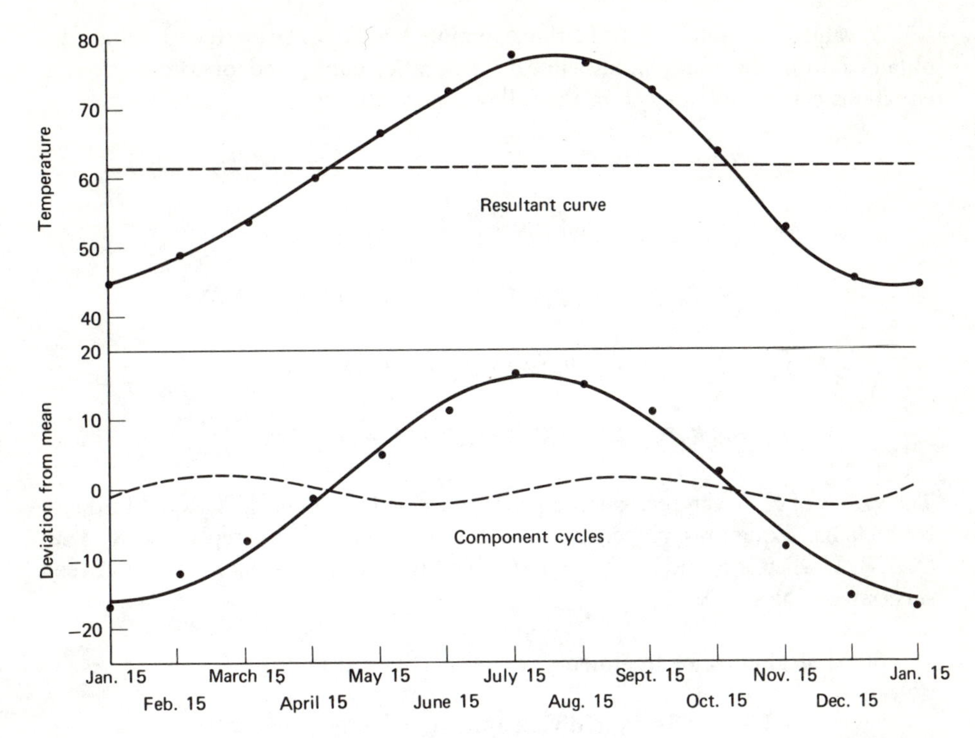

Figure 15.1. Mean monthly temperatures at Stockton, California. Second-degree Fourier curve and its components.

and a phase angle of

$$180° - \tan^{-1}\left(\frac{b_2}{a_2}\right) = 180° - \tan^{-1} 7.1905 = 180° - 82°5' = 97°5'$$

This must be divided by 2, since we are now dealing with a two-cycle curve, so we have a maximum at 48°32.5′ or about 1 month and 18 days after January 15. There is another maximum 6 months later, and a minimum at 3 months after each maximum. This is plotted as the dotted curve at the bottom of Figure 15.1.

Adding these two curves to the mean of 61.34 gives the resultant curve in the upper half of Figure 15.1.

SUMMARY

For equally spaced observations or treatments, a table (Table A.11) is furnished which greatly simplifies the calculations for deriving linear, quadratic, cubic, and quartic regression equations, or partitioning treatment sums of squares into trend components. The table contains three parts under each number of observations from 3 to 25: the c coefficients, the divisors, and the K values.

P values are obtained from the equation $P = \Sigma c_i T_i$. After the P values are obtained from the observations, linear, quadratic, cubic, and quartic regression equations can be obtained from the following equations:

$$\hat{Y}_L = \bar{Y} + (K_2P_1)X'$$

$$\hat{Y}_Q = (\bar{Y} - K_1P_2) + (K_2P_1)X' + (K_4P_2)X'^2$$

$$\hat{Y}_C = (\bar{Y} - K_1P_2) + (K_2P_1 - K_3P_3)X' + (K_4P_2)X'^2 + (K_5P_3)X'^3$$

$$\hat{Y}_4 = (\bar{Y} - K_1P_2 + K_8P_4) + (K_2P_1 - K_3P_3)X'$$

$$+ (K_4P_2 - K_7P_4)X'^2 + (K_5P_3)X'^3 + (K_6P_4)X'^4$$

The values of X' in the regression equations are *coded values of* X, equal to the c_1 coefficients. Equations in terms of X can be obtained by replacing X' with $(X - \bar{X})/L$ when n is odd or $(X - \bar{X})2/L$ when n is even. L is the interval between successive values of X.

Sums of squares for treatments can be partitioned into:

Linear $SS = P_1^2/$(divisor times number of replicates)

Quadratic $SS = P_2^2/$(divisor times number of replicates)

Cubic $SS = P_3^2/$(divisor times number of replicates)

Quartic $SS = P_4^2/$(divisor times number of replicates)

Residual SS = treatment SS − linear SS − quadratic SS − cubic SS − quartic SS

Table A.12 gives sets of coefficients for calculating periodic curves for data equally spaced throughout a time cycle. The table contains two sets of coefficients called U and V for each of the first four degrees (harmonics) for selected values of n.

Two P values are calculated for each degree of fit, from the equations

$$PU_i = \Sigma U_i Y \qquad \text{and} \qquad PV_i = \Sigma V_i Y$$

After the P values are determined, an equation of any desired degree up to the fourth can be written directly from the following equation:

$$\hat{Y} = \bar{Y} + \left(\frac{2PU_1}{n}\right)\cos CX + \left(\frac{2PV_1}{n}\right)\sin CX + \ldots + \left(\frac{2PU_i}{n}\right)\cos iCX + \left(\frac{2PV_i}{n}\right)\sin iCX$$

where X is the number of units of time from the beginning of a cycle, and C is the length of each unit in degrees.

The sum of squares for any degree has 2 degrees of freedom and is found from the relation

$$\text{SS for ith degree} = \frac{2(PU_i^2 + PV_i^2)}{n}$$

and the sum of squares for deviations from the curve can be obtained by subtraction of those regression components from the total SS.

The methods of this chapter are applicable only when the values of X are equally spaced, except for several commonly encountered sets of unequally spaced treatments for which orthogonal coefficients are given in Table A.11a, which can be used for calculating regression sums of squares but not regression equations.

16

MULTIPLE CORRELATION AND REGRESSION

So far, we have discussed only relations between two variables. We are often interested in the relation between a dependent variable and more than one independent variable. The law of supply and demand, for example, implies a relation between price (the dependent variable) and two variables—supply and demand. In livestock, we may be interested in weight gain in relation to various components of feed. In crops, we may want to study the effect on yield as N, P, and K all vary.

CORRELATION COEFFICIENTS

The correlation between two variables, disregarding any other variables that may be varying simultaneously, is called *simple* or *total correlation*. The correlation between two variables, when one or more other variables are held at a constant level, is called *partial correlation*. The combined relation between a variable and two or more other variables varying simultaneously is called *multiple correlation*.

Suppose we have a dependent variable, Y, and for each value of Y there are corresponding values of two other variables, X_1 and X_2. The simple or total correlation between Y and X_1 is the linear correlation coefficient we discussed in Chapter 13. You will recall the formula was:[1]

$$r^2 = \frac{(\Sigma xy)^2}{\Sigma x^2 \Sigma y^2}$$

To show clearly that this is the simple correlation of Y with X_1, it is customary to include explanatory subscripts, so we write the formula as

$$r^2_{YX_1} = \frac{(\Sigma x_1 y)^2}{\Sigma x_1^2 \Sigma y^2}$$

[1]As before, the formulas are expressed in terms of r^2 rather than r. It should be remembered that r, the coefficient of correlation, is the square root of r^2.

Likewise, the simple correlation between Y and X_2 is written

$$r^2_{YX_2} = \frac{(\Sigma x_2 y)^2}{\Sigma x_2^2 \Sigma y^2}$$

Finally, in order to calculate *partial* and *multiple* correlation, we need a third simple correlation, that between X_1 and X_2:

$$r^2_{X_1X_2} = \frac{(\Sigma x_1 x_2)^2}{\Sigma x_1^2 \Sigma x_2^2}$$

The partial correlation between Y and X_1 with a fixed X_2 is designated as $r_{YX_1 \cdot X_2}$ and is calculated from the simple correlations in the following manner:

$$r^2_{YX_1 \cdot X_2} = \frac{(r_{YX_1} - r_{YX_2} r_{X_1X_2})^2}{(1 - r^2_{YX_2})(1 - r^2_{X_1X_2})}$$

Likewise,

$$r^2_{YX_2 \cdot X_1} = \frac{(r_{YX_2} - r_{YX_1} r_{X_1X_2})^2}{(1 - r^2_{YX_1})(1 - r^2_{X_1X_2})}$$

The multiple correlation coefficient, designated as $R_{Y.X_1X_2}$, measures the combined relation of X_1 and X_2 with Y. It is found by taking the square root of:

$$R^2_{Y.X_1X_2} = \frac{r^2_{YX_1} + r^2_{YX_2} - 2 r_{YX_1} r_{YX_2} r_{X_1X_2}}{1 - r^2_{X_1X_2}}$$

Just as r^2 was called the coefficient of determination, R^2 is called the multiple coefficient of determination. It is the proportion of the variation in Y accounted for by the variation in the two or more independent variables.

Notice how the addition of just one more variable has added to the complexity of correlation. With two variables, X and Y, we had only one coefficient of correlation. With three variables, X_1, X_2 and Y, we have three simple coefficients, three partial coefficients, and the multiple coefficient.

The problem of visualizing a three-variable relation is also much more difficult than with two variables. In the two-variable case, we can depict the observations on a two-dimensional graph. The relation is described by a regression line, and with many observations, the scatter diagram of points will appear as an ellipse. The narrower the ellipse, the higher the correlation. With three variables, the relation must be described as a plane in three-dimensional space. The scatter of points around this plane will be in the shape of an ellipsoid. The projection of the ellipsoid on the X_1Y plane shows the simple correlation of X_1 and Y. A section

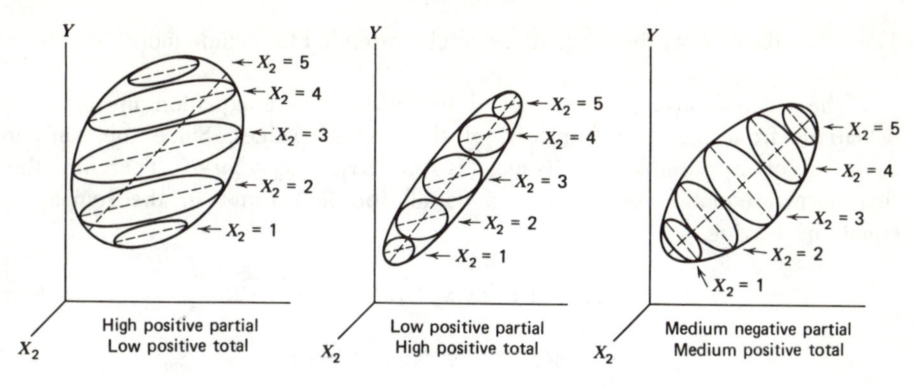

Figure 16.1. Diagram of various combinations of partial and total correlations involving three variables.

through the ellipsoid paralled to the X_1Y plane and projected on the X_1Y plane will show the partial correlation of X_1 and Y with X_2 fixed, written $r_{YX_1.X_2}$

In Figure 16.1, various situations are shown diagrammatically. Note that the simple correlation can be low, but the partial correlation high, or vice versa. They can even be different in sign.

The multiple coefficient of correlation, R, shows how closely the points in the ellipsoid are clustered around the regression plane. The value of R is always positive, ranging from zero to one. Furthermore, it is always at least as large as the largest simple and partial coefficients. This fact serves as a good check on the calculations.

REGRESSION COEFFICIENTS

So far, we have talked only about correlations—the closeness of the relations among the variables. We also want to know the nature of the relations. What change in Y is associated with unit changes in the independent variables? To answer this, we need an equation of the form

$$\hat{Y} = a + b_1X_1 + b_2X_2.$$

The terms b_1 and b_2 are called *partial regression coefficients*. The best-fitting equation of this form will be the one that makes the sum of squares of deviations of the observed Y's from the estimated $\hat{Y}$'s a minimum. To find the values of a, b_1, and b_2 that will meet this requirement, we solve *normal equations* very similar to the ones we solved for curvilinear regression:

$$an + b_1\Sigma X_1 + b_2\Sigma X_2 + \ldots = \Sigma Y$$

$$a\Sigma X_1 + b_1X_1^2 + b_2\Sigma X_1X_2 + \ldots = \Sigma X_1Y$$

$$a\Sigma X_2 + b_1\Sigma X_2X_1 + b_2\Sigma X_2^2 + \ldots = \Sigma X_2Y$$

. . . .

The dots indicate how these equations can be extended to include more than three variables.

The calculations can be reduced by rewriting the equation in terms of deviations from the means instead of the original values. Since the sum of deviations of any variable from its mean is zero, $\Sigma x_1 = \Sigma x_2 = \Sigma y = 0$. Therefore, the first normal equation drops out, as do all the first terms in the remaining equations, leaving

$$b_1 \Sigma x_1^2 + b_2 \Sigma x_1 x_2 + \ldots = \Sigma x_1 y$$

$$b_1 \Sigma x_2 x_1 + b_2 \Sigma x_2^2 + \ldots = \Sigma x_2 y$$

$$\cdots\cdots\cdots\cdots\cdots\cdots$$

Solving these equations for the b's gives a regression equation of the form $\hat{y} = b_1 x_1 + b_2 x_2 + \ldots$. If we wish an equation in terms of the original observations, we can calculate: $a = \bar{Y} - b_1 \bar{X}_1 - b_2 \bar{X}_2 \ldots$. Then, $\hat{Y} = a + b_1 X_1 + b_2 X_2 + \ldots$.

AN EXAMPLE WITH THREE VARIABLES

To illustrate partial and multiple correlation and regression, we will analyze some data on the specific gravity of potatoes (Y), the nitrogen content (X_1) and the phosphorous content (X_2). The observations will be coded to simplify the calculations (see Table 16.1).

First, we calculate the various coefficients of correlation. The simple or total correlations are as follows:

$$r^2_{YX_1} = (\Sigma y x_1)^2 / \Sigma y^2 \Sigma x_1^2 = \frac{-29{,}218.35^2}{(51{,}172.95)(21{,}240.55)} = 0.7854$$

$$r_{YX_1} = \sqrt{r^2_{YX_1}} = -0.8862 \text{ (Note that it is negative because } \Sigma y x_1 \text{ was negative)}$$

$$r^2_{YX_2} = (\Sigma y x_2)^2 / \Sigma y^2 \Sigma x_2^2 = \frac{-6{,}611.8^2}{(51{,}172.95)(1{,}663.2)} = 0.5136$$

$$r_{YX_2} = \sqrt{r^2_{YX_2}} = -0.7167$$

$$r^2_{X_1X_2} = (\Sigma x_1 x_2)^2 / \Sigma x_1^2 \Sigma x_2^2 = \frac{2{,}584.4^2}{(21{,}240.55)(1{,}663.2)} = 0.1891$$

$$r_{X_1X_2} = \sqrt{r^2_{X_1X_2}} = 0.4348$$

TABLE 16.1
Specific gravity, nitrogen and phosphorous content of twenty samples of potatoes

	Y (Sp. Gr. − 1.07)10^4	X_1 (Nitrogen − 1)100	X_2 (Phosphorous)100
	2	96	40
	14	82	36
	15	121	30
	15	88	42
	16	100	28
	27	114	26
	48	71	33
	54	94	26
	58	74	15
	68	36	35
	82	36	25
	83	73	15
	91	58	26
	97	31	25
	98	38	24
	101	56	11
	128	24	22
	140	37	11
	163	10	24
	179	14	10
Totals	1,479	1,253	504
	$\Sigma Y^2 = 160{,}545$	$\Sigma X_1^2 = 99{,}741$	$\Sigma X_2^2 = 14{,}364$
	$(\Sigma Y)^2/20 = 109{,}372.05$	$(\Sigma X_1)^2/20 = 78{,}500.45$	$(\Sigma X_2)^2/20 = 12{,}700.8$
	$\Sigma y^2 = 51{,}172.95$	$\Sigma x_1^2 = 21{,}240.55$	$\Sigma x_2^2 = 1{,}663.2$
	$\Sigma YX_1 = 63{,}441$	$\Sigma YX_2 = 30{,}659$	$\Sigma X_1X_2 = 34{,}160$
	$\Sigma Y\Sigma X_1/20 = 92{,}659.35$	$\Sigma Y\Sigma X_2/20 = 37{,}270.8$	$\Sigma X_1\Sigma X_2/20 = 31{,}575.6$
	$\Sigma yx_1 = -29{,}218.35$	$\Sigma yx_2 = -6{,}611.8$	$\Sigma x_1x_2 = 2{,}584.4$

The partial correlation coefficients are as follows:

$$r^2_{YX_1 \cdot X_2} = \frac{(r_{YX_1} - r_{YX_2} r_{X_1X_2})^2}{(1 - r^2_{YX_2})(1 - r^2_{X_1X_2})} = \frac{[-0.8862 - (-0.7167)(0.4338)]^2}{(1 - 0.5136)(1 - 0.1891)}$$

$$= \frac{-0.5746^2}{(0.4864)(0.8109)} = 0.8371$$

$$r_{YX_1 \cdot X_2} = \sqrt{r^2_{YX_1 \cdot X_2}} = -0.9149$$

$$r^2_{YX_2 \cdot X_1} = \frac{(r_{YX_2} - r_{YX_1} r_{X_1X_2})^2}{(1 - r^2_{YX_1})(1 - r^2_{X_1X_2})} = \frac{[-0.7167 - (-0.8862)(0.4348)]^2}{(1 - 0.7854)(1 - 0.1891)}$$

$$= \frac{-0.3314^2}{(0.2146)(0.8109)} = 0.6310$$

$$r_{YX_2 \cdot X_1} = \sqrt{r^2_{YX_2 \cdot X_1}} = -0.7944$$

Finally, we calculate R, the multiple coefficient:

$$R^2_{Y \cdot X_1X_2} = \frac{r^2_{YX_2} + r^2_{YX_1} - 2r_{YX_1} r_{YX_2} r_{X_1X_2}}{1 - r^2_{X_1X_2}}$$

$$= \frac{0.5136 + 0.7854 - 2(-0.8862)(-0.7167)(0.434)}{1 - 0.1891}$$

$$= \frac{0.7467}{0.8109} = 0.9208$$

$$R_{Y \cdot X_1X_2} = \sqrt{0.9208} = 0.9596$$

The simple correlations of either nitrogen or phosphorus content alone with specific gravity are not very large, but when the two variables are considered simultaneously, the relation with specific gravity is very close. Stated in percentage figures, nitrogen alone accounts for 78.54% of the variability in specific gravity, $(100 \times r^2_{YX_1})$. Phosphorus accounts for 51.36%. Nitrogen and phosphorus jointly account for 92.08%.

We now need to describe the relation by calculating the regression equation.

Using the normal equations based on deviations from means, we have

$$b_1\Sigma x_1^2 + b_2\Sigma x_1 x_2 = \Sigma x_1 y$$

$$b_1\Sigma x_1 x_2 + b_2\Sigma x_2^2 = \Sigma x_2 y$$

Substituting the observed values from the data:

$$21{,}240.55b_1 + 2{,}584.4b_2 = -29{,}218.35$$

$$2{,}584.40b_1 + 1{,}663.2b_2 = -6{,}661.8$$

Multiplying the first equation by 2,584.4, and the second equation by 21,240.55 and subtracting, we get

$$28{,}648{,}159.4b_2 = -64{,}926{,}364.75$$

$$b_2 = \qquad -2.266$$

Substituting this value of b_2 in either of the original equations, and solving for b_1, we find

$$b_1 = -1.100$$

To have a regression equation in terms of the original values, we need to find a:

$$a = \overline{Y} - b_1\overline{X}_1 - b_2\overline{X}_2$$

$$= \frac{1479}{20} - \left(-1.100\frac{1253}{20}\right) - \left(-2.266\frac{504}{20}\right) = 199.968$$

We can now write the regression equation: $\hat{Y} = 199.968 - 1.100X_1 - 2.266X_2$.

From this equation we can calculate values of $\hat{Y}$ and compare them with the observed values (Table 16.2).

The sum of the deviations is zero, as it should be. This furnishes a good check on the computations. The sum of squares of deviations is 4,051.16. This represents the variation in specific gravity (Y) not associated with the variation in nitrogen content (X_1) or phosphorus content (X_2). It can be calculated, without computing each Y, by taking $(1-R^2)\Sigma y^2$, which is

$$(1-0.9208)51{,}172.95 = 4{,}052.90$$

The two answers are in close agreement, the small difference resulting from rounding.

The results we have obtained can be summarized in an analysis of variance table as follows:

Source of variation	Method of computing SS	SS	df	MS	F
Total	Σy^2	51,172.95	19		
Regression due to X_1	$r^2_{YX_1}(\Sigma y^2)$	40,191.23	1	40,191.23	65.9**
Deviation from simple regression	$(1-r^2_{YX_1})\Sigma y^2$	10,981.72	18	610.10	
Additional regression due to X_2	$r^2_{YX_2 \cdot X_1}(1-r^2_{YX_1})\Sigma y^2$	6,929.47	1	6,929.47	29.07**
Deviation from multiple regression	$(1-R^2_{Y \cdot X_1X_2})\Sigma y^2$	4,052.90	17	238.41	

The last sum of squares can be obtained by subtraction: $10{,}981.72 - 6{,}929.47 = 4{,}052.25$. The discrepancy between this value and the one in the table is the result of rounding and will have no important bearing on the F value. The square root of 238.41 or 15.44 is called *standard error of estimate*, and is designated by the symbol $s_{Y \cdot X_1X_2}$.

There is another way in which the analysis of variance table can be set up, giving quite different F values:

Source of variation	Method of computing SS	SS	df	MS	F
Total	Σy^2	51,172.95	19		
Regression due to X_2	$r^2_{YX_2}(\Sigma y^2)$	26,282.43	1	26,282.43	19.01**
Deviation from simple regression	$(1-r^2_{YX_2})\Sigma y^2$	24,890.52	18	1,382.81	
Additional regression due to X_1	$r^2_{YX_1 \cdot X_2}(1-r^2_{YX_2})\Sigma y^2$	20,835.85	1	20,835.85	87.40**
Deviation from multiple regression	$(1-R^2_{Y \cdot X_1X_2})\Sigma y^2$	4,052.90	17	238.41	

In the first of these two tables, we considered the total effect of nitrogen and then the additional effect of phosphorus. In the second table, we considered the total effect of phosphorus and then the additional effect of nitrogen. The fact that the order in which variables are considered makes a marked difference in the outcome of the analysis can be confusing to anyone during first exposure to multiple regression.

A simple example might help clarify some of the confusion. It is well known that the yield of many crops is influenced by both temperature and day length.

TABLE 16.2.
Observed and calculated specific gravity of 20 samples of potatoes

Y	$\hat{Y}$	$d = Y - \hat{Y}$
2	3.7	−1.7
14	28.2	−14.2
15	−1.1	16.1
15	8.0	7.0
16	26.5	−10.5
27	15.7	11.3
48	47.1	0.9
54	37.7	16.3
58	84.6	−26.6
68	81.1	−13.1
82	103.7	−21.7
83	85.7	−2.7
91	77.2	13.8
97	109.2	−12.2
98	103.8	−5.8
101	113.4	−12.4
128	123.7	4.3
140	134.3	5.7
163	134.6	28.4
179	161.9	17.1

Suppose we have numerous crop yield records of a crop grown in different seasons of the year. For each yield record, we have a record of the mean day length and of the mean temepratabove during the growing season. We expect day length and temperature to be closely correlated with each other. Since this is true, we should not be surprised if we found that yield was closely correlated with temperature but that the additional consideration of day length would explain little of the variation in yield not already accounted for. At the same time, day length alone might be closely correlated with yield, while temperature might have little added effect. The conclusion would be that long, warm days are associated with higher yields than are short, cold days. We could tell little about which factor was the more important, temperature or day length. To answer this question, we would need an experiment in which the day length and/or temperature were controlled so that they would be less closely correlated than they are in nature.

In Chapter 13, we gave an example of a spurious correlation between cigarette consumption and hay production. This high correlation was apparently

TABLE 16.3.
Multiple regression analysis of hay production (Y), cigarette consumption (X_1), and time (X_2)

Source of variation	df	SS	MS	F
X_1 considered first				
Total	14	10,094.00		
Regression due to X_1	1	8,855.31	8,855.31	92.94**
Deviation from simple regression	13	1,238.69	95.28	
Additional regression due to X_2	1	918.01	918.01	34.35**
Deviation from multiple regression	12	320.67	26.72	
X_2 considered first				
Total	14	10,094.00		
Regression due to X_2	1	9,723.21	9,723.21	340.90**
Deviation from simple regression	13	370.79	28.52	
Additional regression due to X_1	1	50.11	50.11	1.88ns
Deviation from multiple regression	12	320.67	26.72	

caused by the fact that both variables were closely related to a third variable, time. A multiple regression analysis will show a striking difference between two analyses, depending on which independent variable is considered first (Table 16.3).

In the second analysis, where we removed the regression with time first, we see there is no significant additional regression related to cigarette consumption.

MORE THAN THREE VARIABLES

For the sake of simplicity, most of our discussion and the illustrative examples have been based on three variables, one dependent and two independent. Actually, multiple and partial correlation coefficients and regression equations can be calculated for any number of variables. A recent study at the University of California included 35 variables. We can do no more here than indicate, in a general way, how the methods described can be extended to more than three variables and point out some of the difficulties involved.

We have already shown how the normal equations for calculating the regression coefficients, b_1, b_2, and so on can be extended to include as many variables as we wish. Each new variable requires only the addition of another term on the left-hand side of each equation and the addition of one new equation following the same pattern as the previous ones. For m variables the last normal equation will be

$$b_1\Sigma x_1x_m + b_2\Sigma x_2x_m + b_3\Sigma x_3x_m + \cdots + b_m\Sigma x_m^2 = \Sigma x_m y$$

The algebra does not change, but the arithmetic involved in solving the equations becomes increasingly difficult as we add new variables. For this reason, it is suggested that one of the systematic procedures mentioned in the previous chapter be used, or if possible, use an electronic computer.

We have seen how, with only two variables, there was just 1 coefficient of correlation, but with three variables there were 7, including 1 multiple, 3 simple, and 3 partial coefficients. With four variables, the total increases to 25, and with five to 81. One of the reasons for the big increases is the fact that we have the addition of *high order partial coefficients*. The *order* of a partial correlation coefficient is the number of variables that are fixed. With three variables, we had only first-order partials, such as $r_{YX_1 \cdot X_2}$. With four variables, we have simple and first-order partials and second-order partials, such as $r_{YX_1 \cdot X_2X_3}$, which is read "the correlation of Y and X_1 for fixed values of X_2 and X_3."

There is a general equation that enables us to compute a partial correlation coefficient of any order if we know three partials of one order lower:

$$r^2_{YX_1.X_2X_3\ldots X_m} = \frac{(r_{YX_1.X_3\ldots X_m} - r_{YX_2.X_3\ldots X_m} r_{X_1X_2.X_3\ldots X_m})^2}{(1 - r^2_{YX_2.X_3\ldots X_m})(1 - r^2_{X_1X_2.X_3\ldots X_m})}$$

The equations given for finding the first order partials involving three variables from the three simple correlations were simply special cases of this general equation.

A general equation for finding the multiple coefficient of correlation involving m independent variables is

$$1 - R^2_{Y.X_1\ldots X_m} = (1 - r^2_{YX_1})(1 - r^2_{YX_2.X_1})(1 - r^2_{YX_3.X_1X_2}) \cdots (1 - r^2_{YX_m.X_1\ldots X_{m-1}})$$

In the case of two independent variables, this reduces to the fairly simple form already given for $R^2_{Y \cdot X_1X_2}$.

We have seen that the arithmetic becomes increasingly difficult as we consider more variables, but perhaps the greatest difficulty encountered when one considers more than three variables is in visualizing the relations. The relation between two variables can be pictured on a two-dimensional graph. The relations among three variables can be depicted in a three-dimensional diagram. But how do we draw a picture of the relations among four or more variables? The answer is that we just do not try. We have to learn not to be bothered by our inability to visualize relations involving four or more dimensions. Instead, we need to think in terms of equations rather than diagrams. After all, we have no trouble grasping the idea that the yield of a crop is related to the N, P, and K levels in the soil, the amount of water applied, the weed competition, the amount of disease, the number of injurious insects, the temperature, and the day length. With enough data, we can even write an easily understood equation that describes these relations. Should we worry if we cannot draw a picture descriptive of this complex interplay of factors? One equation may be worth a thousand pictures.

One more thing needs to be said about correlation and regression involving a large number of variables. We showed that, with three variables, two different analyses could be made, depending on which of the independent variables we considered first. With three independent variables, the number of possible analyses increases to six, and with m independent variables there are m! possible ways of ordering the variables. (The symbol "m!" is read *factorial m* and means the product of all the numbers from one to m. Thus $10! = 1 \times 2 \times 3 \times 4 \times 5 \times 6 \times 7 \times 8 \times 9 \times 10 = 3{,}628{,}800$.) What is the *best* order in which to consider the variables? A related question is, "Out of a large number of independent variables, how can we find the best set of a given size?" Finding a direct simple method for obtaining the best set is one of the great unsolved problems of statistics. Programs are available on electronic computers for arriving at the solution, but time is the limiting factor.

RESPONSE SURFACES

The independent variables in multiple regression problems need not be distinct variables. They may be different powers of the same variable such as X, X^2, and X^3, or the products of two or more variables, such as X_1X_2, $X_1{}^2X_2$, and so forth. Thus, polynomial curve fitting is a special case of multiple regression. We pointed out in Chapter 14 that the proportion of the variability in Y accounted for by the linear plus the quadratic sum of squares is designated R^2, or the coefficient of multiple determination.

If we have two variables, each having a significant curvilinear relation with Y, we can find an equation describing this entire relationship. Not only can we find such an equation but we can also show it graphically by one of several kinds of three-dimensional presentations. Such a graph is called a *response surface*.

The sugar beet nitrogen and time of harvest experiment described in Chapter 10 is a good example. We showed in that chapter that the significant components were: nitrogen linear, nitrogen quadratic, harvest dates linear, harvest dates quadratic, nitrogen linear × harvest dates linear, and nitrogen quadratic × harvest dates linear. To include the effects of all these components on yield in a single equation would require an equation of the form:

$$\hat{Y} = a + bH + cN + dH^2 + eN^2 + fNH + gN^2H$$

To find this equation requires the solving of seven simultaneous equations in seven unknowns. The normal equations are

$$an + b\Sigma H + c\Sigma N + d\Sigma H^2 + e\Sigma N^2 + f\Sigma NH + g\Sigma N^2H = \Sigma Y$$

$$a\Sigma H + b\Sigma H^2 + c\Sigma NH + d\Sigma H^3 + e\Sigma N^2H + f\Sigma NH^2 + g\Sigma N^2H^2 = \Sigma HY$$

$$a\Sigma N + b\Sigma NH + c\Sigma N^2 + d\Sigma NH^2 + e\Sigma N^3 + f\Sigma N^2H + g\Sigma N^3H = \Sigma NY$$

$$a\Sigma H^2 + b\Sigma H^3 + c\Sigma NH^2 + d\Sigma H^4 + e\Sigma N^2H^2 + f\Sigma NH^3 + g\Sigma N^2H^3 = \Sigma H^2Y$$

$$a\Sigma N^2 + b\Sigma N^2H + c\Sigma N^3 + d\Sigma N^2H^2 + e\Sigma N^4 + f\Sigma N^3H + g\Sigma N^4H = \Sigma N^2Y$$

$$a\Sigma NH + b\Sigma NH^2 + c\Sigma N^2H + d\Sigma NH^3 + e\Sigma N^3H + f\Sigma N^2H^2 + g\Sigma N^3H^2 = \Sigma NHY$$

$$a\Sigma N^2H + b\Sigma N^2H^2 + c\Sigma N^3H + d\Sigma N^2H^3 + e\Sigma N^4H + f\Sigma N^3H^2 + g\Sigma N^4H^2 = \Sigma N^2HY$$

Solving these seven simultaneous equations appears at first to be a formidable task, but if we code the values of N and H properly, many of the sums will be zero, and the equations will be greatly simplified. H, since it consists of five equally spaced dates, can be coded by using the c_1 coefficients under $n=5$ in Table A11. These are -2, -1, 0, 1 and 2. To code the nitrogen levels, we observe that dividing by 80 gives the series: 0, 1, 2, 4, and the linear coefficients for this series in Table A.11a are -7, -3, -1, and 9.

Using these coded values, the following terms in the normal equations are equal to zero: ΣH, ΣN, ΣNH, ΣN^2H, ΣH^3, ΣNH^2, ΣN^3H, ΣNH^3, ΣN^2H^3, ΣN^4H. This leaves the following sums that are needed for the normal equations:

$\Sigma H^2 = 40$ $\quad$ $n = 20$

$\Sigma N^2 = 700$ $\quad$ $\Sigma Y = 1600$

$\Sigma N^2H^2 = 1400$ $\quad$ $\Sigma HY = 751$

$\Sigma N^3 = 1800$ $\quad$ $\Sigma NY = 1430.4$

$\Sigma H^4 = 136$ $\quad$ $\Sigma H^2Y = 3006.6$

$\Sigma N^4 = 45{,}220$ $\quad$ $\Sigma N^2Y = 54{,}867.2$

$\Sigma N^3H^2 = 3600$ $\quad$ $\Sigma NHY = 744.2$

$\Sigma N^4H^2 = 90{,}440$ $\quad$ $\Sigma N^2HY = 25{,}967.8$

The normal equations are now:

$$20a \qquad\qquad + \ 40d + \ 700e \qquad\qquad = 1{,}600.0 \qquad (1)$$

$$40b \qquad\qquad\qquad\qquad 1400g = \ 751.0 \qquad (2)$$

$$700c \qquad + \ 1800e \qquad\qquad = 1{,}430.4 \qquad (3)$$

$$40a \qquad\qquad + \ 136d + \ 1400e \qquad\qquad = 3{,}006.6 \qquad (4)$$

$$700a \qquad + 1800c + 1400d + 45220e \qquad\qquad = 54{,}867.2 \qquad (5)$$

$$1400f + \ 3600g = \ 744.2 \qquad (6)$$

$$1400b \qquad\qquad + 3600f + 90440g = 25{,}967.8 \qquad (7)$$

Multiplying equation (2) by 35 and subtracting from equation (7) gives

$$3600f + 41440g = -317.2 \qquad (8)$$

Multiplying equation (8) by 7 and subtracting equation (6) times 18 gives:

$$225{,}280g = -15.616$$

$$g = -0.069318$$

Substituting g in equation (6) gives $f = 0.709818$.
Substituting g in equation (2) gives $b = 21.201136$.
Multiplying equation (1) by 2 and subtracting from equation (4) gives

$$56d = -193.4$$

$$d = -3.453571$$

Multiplying equation (1) by 35 and subtracting from equation (5) gives

$$1800c + 20720e = -1132.8 \qquad (9)$$

Multiplying equation (9) by 7 and subtracting equation (3) times 18 leaves:

$$112{,}640e = -33{,}676.8$$

$$e = -0.298977$$

Substituting d and e in equation (1) gives

$$20a = 1947.4269$$

$$a = 97.371345$$

Substituting e in equation (3) gives

$$700c = 1968.559$$

$$c = 2.812227$$

We now have all the terms for the equation in terms of coded values of N and H:

$$\hat{Y} = 97.371345 + 21.201136H' + 2.812227N' - 3.453571H'^2$$
$$- 0.298977N'^2 + 0.709818N'H' - 0.069318N'^2H'$$

The original harvest date levels were 0, 3, 6, 9, and 12 weeks, and the N rates were

TABLE 16.4.
Observed and predicted sugar beet yields
for each treatment combination, based on totals of four replicates

N	H	Y	$\hat{Y}$	$(Y-\hat{Y})$	$(Y-\hat{Y})^2$
0.0	0	22.0	23.55	−1.55	2.4025
0.0	3	47.4	46.75	0.65	0.4225
0.0	6	61.1	63.04	−1.94	3.7636
0.0	9	69.8	72.42	−2.62	6.8644
0.0	12	76.1	74.89	1.21	1.4641
0.8	0	39.4	35.53	3.87	14.9769
0.8	3	67.9	64.34	3.56	12.6736
0.8	6	85.6	86.24	−0.64	0.4096
0.8	9	105.0	101.24	3.76	14.1376
0.8	12	110.1	109.33	0.77	0.5929
1.6	0	40.7	42.39	−1.69	2.8561
1.6	3	74.4	74.59	−0.19	0.0361
1.6	6	91.9	99.88	−7.98	63.6804
1.6	9	120.1	118.27	1.83	3.3489
1.6	12	129.3	129.75	−0.45	0.2025
3.2	0	37.9	40.70	2.80	7.8400
3.2	3	77.5	73.04	4.46	19.8916
3.2	6	96.6	98.46	−1.86	3.4596
3.2	9	122.1	116.98	5.12	26.2144
3.2	12	125.1	128.60	−3.50	12.2500
Total		1600.0	1599.99	0.01	197.4873

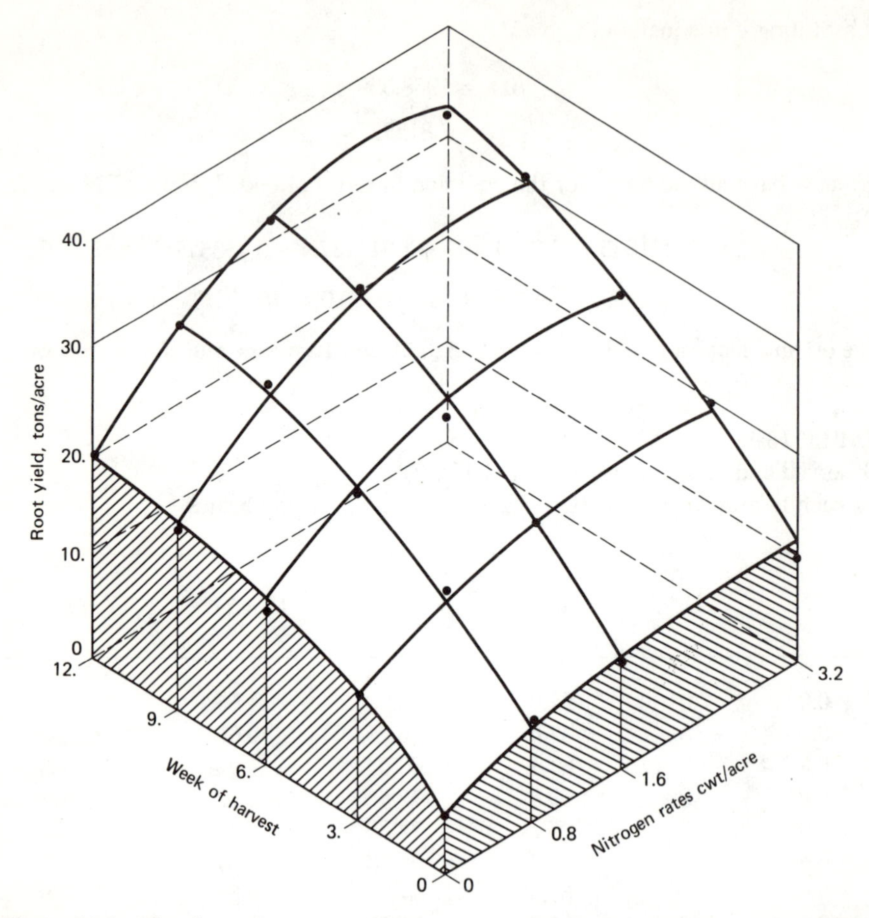

Figure 16.2. The three-dimensional response curve fitted to the observed response (solid points) of sugar beet to rates of nitrogen fertilizer and week of harvest.

0, 0.8, 1.6, and 3.2 cwt. To convert the above equation to these units we must substitute $(H/3)-2$ for H′ and $5N-7$ for N′. (See the summary of Chapter 15 for the equation for changing H′ and N′ to the original values of H and N.) The resultant equation is

$$\hat{Y} = 23.55 + 8.8834H + 18.1868N - 0.38373H^2$$

$$-4.00853N^2 + 2.80045NH - 0.577652N^2H$$

Substituting the values of N and H in this equation gives the calculated values shown in Table 16.4.

The sum of squares of deviations of observed from predicted has to be divided by 4 to put it on a per-plot basis, since there were four replicates. This

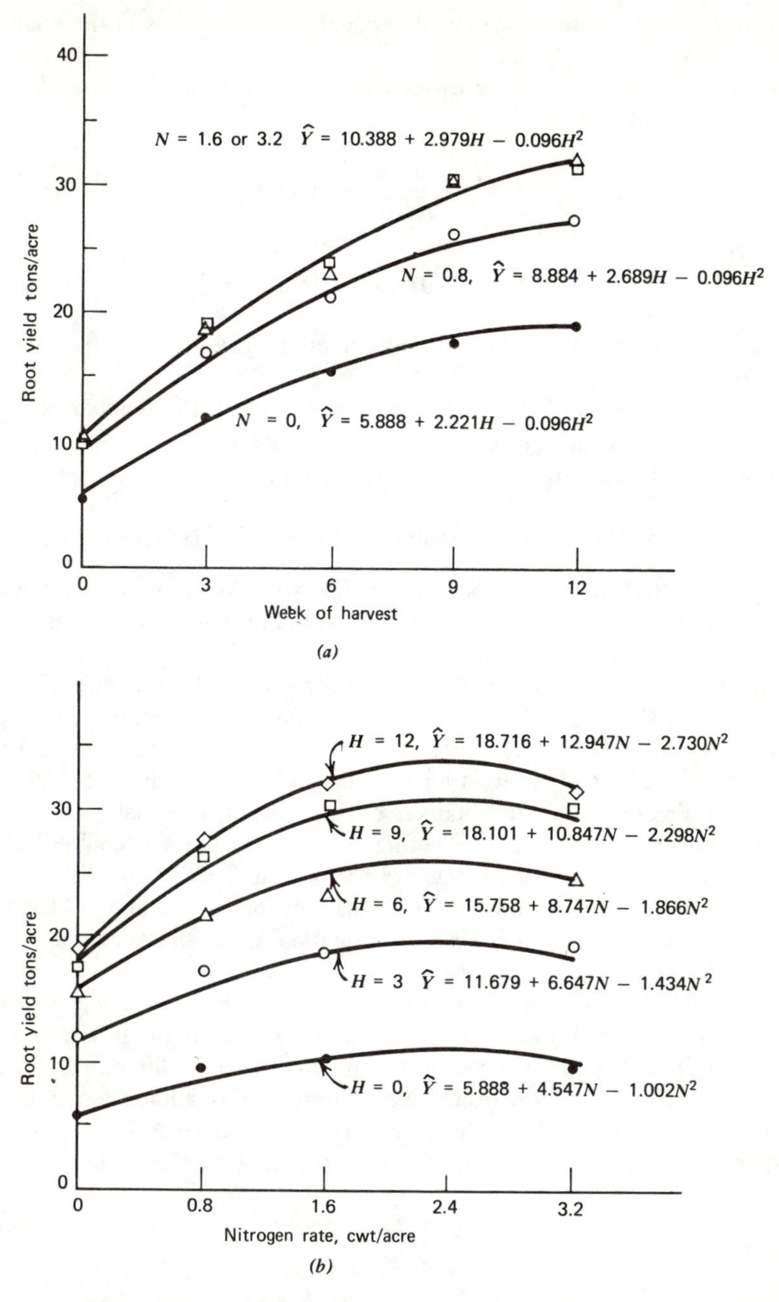

Figure 16.3. "Slices" through the response surface of Figure 16.2. The effect of date of harvest for each N level (*a*, equations for $N_{1.6}$ and $N_{3.2}$ have been averaged) and the effect of nitrogen rate at each harvest date (*b*).

gives 49.372, which is exactly the total residual sum of squares in the analysis of variance in Chapter 10.

The residual sum of squares divided by the total sum of squares for treatments is equal to $(1-R^2)$, so

$$(1-R^2)=\frac{49.372}{4969.24}=0.0099$$

and

$$R^2=1-0.0099=0.9901$$

Thus 99% of the variability in Y is accounted for by the equation we have calculated.

Dividing each term of the regression equation by 4 (the number of replications of each treatment) gives $\hat{Y}$ in tons of roots per acre, the units most appropriate for publication. In this form the equation is

$$\hat{Y}=5.888+2.221H+4.547N-0.096H^2-1.002N^2+0.700NH-0.144N^2H$$

where N = hundred-weights of N/acre, and H = weeks from the first date of harvest (H date 1 = 0). The results can be presented in several ways depending on the researcher's objectives and the points to be stressed.

Figure 16.2 shows the three-dimensional response surface. Figure 16.3 shows "slices" through the response surface; Figure 16.3*a* gives equations and response curves for the effect of time of harvest for each N level, while Figure 16.3*b* does the same for the effect of N levels for each date of harvest. The two-dimensional equations of Figure 16.3*a* are obtained by first setting N equal to zero in the multiple regression equation and collecting like terms to give $\hat{Y}=5.888+2.221H-0.096H^2$. The other equations of Figure 16.3*a* are similarly obtained by, in turn, setting N equal to 0.8, 1.6, and 3.2. The equations for N = 1.6 and 3.2 have been averaged, as they are nearly identical. The equations of 16.3*b* are computed by, in turn, setting H equal to 0, 3, 6, and 12.

Figure 16.3 illustrates the nature of the interaction terms N linear × H linear and N quadratic × H linear. The NH term of the multiple regression equation results in a different N linear for each date of harvest and a different H linear for each N level. The N^2H term results in a different N quadratic effect for each harvest date (Fig. 16.3*b*). In contrast, note that there is no NH^2 term, and as a consequence the same H quadratic effect is present at each N level (Fig. 16.3*a*).

SUMMARY

When we are considering more than two variables, there are three types of correlation coefficients.

> **Simple** or **total correlation** is the linear correlation between any pair of variables, disregarding the values of the remaining variables.

Partial correlation is the relation between two variables when one or more of the remaining variables are held constant.

Multiple correlation is the joint relation between the dependent variable and all of the independent variables.

The equation for the simple correlation coefficient squared is

$$r^2_{YX_i} = \frac{(\Sigma x_i y)^2}{\Sigma x_i^2 \Sigma y^2}$$

The general equation for a first-order partial coefficient of correlation squared is

$$r^2_{YX_i \cdot X_j} = \frac{(r_{YX_i} - r_{YX_j} r_{X_i X_j})^2}{(1 - r^2_{YX_j})(1 - r^2_{X_i X_j})}$$

The *order* of a partial correlation coefficient is the number of variables held constant, shown symbolically by the number of subscripts following the dot. With three variables, we can have only first order partial coefficients.

The *multiple correlation coefficient* among three variables is found from

$$R^2_{Y \cdot X_1 X_2} = \frac{r^2_{YX_1} + r^2_{YX_2} - 2 r_{YX_1} r_{YX_2} r_{X_1 X_2}}{1 - r^2_{X_1 X_2}}$$

The multiple coefficient is always positive and at least as large as the largest simple and partial coefficients.

A *regression equation* describes the relation between the dependent variable and all of the independent variables. It is of the form:

$$\hat{Y} = a + b_1 X_1 + b_2 X_2 + \ldots$$

The symbols b_1, b_2, and so on are called *partial regression coefficients*. To find the regression equation that best fits the observed data, we solve the following *normal equations* for the partial regression coefficients:

$$b_1 \Sigma x_1{}^2 + b_2 \Sigma x_1 x_2 + \ldots + b_m \Sigma x_1 x_m = \Sigma x_1 y$$

$$b_1 \Sigma x_2 x_1 + b_2 \Sigma x_2{}^2 + \ldots + b_m \Sigma x_2 x_m = \Sigma x_2 y$$

$$\cdots\cdots\cdots\cdots\cdots\cdots\cdots\cdots$$

$$b_1 \Sigma x_m x_1 + b_2 \Sigma x_m x_2 + \ldots + b_m \Sigma x_m{}^2 = \Sigma x_m y$$

where m is the number of independent variables. To solve, we need m equations with m terms on the left-hand side of the each equation.

The equation for finding a in the regression equation is

$$a=\bar{Y}-b_1\bar{X}_1-b_2\bar{X}_2-\ldots-b_m\bar{X}_m$$

(The symbol $\bar{Y}$ denotes the mean of Y and is $\Sigma Y/n$, where n is the number of observations. Similarly $\bar{X}_i=\Sigma X_i/n$ and is the mean of X_i.)

The symbol $\hat{Y}$ is the estimated value of Y from the regression equation. The difference $Y-\hat{Y}$ represents the deviation of an observed value from its estimate, and $\Sigma(Y-\hat{Y})=0$. If this sum fails to equal zero (except for small errors due to rounding), an error has been made in the calculations.

$$\frac{\Sigma(Y-\hat{Y})^2}{\text{total sum of squares of Y}}=1-R^2$$

As we consider more variables, three difficulties arise:

The arithmetic increases at an exponential rate.

Visualization of relations becomes difficult.

Determining the best order for adding or eliminating variables is a difficult problem, and no practical method is known for finding the *best* set of a given size out of a large number of variables.

Powers and products of variables can be considered as additional variables. When the powers and products of two independent variables are used in calculating a multiple regression equation, the results can be graphically plotted as a three-dimensional response surface.

17

ANALYSIS OF COUNTS

Most of the discussion in this book has dealt with the analysis of measurements such as weight, yield, or height. However, we do not always measure some characteristic of an individual. At times we may simply classify individuals into two or more groups, such as dead or alive; healthy or diseased; male or female; red, pink, or white; freshman, sophomore, junior, or senior. Even with characteristics that can be measured, it is sometimes more convenient to classify individuals into broad groups. For example, we might wish to conduct a study that included a measure of people's incomes. Many people in our sample might resent being asked the exact amount of their income but would not hesitate if asked in which one of three or four categories of income they belonged, and such a classification might suffice for the purposes of our study.

Data based on counts of individuals belonging to each of several classes generally require a different kind of statistical analysis than that commonly used for measurements. Consider, for example, a study to determine something about the characteristics of eggs laid by a flock of hens. We could weigh each egg in a sample and determine that the mean or average weight per egg was, say, 21 gm. We could also classify each egg as cracked or sound and find that 5% of the eggs were cracked. It would not make sense to say that the average egg was 5% cracked. Our average applies to the proportion of units in the sample possessing this characteristic.

In the chapter on transformations, we showed how data based on counts can sometimes be transformed and analyzed validly as though they were measurement data. In this chapter, we describe a method called *chi-square* (represented by the symbol χ^2) for analyzing enumeration data.

Before discussing this method, we should first consider what we would like to learn by classifying and counting individuals. The purposes of collecting such data generally fall into one or more of three objectives: (1) to test one or more hypotheses not suggested by the data, (2) to determine whether different characteristics are interrelated, and (3) to test whether samples are drawn from different populations.

CHI-SQUARE

The general formula for chi-square used in solving all these problems is

$$\chi^2 = \sum \frac{(\text{Ob} - \text{Ex})^2}{\text{Ex}}$$

where Ob is the observed value for each of two or more classes, and Ex is the corresponding expected value.

To evaluate this expression, we must first determine the expected value for each class of individuals, according to our hypothesis. The expected value is then subtracted from the observed value and the resulting difference is squared and divided by the expected value. These quotients are summed over all classes. The sum is then compared with values in a χ^2 table at the appropriate degrees of freedom. This tells us the approximate probability of obtaining deviations from expectancies, as large or larger than those observed, by chance alone.

The arithmetic is fairly simple and for certain special cases there are computational shortcuts available. However, there are several things we need to consider in order to use chi-square tests properly.

1. We must exercise care in selecting the hypothesis to be tested. This hypothesis should be a reasonable one based on previously known facts or principles.

2. We need to be aware of the fact that a chi-square distribution is a continuous distribution and is in fact related to the *normal distribution*. On the other hand, the distribution of samples based on counts is a discrete or discontinuous distribution. If the individuals are classified into one of two classes, we are dealing with what is called binomial distribution. Normal and binomial distributions are similar but not identical. That is why it was stated above that reference to a chi-square table gives an *approximate* probability. We need to know what situations result in poor approximations so that we can either avoid these situations or perhaps make adjustments to get closer approximations to the true probability.

3. Given an hypothesis, we need to know how to calculate the expected values for each class correctly.

4. The number of degrees of freedom for entering the chi-square table is not always obvious. We need to learn certain rules for determining this.

5. Interpreting the results of a chi-square test requires caution and good judgment. Even though our observations do not differ significantly from our hypothesis, we may not be justified in accepting the hypothesis if the data also fit other equally logical hypotheses.

Let us illustrate these various points with an example. Suppose we are working with some plant that has red and white flowered forms. We have crossed plants from true-breeding lines of the two forms and the F_1 generation was all red. We grow an F_2 generation of eight plants and find that four are red and four are white flowered. On the basis of what we have already learned, we feel quite certain that red is dominant over white, and we further suspect that it is determined by a single gene. Our knowledge of genetics leads us to adopt the hypothesis that the F_2 will segregate in a 3:1 ratio of reds to whites.

On the basis of this hypothesis, we expect out of eight plants to obtain six reds and two whites, so our observed numbers deviated by two from expected. We ask, "What is the probability that we could have obtained a deviation from expected as large or larger than we observed, by chance alone?" If this probability is very small, we will reject our hypothesis.

Recognizing that chi-square will give us only an approximation of the desired probability, we will calculate the exact probability based on the binomial distribution. To do this, we must find the probability of each possible outcome and pool all of the cases which equal or exceed the observed deviation from expected.

First we must define some symbols. We call the hypothetical ratio $r_1:r_2$. The probability of an individual belonging to the first class is called p and is equal to $r_1/(r_1+r_2)$. The probability of being in the second class is called q and is equal to $r_2/(r_1+r_2)$ or $1-p$. The numbers observed in each class are called n_1 and n_2, and $n_1+n_2=n$, the total number in our sample. The symbol n! is called *factorial n* and is obtained by taking the product of all the integers from 1 to n. Factorial zero is defined as 1.

In a binomial distribution, the probability of obtaining a sample with n_1 in the first class and n_2 in the second is

$$\frac{p^{n_1}q^{n_2}n!}{n_1!n_2!}$$

In our example, $r_1=3, r_2=1, p=r_1/(r_1+r_2)=3/4, q=r_2/(r_1+r_2)=1/4$. The probability of obtaining a sample in which $n_1=4$ and $n_2=4$ is

$$\frac{p^{n_1}q^{n_2}n!}{n_1!n_2!}=\left(\frac{3}{4}\right)^4\cdot\left(\frac{1}{4}\right)^4\cdot\frac{1.2.3.4.5.6.7.8}{1.2.3.4.1.2.3.4}=\frac{81}{256}\times\frac{1}{256}\times 70=.0865$$

Likewise, we can calculate the probability of every other outcome and construct the first three columns of Table 17.1.

The last probability is not actually zero, but is less than .00005.

Notice that the sum of all the probabilities is 1, which furnishes us with a check on the calculations.

The expected value of n_1 is $np=8\times 3/4=6$, so we make a third column in the table showing the differences between the observed values of n_1 and this expected value.

TABLE 17.1

Outcome	Probability	Deviation of n_1 from Expected ($n_1 - 6$)	Class Interval	Probability Based on Normal Curve
8:0	.1001	2	>1.5	.1104
7:1	.2670	1	0.5 to 1.5	.2312
6:2	.3115	0	−0.5 to 0.5	.3168
5:3	.2076	−1	−0.5 to −1.5	.2312
4:4	.0865	−2	−1.5 to −2.5	.0897
3:5	.0231	−3	−2.5 to −3.5	.0186
2:6	.0038	−4	−3.5 to −4.5	.0020
1:7	.0004	−5	−4.5 to −5.5	.0001
0:8	.0000	−6	<−5.5	.0000
Total	1.0000			

We can now answer our original question. The probability of obtaining a deviation of two or more from expected is the sum of the probabilities in the first and last five of the nine cases in the table. This is .1001+.0865+ ... +.0000 =.2139.

Let us see how this result compares with the chi-square test. Our formula is

$$\chi^2 = \sum \frac{(\text{Ob} - \text{Ex})^2}{\text{Ex}} = \frac{(4-6)^2}{6} + \frac{(4-2)^2}{2} = \frac{4}{6} + \frac{4}{2} = 0.67 + 2 = 2.67$$

Looking this value up in a chi-square Table A.6, at 1 degree of freedom, we see that our observed chi-square is very close to the value 2.706 found at the 10% point, indicating that the probability is .10 of getting a deviation at least as large as we observed by chance. (A more precise value from more extensive tables is .1025). This is considerably lower than the exact probability of .2139 that we found.

Yates Correction for Continuity

There is a correction called *Yates correction for continuity* that will greatly reduce the discrepancy between the two methods. Suppose we used the normal distribution to obtain an estimate of the probability of each outcome. To do this, we first must find the variance and standard deviation of the distribution. This can be found by squaring the deviation from the mean (expected value) for each outcome and multiplying by the corresponding probability. These products are summed over all outcomes.

Variance $= 2^2 \times .1001 + 1^2 \times .2670 + \cdots + (5)^2 \times .0004 = 1.4997$. Since we are dealing with the binomial distribution, there is a much simpler formula for obtaining the variance: $\sigma^2 = npq$. Thus, in this example,

$$\sigma^2 = 8 \times 3/4 \times 1/4 = 1.5$$

$$\text{Standard deviation}, \sigma = \sqrt{\sigma^2} = \sqrt{1.5} = 1.225$$

The class intervals can now be expressed in terms of z values by dividing the limits of each interval by the standard deviation. The area under a normal curve for each interval can then be found by reference to a table of probability functions found in most books of mathematical tables.

These estimates are shown in Table 17.1 to point out how the normal and binomial distributions differ. Since the normal is a continuous distribution, we have to lump together all of the portion of the normal curve from $n_1 - Ex = -1.5$ to $n_1 - Ex = -2.5$ and determine the area of this portion to find the probability of $n_1 - Ex$ being -2. Likewise, the probability of $n_1 - Ex$ being 2 is the area under the normal curve from $n_1 - Ex = 1.5$ to infinity. Thus our question regarding the probability of obtaining a deviation of 2 or greater from expected, when using a normal curve, must be reworded to ask, "What is the probability that the deviation from expected will exceed 1.5?" Yates correction takes this into account and consists simply of subtracting 0.5 from the absolute value (disregarding sign) of the differences between observed and expected.
Using this correction, we calculate an adjusted chi-square as follows:

$$\chi^2 = \sum \frac{(|Ob - Ex| - 0.5)^2}{Ex} = \frac{(2-0.5)^2}{6} + \frac{(2-0.5)^2}{2}$$

$$= \frac{(1.5)^2}{6} + \frac{(1.5)^2}{2} = 0.375 + 1.125 = 1.50$$

(Note: The symbol $|x|$ means the absolute value of x.)

Looking this value up in a chi-square (Table A.6) shows that the probability is between .10 and .50 but much higher than it was before. More extensive tables give a P value of .2207, very close to the calculated exact probability of .2139. The probability based on a normal distribution can also be obtained in the same way as the binomial, by adding the probabilities of the first line and last five lines of Table 17.1. This gives .2208 which is, as it should be, equal (within rounding errors) to the result obtained by the chi-square test.

GUIDES FOR USING CHI-SQUARE

We have seen that even with a sample as small as eight the difference between the normal distribution on which chi-square is based and the exact binomial distribution is not very great. The following rules will help in deciding whether chi-square will give a sufficiently close approximation of the correct answer:

1. The larger the sample size, the closer the agreement between the two distributions.

2. The larger the ratio between r_1 and r_2 in our hypothesis, the greater the discrepancy between the two distributions for a given sample size. Thus, if we hypothesize a 1:1 ratio, the agreement will be close even for small samples, but if we hypothesize a 15:1 ratio, a much larger sample size is necessary.

3. A good rule of thumb is to avoid using chi-square if the smallest expected class is less than five. If we have more than two classes, we can pool classes whose expected values are less than five. Increasing sample size can also be used to increase the size of the smallest expected value.

4. Always use Yates correction for determining chi-square with only 1 degree of freedom. *Never* use it for problems in which more than 1 degree of freedom is involved.

Degrees of freedom can be defined in general as the number of classes that can be assigned an arbitrary value. Thus, if we have two classes, as in the example we have been using, we can assign any value to n_1, but n_2 is then fixed because it must include the remaining members of the sample, since $n_2 = n - n_1$. Chi-square therefore has one degree of freedom. *In testing any hypothesis exterior to the data, degrees of freedom is always one less than the number of classes.* Other situations will be discussed later.

INTERPRETING RESULTS

Interpretation is the last and most important step in our analysis of the data. We have seen that the discrepancy between what was observed and what was expected could easily have been due to chance alone. We therefore have no evidence for rejecting our hypothesis. Does this mean that we have strong evidence to *support* our hypothesis? Not necessarily, and this is a point often misunderstood. Look at it this way. There are many other hypotheses we could set up from which this sample would not represent a significant deviation. If we have strong evidence that red and white are determined by a single pair of genes, then a 3:1 ratio is the most reasonable hypothesis, and our sample could be considered as furnishing good supporting evidence. On the other hand, the evidence we have for postulating a single pair of genes may be very weak. We then must consider such possibilities as two pairs of genes giving rise to a 9:7 or 13:3 ratio. Our observed sample of 4 red:4 white would give a "good fit" to either of these ratios. Further tests or much larger F_2 samples will have to be used to distinguish among the various plausible hypotheses.

Table 17.2 shows the sample sizes needed to distinguish between various common ratios. For example, the table shows that a sample of 105 is necessary to ensure that either a 3:1 or a 9:7 ratio will be rejected at the 5% level. The rejection value in the chi-square table is 3.84. If we observed a 70:35 ratio, the chi-square value in testing the 3:1 hypothesis would be 3.46, not large enough to reject at the 5% level. Tested against the 9:7 hypothesis we get a chi-square value of 4.22, large enough to reject the hypothesis at the 5% level. On the other hand, an observed ratio of 69:36 would give chi-square values of 4.34 and 3.45 for the 3:1 and 9:7 hypotheses respectively. We would therefore reject the 3:1 hypothesis. Verifying these chi-square values is left as an exercise. Be sure to use the correction for continuity.

TABLE 17.2.
Sample size to ensure that at least one of two alternative hypotheses will be rejected (Top number at 5% level, bottom at 1%)

	15:1	7:1	13:3	3:1	11:5	5:3	9:7
1:1	16	24	38	62	112	254	1008
	24	38	61	101	186	428	1718
9:7	20	33	56	105	243	977	
	31	53	92	174	407	1664	
5:3	27	49	94	223	915		
	42	79	155	374	1558		
11:5	39	80	195	823			
	61	130	326	1398			
3:1	60	159	699				
	97	264	1184				
13:3	114	543					
	186	915					
7:1	354						
	589						

TESTING FOR INDEPENDENCE

One of the things we often want to learn about counted data is whether two variables are related. For example, one variable used to classify individuals might be level of education and another level of income. We could test to see if education and income are related.

We might deliberately impose two levels of a variable such as inoculation on two groups, treating one group and leaving the other untreated. We could then classify each group into healthy and diseased after a certain period of time, and test for any relation between treatment and disease incidence. In genetic research, it is often desired to find if two traits are inherited independently or show evidence of linkage. All these problems are analogous to correlation analysis with measurement data.

In analyzing for a relation between two variables, it is most convenient to set up a null hypothesis that they are independent. If the deviation from independence is much greater than we would expect by chance, we reject the hypothesis that the two variables are independent and accept the alternate hypothesis that they are related.

To find the expected values for applying the chi-square formula, we use a principle in the theory of probability that states: *If two events are independent*, the probability of simultaneous occurrence of the two events is the product of the probabilities of their individual occurrence. Let us illustrate this principle with an example and show how the chi-square test is performed.

One hundred animals were treated with an antibiotic and after a period of time examined for symptoms of disease. There were 88 animals that were healthy and 12 that showed disease symptoms. Another group of 200 animals was given no antibiotic, and when examined later, 143 were found to be healthy and 57 diseased. These results can be summarized in what is called a 2×2 contingency table, Table 17.3.

TABLE 17.3.
Disease incidence in treated and untreated cattle

	Disease Categories		
Treatment	Healthy	Diseased	Total
Treated	88	12	100
Expected	(77)	(23)	
Untreated	143	57	200
Expected	(154)	(46)	
Totals	231	69	300

We will test the hypothesis that there is *no* relation between treatment with antibiotic and the incidence of disease. If these two variables are independent, the expected proportion of healthy treated animals will be the proportion of healthy times the proportion of treated. This is $231/300 \times 100/300 = 77/300$. Since there are 300 animals altogether, $77/300 \times 300 = 77$ is the number of animals that we expect to be treated and healthy. The computation can be considerably shortened by noting that the grand total appears as the denominator of both fractions which are multiplied to give the joint probability. The resulting *proportion* was then multiplied by the grand total to get the expected *number*. We can cancel one of the grand totals in our calculation and find the expected number from $(100 \times 231)/300 = 77$. In words, this can be stated: The expected number of treated healthy animals is the total number of treated times the total number of healthy divided by the grand total. Similarly, every other expected class can be calculated. Actually, in a 2×2 table, only one expected value needs to be calculated. Since we expect 77 of the treated animals to be healthy, we expect the remaining 23 to be diseased. Likewise, we expect 77 of the healthy animals to be in the treated class; we expect the remainder of the 231 healthy animals or 154, to be in the untreated class. Then, of the 200 untreated animals, since we expect 154 to be healthy, we expect the remaining 46 to be diseased. Notice that once a number is assigned to one of the classes, the remaining three classes are fixed. Thus, we have only *one* degree of freedom in a 2×2 table. The general rule for an $r \times c$ (r rows and c columns) contingency table is that the degrees of freedom equals $(r-1) \times (c-1)$.

One feature of a 2×2 table to notice is that the difference between observed and expected is the same for every cell of the table except that two of the differences are positive and the other two are negative. This common difference in our example is 11 (e.g., $88 - 77 = 11$, etc.), and since we are dealing with 1 degree of freedom, we should apply Yates correction and consider the differences as 10.5.

Applying our chi-square formula, we get

$$\chi^2 = \sum \frac{(|\text{Ob} - \text{Ex}| - 0.5)^2}{\text{Ex}} = \frac{(10.5)^2}{77} + \frac{(10.5)^2}{23} + \frac{(10.5)^2}{154} + \frac{(10.5)^2}{46} = 9.34$$

Referring to the chi-square Table A.6 under 1 degree of freedom, we see that we would expect a chi-square value of 6.635 1% of the time by chance alone, and 10.827 only 0.1% of the time. Therefore, we can say that the probability of obtaining a chi-square value as large as 9.34 is only slightly more than 1 in a 1000, so we reject the hypothesis of independence and say that there is a relation between antibiotic and the incidence of disease.

To show how chi-square is used to test independence between two pairs of genes, we will analyze some data from a large progeny of marigolds, segregating for two factors, earliness and virescence (a mild chlorophyll deficiency). It was known that earliness is recessive to late development and determined in this genetic material by a single pair of genes. Virescence is recessive to normal and also controlled by a single gene pair. Three questions need to be answered. Does the ratio of late:early fit a 3:1 ratio? Does the ratio of normal:virescent fit a 3:1

ratio? Are the two pairs of traits inherited independently, or is there evidence of linkage?The data arranged in a contingency Table 17.4 were as follows:

TABLE 17.4.
Segregation of two traits in a progeny of marigolds

	Normal	Virescent	Total	Ex 3:1
Late	3470	910	4380	4275
Expected	(3457.9)	(922.1)		
Early	1030	290	1320	1425
Expected	(1042.1)	(277.9)		
Totals	4500	1200	5700	
Ex(3:1)	4275	1425		

To answer the first question regarding the ratio of late:early, we calculate chi-square:

$$\chi^2 = \frac{(|4380-4275|-0.5)^2}{4275} + \frac{(|1320-1425|-0.5)^2}{1425}$$

$$= \frac{(104.5)^2}{4275} + \frac{(104.5)^2}{1425}$$

$$= 10.22$$

This is almost equal to the required chi-square value of 10.827 at the 0.1% level. This means that if 3:1 were the true ratio, the probability of finding a deviation as great as we observed was only about 1 in a 1000. We therefore reject the hypothesis that 3:1 is the true ratio. Actually, the hypothesis that late flowering was a simple dominant over early flowering was not rejected, because it was observed that (as with many recessive traits) the early plants were somewhat weaker than the late ones. The small but significant deviation from a 3:1 ratio was therefore attributed to differential survival rates. It is worth noting that this was an unusually large progeny. If it had been one-tenth as large (570 plants) and the ratio of late: early had been the same, the chi-square value would have been only 0.94, not approaching significance.

The question about the ratio of normal: virescent is answered in the same way, and the chi-square value turns out to be 47.16, again very highly significant. Virescent plants, being partially lacking in chlorophyll, show an even greater loss in vigor compared to normal than do early plants compared to late.

In testing for independence, we accept the observed ratios rather than assuming a 3:1 ratio, and calculate the expected values on the assumption of independence. Thus the expected number of late normal plants is

$$\frac{\text{total normal} \times \text{total late}}{\text{grand total}} = \frac{4500 \times 4380}{5700} = 3457.9.$$

The expected values for the remaining three cells in Table 17.4 can be calculated in a similar fashion or obtained by subtraction from the marginal totals. Using both methods furnishes a check on the accuracy of the computations. Note that (Ob − Ex) is 12.1 in the upper left and lower right cells of the table and −12.1 in the other two cells. The numerators of the terms for determining chi-square will be the same for each class. Applying Yates correction for each cell of the table gives $(12.1-.5)^2=(11.6)^2$. Chi-square is therefore

$$\frac{11.6^2}{3456.9} + \frac{11.6^2}{922.1} + \frac{11.6^2}{1042.1} + \frac{11.6^2}{277.9} = .80$$

The probability of obtaining a value of this magnitude by chance alone is between 10% and 50%, so we do not have any evidence to justify rejecting the hypothesis of independence.

Another example will show how to calculate chi-square when more than 1 degree of freedom is involved, and how a contingency table may be "collapsed." Three groups of 39 cattle were each fed a different ration. The condition of health of each animal was measured by recording the number of times it had to be treated for sickness. The results shown in Table 17.5 were obtained.

TABLE 17.5.
Health condition of cattle fed with three rations. Expected values in parentheses

Number of Times Treated	Ration 1	Ration 2	Ration 3	Total
0	19(17.3)	16(17.3)	17(17.3)	52
1	1 (0.3)	0 (0.3)	0 (0.3)	1
2	0 (1.3)	3 (1.3)	1 (1.3)	4
3	7 (5.7)	9 (5.7)	1 (5.7)	17
4	3 (4.7)	5 (4.7)	6 (4.7)	14
5	4 (3.3)	1 (3.3)	5 (3.3)	10
6	2 (2.0)	1 (2.0)	3 (2.0)	6
7	0 (1.3)	2 (1.3)	2 (1.3)	4
8	1 (2.3)	2 (2.3)	4 (2.3)	7
10	2 (0.7)	0 (0.7)	0 (0.7)	2
Totals	39	39	39	117

In this case, the expected values are very easy to calculate, since exactly one-third of all the cattle were in each ration class. This means that we would expect one-third of the animals in each treatment frequency class to fall in each ration class if ration and treatment frequency are independent. We note that many of the expected values are less than five, so we are not really justified in applying the chi-square formula to the data as it stands. However, we will go through the calculations and see how the results compare with those obtained from a collapsed table.

$$\chi^2 = \sum \frac{(\text{Ob}-\text{Ex})^2}{\text{Ex}} = \frac{(19-17.3)^2}{17.3} + \frac{(16-17.3)^2}{17.3} + \ldots + \frac{(0-.7)^2}{.7} = 24.5$$

$$\text{Degrees of freedom} = (r-1)(c-1) = (10-1)(3-1) = 18$$

Looking up our calculated chi-square value of 24.5, Table A.6, opposite 18 degrees of freedom shows that the probability of obtaining the results observed by chance alone is slightly over 10%. We therefore have insufficient evidence to reject the hypothesis that animal health was *not* related to ration.

In order to satisfy the rule that no expected class should be less than 5, we can collapse the table by combining frequency classes 1, 2 and 3; 4 and 5; and 6, 7, 8 and 10. This gives a new table (Table 17.6).

Calculating chi-square gives us a value of 10.61, which we look up in the table opposite 6 degrees of freedom. We find it is almost exactly equal to the tabular value at 10% probability. Our conclusions will therefore be the same as those we reached with the original table, though this will not always be the case. It is always safer to collapse a table to avoid too small expected classes. Furthermore, it reduces the number of calculations needed to compute chi-square. Note that the correction for continuity was not used in this example, because we were dealing with more than a single degree of freedom.

TABLE 17.6.
Collapsed version of Table 17.5

Number of Times Treated	Ration 1	Ration 2	Ration 3	Total
0	19 (17.3)	16 (17.3)	17 (17.3)	52
1–3	8 (7.3)	12 (7.3)	2 (7.3)	22
4–5	7 (8.0)	6 (8.0)	11 (8.0)	24
6–10	5 (6.3)	5 (6.3)	9 (6.3)	19
Totals	39	39	39	117

HETEROGENEITY

The third and final use we will consider in connection with chi-square is that of testing whether a group of samples could have been drawn from the same population. Consider eight progenies of marigolds each segregating for normal and virescence as shown in Table 17.7.

TABLE 17.7.
Normal and virescent marigolds in eight progenies

Progeny	Normal	Virescent	$\chi^2(3:1)$	$\chi^2(3106:854)$
1	315	85	3.00	0.023
2	602	170	3.65	0.094
3	868	252	3.73	0.578
4	174	42	3.56	0.575
5	192	48	3.20	0.348
6	165	39	3.76	0.723
7	161	43	1.67	0.028
8	629	175	4.48	0.019
Totals			27.05	2.388
Pooled	3106	854	24.91	0.000
Heterogeneity			2.14	2.388

We will carry out two kinds of analyses. First we will test each progeny and the pooled data from all progenies for deviation from a hypothetical 3:1 ratio.

The chi-square calculated for each progeny is shown in column four. These were calculated without the correction for continuity, because we will want to add them, and only unadjusted chi-squares are additive. Note that only one of these exceeds the required value of 3.84 for significance at the 5% level. We therefore have very little evidence from the individual progenies for rejection of our hypothesis. Still we are *not* justified in concluding that, since seven out of eight progenies gave a "good fit" (i.e., did not deviate significantly from 3:1), there is overwhelming evidence to support our hypothesis. We must carry the analysis further. Adding the eight individual chi-squares, each with 1 degree of freedom, gives a total chi-square of 27.05 with 8 degrees of freedom. This exceeds the tabular chi-square value of 26.125 at the 0.001 level. In other words, the probability is less than 1 in 1000 that such a large value could simply be the result of chance. Another test can be applied to the total of 3106 normal and 854 virescent.

The expected numbers are: $3960 \times 3/4 = 2970$ and $3960 \times 1/4 = 990$,

$$\text{so } \chi^2 = \frac{(3106-2970)^2}{2970} + \frac{(854-990)^2}{990} = 24.91.$$

This far exceeds the tabular chi-square value for 1 degree of freedom at the 0.001 level, so we now definitely reject the hypothesis that all of the progeny are samples from a population with 3 : 1 ratio. We still would like to know whether all of these progenies might represent samples from a single population. To test this hypothesis, we calculate what is called *heterogeneity chi-square*.

$$\text{heterogeneity chi-square} = \text{total chi-square} - \text{pooled chi-square.}$$

Since total chi-square was 27.05 and pooled chi-square was 24.91, heterogeneity chi-square is 2.14 with 7 degrees of freedom. Reference to the table shows this to be even less than the 2.167 required at the 0.95 level. The probability is about 95% that a chi-square of this size or larger could come from a homogeneous set of samples just by chance. All of these tests can be summarized in a table similar to an analysis of variance table.(Table 17.8).

TABLE 17.8.
Summary of data from eight marigold progenies based on 3 : 1 ratio

Source	df	Chi-square
Total	8	27.05***
Pooled	1	24.91***
Heterogeneity	7	2.14 ns

Instead of testing each progeny against a hypothetical ratio, we might test the observed ratio of the totals. This is done in the last column of Table 17.7. The pooled chi-square of course has a value of zero, since the observed ratio is the one which we are testing. A table analogous to the one above is given as Table 17.9.

We still have no evidence of heterogeneity, conclude that we are dealing with a homogeneous set of progenies, and that our best estimate of the true ratio is 3106 : 854.

Notice that in this last test, the calculations were exactly the same as for testing independence. In other words, when testing each sample against the observed total ratio, *heterogeneity chi-square = independence chi-square*. It is only when the samples and totals are being tested against a hypothetical ratio that we need to partition the total chi-square into two components.

TABLE 17.9.
Summary of marigold data, based on observed totals

Source	df	Chi-square
Total	8	2.388
Pooled	1	0.000
Heterogeneity	7	2.388

Table 17.10 indicates what the analysis would have looked like if the first four progenies had shown the same deviation from a 3:1 ratio, but in the opposite direction.

TABLE 17.10.
Hypothetical set of marigold data showing heterogeneity

Progeny	Normal	Virescent	$\chi^2(3:1)$	$\chi^2(2950:1010)$
1	285	115	3.00	2.05
2	556	216	3.65	2.49
3	812	308	3.73	2.35
4	150	66	3.56	2.90
5	192	48	3.20	3.82
6	165	39	3.76	4.38
7	161	43	1.67	2.10
8	629	175	4.48	5.92
Totals			27.05	26.01
Pooled	2950	1010	.54	.00
Heterogeneity			26.51	26.01

Note that the pooled data now came very close to fitting a 3:1 ratio, but the heterogeneity chi-square is highly significant. Again we reject the hypothesis that all of the progenies are samples from a population in which the ratio is 3 normal to 1 virescent. The rejection in this case is because there is strong evidence that the samples are not a homogeneous set, so that pooling of the data is not justified.

Throughout this discussion we have used a single formula:

$$\chi^2 = \sum \frac{(Ob - Ex)^2}{Ex}$$

With only one slight modification for cases where the correction for continuity is required. There are many modifications of this formula that provide computational shortcuts for special cases. A person who has a great many chi-squares to calculate would be well advised to refer to a more advanced text for the appropriate shortcut formula. For the reader who only occasionally encounters problems requiring chi-square analysis, we feel it is preferable to learn this single basic formula.

SUMMARY

The general formula for calculating chi-square is

$$\chi^2 = \sum \frac{(Ob - Ex)^2}{Ex}$$

Individuals classified in one way into two or more classes may be compared to a hypothetical ratio. Degrees of freedom are one less than the number of classes.

By comparing the calculated chi-square with a table, we can find the probability of the occurrence of a deviation at least as great as that observed by chance alone.

Individuals classified in two ways, into r and c classes, can be tested for independence between the two criteria of classification. Degrees of freedom are $(r-1)\times(c-1)$.

If two or more samples are each tested against a common hypothetical ratio, the sum of the resulting chi-squares can be partitioned into two components as follows.

Source	df
Total	$r(c-1)$
Pooled	$(c-1)$
Heterogeneity	$(r-1)(c-1)$

The number of classes into which each sample is classified is c, and r is the number of samples.

18

IMPROVING PRECISION

The precision of an experiment refers to its ability to detect true treatment effects. In general, the more precise the experiment, the smaller the treatment difference that the experiment is capable of detecting. The greater the variability among experimental units treated alike, the greater will be the error associated with the difference between two means and the less precise the experiment will be in detecting differences resulting from treatments. The standard error of the difference between two means decreases as s decreases and n increases, $s_{\bar{d}} = \sqrt{2s^2/n}$ (where n is the number of replications). Thus, methods to increase the precision of an experiment are designed to lower the unaccounted variability per plot or to increase the effective number of replications.

Precision may be improved by (1) increased replication, (2) careful selection of treatments, (3) refinement of technique, (4) selection of experimental material, (5) selection of the experimental unit, (6) taking additional measurements, and (7) planned grouping of experimental units.

INCREASED REPLICATION

The precision of an experiment can always be increased by additional replications, but the degree of improvement falls off rapidly as the number of replications increases. For example, compared to an experiment with four replications, to double the degree of precision with which two means can be separated requres 16 replications. This follows from the effect of the number of replications (n) on the difference required to separate two means at a given level of significance, $\text{LSD} = t\sqrt{2s^2/n}$. This is not exactly so because, as n increases, t becomes slightly smaller, but it is close enough to use as a rule of thumb.

In general, in field and vegetable crop research, from four to eight replications are required for reasonable precision. In planning an experiment, you should be reasonably sure that you will be able to detect a true difference of the magnitude in which you are interested. If the probability is poor that you can accomplish your objective with the number of replications you are willing to employ, and there are no other reasonable means for improving precision, you would be well advised not to do the experiment—or at least to postpone it until you have sufficient resources to conduct it in a way that does have a good chance of accomplishing your objective.

Table 2.1 of Cochran and Cox (1964) is convenient for estimating the number of replications required to detect a specified difference. Their table is based on the formula $r \geqslant 2[(CV)^2/D^2](t_1+t_2)^2$, where CV is the coefficient of variation [$CV = s(100)/\bar{Y}_{..}$]; D is the difference you desire to detect expressed as a percent of the mean of the experiment; t_1 is a tabular t value for a specified level of significance (say 5%) and the degrees of freedom for experimental error; and t_2 is a tabular t value for degrees of freedom for error and a probability of $(1-P)2$, where P is the probability of detecting a significant result in a given run of the experiment. If $P=0.80$, then $(1-P)2=0.40$, a two-tailed area for a t distribution based on degrees of freedom for experimental error.

To use the equation, start by specifying the number of replications you think may be needed and then work it to approximate r. Based on this r, solve the equation again and then take the next larger value of r as the number of replications required.

For example, suppose we wish to conduct an experiment involving six treatments in a randomized complete block design. We want an 80% chance of detecting a mean difference as small as 10% of the experimental mean at the 5% level of significance. Other experiments with the experimental units we will use indicate that a well-conducted experiment should have a coefficient of variation of about 5%. We think six replications may be enough. Thus, for the first run of the equation, $r=6$, treatments$=6(n=6)$, df error$=(r-1)(n-1)=25$, $t_1=2.060$, $t_2=0.856$ (see Table A.2), and $r \geqslant 2(5/10)^2(2.060+0.856)^2=4.25$.

Now let $r=5$, then df error$=(4\text{-}1)(5-1)=20$, $t_1=2.086$, and $t_2=0.860$. Solving again for r gives $r \geqslant 2(5/10)^2(2.086+0.860)^2=4.34$; therefore we take 5 as our estimate of the number of replications required. When we conduct the experiment, we will have an 80% chance of detecting a 10% difference at the 5% level with five replications unless the coefficient of variation turns out to be larger than expected.

SELECTION OF TREATMENTS

Careful selection of treatments is not only important in achieving the experimenter's objectives but it also can increase the precision of the experiment. For example, in studying the effect of an herbicide, fungicide, fertilizer, or insecticide, it is more useful to determine how the experimental units respond to increasing doses of your treatment material, than to decide whether or not two succeeding doses are significantly different. Thus, a proper series of doses will make it possible to plan tests of significance that are more sensitive than merely comparing adjacent means in an array. As mentioned before, doses in equal increments covering the range of the response expected are most efficient in establishing a dose-response curve and facilitate the computation of sums of squares and equations for responses. Also, as pointed out in Chapter 3, factorial experiments, where two or more types of treatments are tested simultaneously, can result in considerable improvement in the precision of main factor comparisons.

REFINEMENT OF TECHNIQUE

Faulty technique may increase experimental error and bias treatment effects. A good technique should (1) uniformly apply treatments, (2) devise suitable and unbiased measure of treatment effects, (3) prevent gross errors, and (4) control external influences so that all treatments are comparably affected.

SELECTION OF EXPERIMENTAL MATERIAL

For certain kinds of studies, carefully selected, uniform material is desirable. In selecting experimental material, however, you must keep in mind the population about which you wish to make inferences. Thus, for most applied research in agriculture, it is important to use the kinds of experimental materials that will be used in actual production.

SELECTION OF THE EXPERIMENTAL UNIT

The size and shape of the field plot affects precision. In general, variability decreases with an increase in plot size, but once a certain size has been reached, the increase in precision falls off rapidly with larger sizes. For determining yield, there is usually little gain in precision by using plots larger than 0.1 acre. For most crops, harvested areas of 0.01 to 0.02 acres result in good precision. LeClerg et al. (1962) discuss size and shape of field plots for various crops and cite many useful references. Rectangular plots are most efficient in overcoming soil heterogeneity when their long axes are in the direction of greatest soil variation.

Increasing the number of animals or the number of trees per experimental unit also increases precision. However if animals or trees can be handled individually, precision will be increased more by using individuals as experimental units and having more replications rather than using the same number of animals or trees with more than one per experimental unit.

TAKING ADDITIONAL MEASUREMENTS—COVARIANCE

One of the techniques for reducing error in an experiment is to remove the variability in Y associated with some independent variable X. This techniques is called *covariance.*

Suppose that in a crop experiment there was a considerable amount of variation in stand from plot to plot. If we can make a reasonable estimate of what the plot yields would have been if all plots had the same stand, the precision with which we measure treatment effects can be improved. An estimate based on the assumption that yield is directly proportional to stand is not reasonable for example, for it nearly always introduces a bias favoring the plots with the thinner stands.

Another example of the usefulness of covariance analysis is in animal feeding experiments in which there is variation in the initial weights of the animals. If weight gain is found to be related to initial weights, adjustments can be made to increase the precision of measuring treatment effects.

The whole subject of covariance is a fairly complicated one, both from the point of view of the calculations involved and in the interpretation of results. Many of the texts in our list of references deal with covariance in great detail. In our experience, few agricultural research workers become involved in covariance analysis except in a minor way, so that a discussion of all the intricacies of the technique may not be very fruitful. We will therefore describe only the general method of the analysis and some of the simpler aspects of interpretation.

Table 18.1 consists of some hypothetical data contrived for easy calculation to illustrate the procedures in covariance analysis. You can think of X and Y as representing stand and yield, initial weight and weight gain, or any other pair of variables that you might encounter.

TABLE 18.1.
Hypothetical data representing the values of two variables, X and Y, in a randomized complete block experiment with four replicates and five treatments

	X					Y				
Block:	1	2	3	4	Total	1	2	3	4	Total
Treatment										
1	8	6	7	7	28	7	5	6	6	24
2	8	4	12	12	36	9	5	9	9	32
3	4	10	10	8	32	6	12	10	12	40
4	1	7	4	12	24	9	11	10	18	48
5	9	8	12	11	40	14	7	15	20	56
Totals	30	35	45	50	160	45	40	50	65	200

The regular analysis for both X and Y can be carried out in the usual way, with the results shown in Table 18.2.

We note that the treatments had no significant effect on the X variable, but their effect on Y was significant at the 5% level.

To carry out the analysis of covariance, we need, in addition to the sums of squares of X and Y, the sums of cross-products, which we will designate as SXY. First, we need a correction term:

$$C = \frac{\left(\sum X\right)\left(\sum Y\right)}{rn} = \frac{(160)(200)}{20} = 1600$$

TABLE 18.2.
Separate analyses of variance for X and Y from Table 18.1

Source of Variation	df	SSX	MSX	F	SSY	MSY	F
Total	19	186			334		
Blocks	3	50	16.67		70	23.33	
Treatments	4	40	10.00	1.25	160	40.00	4.62*
Error	12	96	8.00		104	8.67	

The sum of cross-products for blocks is

$$SXYB = \frac{\sum T_{bx}T_{by}}{n} - C = \frac{(30)(45) + \ldots + (50)(65)}{5} - 1600$$

$$= 50$$

For treatments, it is

$$SXYT = \frac{\sum T_{tx}T_{ty}}{b} - C = \frac{(28)(24) + \ldots + (40)(56)}{4} - 1600$$

$$= 24$$

The total sum of cross-products is

$$SXY = \sum XY - C = (8)(7) + \ldots + (11)(20) - 1600$$

$$= 142$$

The sum of cross-products for error can be obtained by subtraction:

$$SXYE = SXY - SXYB - SXYT = 142 - 50 - 24 = 68$$

To show where the error sums of squares and cross-products come from and ultimately how we arrive at the regression equation, we remove the block and treatment effects and the general mean from each variate, leaving only the residual error components, as we did in Chapter 5 (Table 18.3).

It is easy to verify that the sums of squares of these components are the same as the error sums of squares in the analyses of variance in Table 18.2. Also the sum of products of corresponding components of X and Y is the same as the value of SXYE obtained indirectly by subtraction above. It is these 20 pairs of error components that are used to calculate the regression of Y and X free from treatment and block effects.

TABLE 18.3.
Error components of X and Y after removal of block and treatment effects and general mean

	X					Y				
Block:	1	2	3	4	Total	1	2	3	4	Total
Treatment										
1	3	0	−1	−2	0	2	1	0	−3	0
2	1	−4	2	1	0	2	−1	1	−2	0
3	−2	3	1	−2	0	−3	4	0	−1	0
4	−3	2	−3	4	0	−2	1	−2	3	0
5	1	−1	1	−1	0	1	−5	1	3	0
Totals	0	0	0	0	0	0	0	0	0	0

We learned in the chapter on linear regression and correlation that a sum of squares for deviation from regression could be found by taking $(1-r^2)$ SSY. This can be rewritten as

$$(1-r^2)\text{SSY}=\left[1-\frac{(\text{SXY})^2}{(\text{SSX})(\text{SSY})}\right]\text{SSY}=\text{SSY}-\frac{\text{SXY}^2}{\text{SSX}}$$

This sum of squares for deviation from regression can be considered as a sum of squares of Y after removing the effect of X on Y. It is therefore called "Y adjusted for X."

We now have all the information we need to make a complete analysis of covariance table (Table 18.4).

The error sum of squares of Y adjusted for X is

$$\text{SSY}-\frac{\text{SXY}^2}{\text{SSX}}=104-\frac{68^2}{96}=55.833$$

This has 11 degrees of freedom, 1 less than the 12 for unadjusted error.

The degrees of freedom and sums of squares and products in the row called "treatments + error" are simply obtained by adding the numbers in the "treatments" row to those in the "error" row. We then obtain a sum of squares of Y adjusted for X in the same way on this row as we did for error:

$$(\text{treatment}+\text{error})\text{adjusted SS}=264-\frac{92^2}{136}=201.765$$

TABLE 18.4.
Analysis of covariance of data from Table 18.2

Source of Variation	df	Sums of Squares and Products SSX	SXY	SSY	Y Adjusted for X df	SS	MS	F
Total	19	186	142	334				
Blocks	3	50	50	70				
Treatments	4	40	24	160				
Error	12	96	68	104	11	55.833	5.076	
Treatments + error	16	136	92	264	15	201.765		
Treatments adjusted					4	145.932	36.483	7.19**

The treatment sum of squares of Y adjusted for X is now obtained by subtraction: $201.765 - 55.833 = 145.932$. It is important to note that the adjusted sum of squares for treatment *cannot* be obtained directly by applying the formula $SSY - SXY^2/SSX$ to the treatment line. In this case, we would get $160 - 24^2/40 = 145.6$. The fact that this is fairly close to the correct value is merely coincidence. The two values will not generally be this close.

The regression coefficient is found from the error line by the usual relation: $b = SXY/SSX = 68/96 = 0.70833$. It is informative to see what happens when we adjust the error terms of Y in Table 18.3 for the corresponding error terms of X. This can be done by applying the equation: $Y_{ij}\text{adjusted} = Y_{ij} - bX_{ij}$ to each value of Y in the table, as shown in Table 18.5.

TABLE 18.5.
Error terms of Y Adjusted for X

Block	1	2	3	4
Treatment				
1	−0.12500	1.00000	0.70833	−1.58333
2	1.29167	1.83333	−0.41667	−2.70833
3	−1.58333	1.87500	−0.70833	0.41667
4	0.12500	−0.41667	0.12500	0.16667
5	0.29167	−4.29167	0.29167	3.70833

Not only are the sums for blocks and treatments still zero as they should be but also the sum of squares of these adjusted error terms is 55.833, exactly the same as in the analysis of covariance.

Adjusting More than One Source of Variation

Regardless of the design of the experiment or the number of factors being studied, the general pattern of the analysis of covariance table (Table 18.4) can be followed. The important point to remember is that for each source of variation to be adjusted, the sums of squares and cross-products for that source must be added to the corresponding error sums of squares and cross-products. The resulting "source + error" line is used to calculate a sum of squares of Y adjusted for X, and from this we subtract the adjusted error sum of squares to find the adjusted sum of squares for the source of variation being studied. We illustrate this procedure by partitioning the treatment sum of squares in our example into four sources of variation or components, each with a single degree of freedom:

Component	Coefficients					$\Sigma(c_iT_i)_X$	$\Sigma(c_iT_i)_Y$	Σc_i^2
I	4	−1	−1	−1	−1	−20	−80	20
II	0	3	−1	−1	−1	12	−48	12
III	0	0	2	−1	−1	0	−24	6
IV	0	0	0	1	−1	−16	−8	2

The sums of squares for each component is obtained by the usual formula: $[\Sigma(c_iT_i)]^2/r(\Sigma c_i^2)$. The sum of cross-products requires a slight modification of this formula: $SXY = \Sigma(c_iT_i)_X\Sigma(c_iT_i)_Y/r(\Sigma c_i^2)$. The analysis of covariance of the partitioned treatment effects is given in Table 18.6.

There is a very important feature of this table to notice. The unadjusted sums of squares and cross-products are additive. That is, the sums of the four components equal the total treatment sums of squares and cross-products. On the other hand, the adjusted sums of squares are *not* additive. The sum for the four components is 141.026 compared to the value of 145.932 for the total adjusted treatment sum of squares. This means that we cannot find an adjusted component sum of squares by subtracting all the remaining components from the total adjusted treatment sum of squares.

Adjusting the Treatment Means

It is often desirable to estimate what the treatment means of the dependent variable would be if the means of the independent variable were the same for all

TABLE 18.6.
Analysis of covariance of partitioned treatment effects

Source of Variation	df	Sums of Squares and Products SSX	SXY	SSY	Y Adjusted for X df	SS	MS	F
Total	19	186	142	334				
Blocks	3	50	50	70				
Treatments	4	40	24	160				
Comp. I	1	5	20	80				
Comp. II	1	3	−12	48				
Comp. III	1	0	0	24				
Comp. IV	1	32	16	8				
Error	12	96	68	104	11	55.833	5.076	
CI + Error	13	101	88	184	12	107.327		
CI Adj.					1	51.494	51.494	10.14**
CII + Error	13	99	56	152	12	120.323		
CII Adj.					1	64.490	64.490	12.70**
CIII + Error	13	96	68	128	12	79.83		
CIII Adj.					1	24.000	24.000	4.73 NS
CIV + Error	13	128	84	112	12	56.875		
CIV Adj.					1	1.042	1.042	0.21 NS
Total for 4 components							141.026	

treatments. These adjusted means are found from the equation:

$$\hat{Y}_i = \bar{Y}_i - b(\bar{X}_i - \bar{X})$$

where b = error SXY/error SSX. In our example, b = 68/96 = 0.7083, and the adjusted means are:

$\bar{Y}_i$	$(\bar{X}_i - \bar{X})$	$b(\bar{X}_i - \bar{X})$	$\hat{\bar{Y}}$
6	−1	−0.7083	6.7083
8	1	0.7083	7.2917
10	0	0.0000	10.0000
12	−2	−1.4166	13.4166
14	2	1.4166	12.5834

One might expect that the adjusted treatment sum of squares could be found directly from the adjusted treatment means. In fact, this is sometimes suggested as an approximate method of covariance analysis when there is no significant treatment effect on X, the independent variable. However, there is a fact seldom explicitly stated in statistics texts: *The sum of squares of adjusted treatment means is always greater than the adjusted treatment sum of squares*. The difference is

$$\frac{\left[\text{SSXE}(\text{SXYT}) - \text{SXYE}(\text{SXXT})\right]^2}{(\text{SSXE})^2(\text{SXXT} + \text{SXXE})}$$

In our example, this is

$$\frac{\left[96(24) - 68(40)\right]^2}{(96)^2(40 + 96)} = 0.1381$$

The sum of squares of adjusted treatments is

$$4(6.7083^2 + \ldots + 12.5834^2) - \frac{200^2}{20} = 146.0694$$

(Note that since we are working with means, we multiply rather than divide by the number of replicates before subtracting the correction term.) The adjusted treatment sum of squares from the analysis of covariance was 145.932, and the difference between these two sums of squares is 0.1374, the same as calculated from the formula except for rounding.

Since the treatment sum of squares obtained from adjusted treatment means always overstimates the correct sum of squares, the resulting F values are likewise too high. Therefore, if one uses the approximate method and finds F values that are only slightly above the significance level, the exact procedure should be used. On the other hand, if the F values found by the approximate method are not significant, we can be sure that they will not be significant by the exact method.

Comparing Two Adjusted Treatment Means

Since the variance of adjusted treatment means is larger than the correct adjusted treatment mean square, the usual LSD is not appropriate for comparing adjusted treatment means. Technically a different standard error of difference must be calculated for each pair of means. The formula is

$$s_{\bar{d}}^2 = \text{Adj. EMS}\left[\frac{2}{r} + \frac{(\bar{x}_p - \bar{x}_q)^2}{\text{SSXE}}\right]$$

If the degrees of freedom for error are 20 or more, and if there is no significant treatment effect on X, an approximation that can be used for all pairs of means is:

$$s_{\bar{d}}^2 = \text{Adj. EMS}\left[\frac{2}{r} + \frac{2SSXT}{r(t-1)SSXE}\right]$$

Interpretation of Covariance Analysis

The error mean square is nearly always reduced considerably by covariance analysis, and the adjusted treatment mean square is usually reduced also. For this reason, the F value for treatments after adjustment may be greater or less than before adjustment. The interpretation of the results depends on whether there was a significant effect on X, the independent variable.

If there was no significant treatment effect on X, and the treatment effects on Y were significant before but not after adjustment, this would indicate that the apparent treatment effects on Y were exaggerated by chance variation in X and should be interpreted with considerable caution.

If X was not significant, and the treatment effects on Y were significant after but not before adjustment, it is likely that the true treatment effects were obscured by variation in X.

If the treatment did have a significant effect on X, then the F value after adjustment is usually less than before adjustment. If it is still significant, then we can conclude that the treatments had a significant effect on Y over and above that associated with the variation in X.

We have seen that the techniques of covariance analysis are considerably more cumbersome than ordinary analysis of variance, and interpretation of results is often difficult. Our best advice is to avoid random distribution of a known independent variable if possible. This can be done by careful grouping of experimental units into blocks, thereby making it possible to remove most of the variability in X along with the block effects.

PLANNED GROUPING OF EXPERIMENTAL UNITS—DESIGN

We have devoted a considerable portion of this book to a discussion of experimental designs and their role in improving precision. There are many other designs we have not discussed. In our experience, however, the designs presented here are used in the great majority of agricultural experiments. The reader interested in other designs should consult more advanced texts, such as that by Cochran and Cox (1964).

SUMMARY

Precision is the ability of an experiment to detect a true treatment effect. It can be improved by increased replication, treatment selection, improved technique to reduce the variability among units treated alike, increasing the size of experimental units (within limits), the use of covariance, and the employment of a more efficient experimental design.

SELECTED REFERENCES

Alder, H. L. and E. V. Roessler. 1968, 5th Ed. *Introduction to Probability and Statistics* W. H. Freeman & Co., San Francisco. 333 p.

Bancroft, T. A. 1968. *Topics in Intermediate Statistical Methods*, Volume One. The Iowa State University Press, Ames, Iowa. 129 p.

Cochran, W. G. and G. M. Cox. 1964, 2nd Ed. *Experimental Design*. John Wiley & Sons, Inc., New York. 617 p.

Finney, D. J. 1962, 2nd Ed. *An Introduction to Statistical Science in Agriculture*. Munksgaard, Copenhagen, Denmark, and Oliver & Boyd Ltd., Edinburgh, Scotland. 216 p.

LeClerg, E. L., W. H. Leonard, and A. G. Clark. 1962, 2nd Ed. *Field Plot Technique*. Burgess Publishing Co., Minneapolis, Minnesota. 373 p.

Snedecor, G. W. and W. G. Cochran. 1967, 6th Ed. *Statistical Methods*. The Iowa State University Press, Ames, Iowa. 593 p.

Sokal, R. R. and R. J. Rohlf. 1969. *Biometry, The Principles and Practice of Statistics in Biological Research*. W. H. Freeman & Co., San Francisco. 776 p.

Steel, R. G. D. and J. H. Torrie. 1960. *Principles and Procedures of Statistics with Special Reference to the Biological Sciences*. McGraw-Hill Book Co., Inc., New York. 481 p.

APPENDIX TABLES

A.1.	Random Numbers	296
A.2.	Distribution of t	297
A.3.	10%, 5%, and 1% Points for the F Distribution	299
A.4.	Significant Studentized Factors (R) to Multiply by LSD for Testing Means at Various Ranges, 5% Level	307
A.5.	Significant Studentized Factors (R) to multiply by LSD for Testing Means at Various Ranges, 1% Level	308
A.6.	Distribution of χ^2 (Chi-Square)	309
A.7.	Values of the Correlation Coefficient, r, for Certain Levels of Significance	310
A.8.	The Angular Transformation of Percentages to Degrees	311
A.9.	Logarithms	312
A.10.	Squares and Square Roots	316
A.11.	Coefficients, Divisors, and K Values for Fitting Up to Quartic Curves to Equally Spaced Data, and Partitioning the Sum of Squares	331
A.11a.	Coefficients and Divisors for Some Selected Sets of Unequally Spaced Treatments	341
A.12.	Coefficients for Fitting Periodic Curves and Partitioning Sums of Squares for Data Taken at Equal Time Intervals Throughout a Complete Cycle	342

TABLE A.1.
Random Numbers

To randomize any set of 10 items or less, begin at a random point on the table and follow either rows, columns or diagonals in either direction. Write down the numbers in the order they appear, disregarding those that are higher than the number being randomized and those that have appeared before in the series. If you wish to randomize more than 10 numbers, pairs of columns or rows can be combined to form two digit numbers and the same process followed as that described above.

8	2	0	3	1	4	5	8	2	1	7	2	7	3	8	5	5	2	9	0	6	3	1	6	4
0	8	7	3	3	1	9	7	5	2	5	7	6	9	8	0	3	6	2	5	1	2	7	5	2
2	3	3	8	6	1	4	2	4	0	2	6	1	8	9	5	2	6	9	8	3	4	0	1	0
4	7	5	5	6	3	0	7	7	1	9	1	6	1	7	4	1	7	1	3	7	9	3	3	7
1	9	3	9	5	3	4	9	5	5	2	7	5	8	0	3	4	8	8	1	2	7	5	3	4
2	8	7	8	1	4	1	4	9	4	2	4	1	5	2	9	4	6	2	1	5	2	8	1	9
8	4	8	5	1	3	9	6	6	0	7	2	1	9	0	2	0	6	7	0	6	0	1	3	0
0	3	8	8	4	7	5	1	5	1	7	3	4	5	2	0	7	4	7	9	6	6	7	7	4
3	5	3	1	9	3	7	4	9	5	0	2	0	1	4	6	2	5	4	5	8	5	0	9	2
3	4	5	9	5	2	7	9	8	9	0	5	5	8	5	1	7	7	3	5	5	4	7	7	2
4	1	5	3	0	9	1	3	7	2	5	8	7	7	1	3	6	3	9	7	8	7	9	1	7
7	2	9	5	6	7	8	5	4	5	3	4	5	4	1	9	8	6	7	5	7	9	3	1	8
5	9	2	8	9	8	6	4	4	1	5	3	7	7	0	8	0	2	5	6	0	6	1	2	0
1	3	3	3	9	0	5	2	8	7	4	0	9	0	3	7	3	1	7	9	4	5	5	2	8
4	6	0	1	0	8	6	2	1	0	0	5	0	3	1	5	4	9	0	3	7	4	7	0	1
7	7	0	6	6	3	2	8	8	5	8	9	5	6	4	0	5	9	1	8	0	5	4	9	4
3	3	8	5	7	5	7	4	3	4	5	7	9	6	9	5	0	7	7	6	6	8	8	5	9
9	1	7	1	3	6	9	2	9	1	9	4	2	3	3	0	8	1	8	7	7	6	4	7	2
6	2	2	8	0	9	4	5	3	7	2	5	4	6	6	5	6	6	5	0	4	6	5	6	8
1	7	5	9	0	0	2	0	5	6	5	8	5	1	9	5	3	3	7	4	0	5	8	2	4
0	3	9	6	9	4	7	3	5	7	0	6	5	4	7	1	1	8	5	3	2	8	0	9	8
3	0	8	2	8	1	4	4	1	6	7	6	6	9	9	9	7	5	8	9	6	4	5	9	0
9	4	9	1	2	2	0	1	3	2	4	6	7	9	1	8	8	2	9	8	3	2	6	2	9
7	2	5	1	4	4	9	6	5	2	8	5	5	1	0	8	2	6	2	0	6	9	2	2	3
9	9	2	5	7	4	3	1	2	3	6	4	1	5	2	4	0	4	2	2	8	7	1	8	2
2	0	9	1	8	9	4	4	6	1	4	8	6	7	9	2	5	0	6	9	3	3	0	1	2
6	5	2	6	1	2	1	7	7	1	4	7	8	1	4	2	7	3	7	4	0	0	1	2	9
1	2	9	9	6	4	2	5	3	2	7	4	3	2	3	3	8	5	3	3	6	5	5	3	2
3	2	8	3	7	9	6	0	4	8	6	0	5	4	1	1	4	9	0	5	0	9	4	4	1
0	9	3	4	1	1	9	5	8	3	2	4	6	7	3	4	4	9	2	3	7	2	5	7	8
6	7	5	3	4	2	1	5	5	0	1	2	4	7	5	5	2	6	8	7	8	2	8	0	3
9	6	0	1	3	0	5	3	6	6	2	9	6	0	3	4	7	6	1	1	9	1	6	5	3
4	6	9	9	6	7	8	5	8	1	2	9	2	6	2	4	4	9	0	5	5	4	5	2	0
9	7	7	1	9	2	6	5	6	3	3	6	3	6	8	3	9	9	8	7	7	2	7	9	7
7	5	3	3	3	3	7	3	7	6	7	3	9	1	1	2	3	9	0	9	5	9	6	5	7
2	8	1	3	1	3	4	2	1	0	3	1	2	3	2	0	2	3	9	7	7	5	0	6	9
6	0	9	4	8	8	5	5	3	7	9	0	0	0	0	1	9	2	0	6	1	5	8	4	2
3	5	9	0	7	7	0	1	8	1	2	9	3	4	6	9	2	8	9	8	9	8	6	5	5
4	4	8	1	1	7	4	4	7	4	4	4	1	6	5	9	3	6	5	9	8	3	2	4	3
6	3	9	7	0	6	2	5	3	3	2	6	0	5	1	2	4	3	7	1	0	7	8	2	1

TABLE A.2.
Distribution of t[a]

Degrees of Freedom	Probability of Obtaining a Value as Large or Larger					
	0.400	0.200	0.100	0.050	0.010	0.001
1	1.376	3.078	6.314	12.706	63.657	
2	1.061	1.886	2.920	4.303	9.925	31.598
3	0.978	1.638	2.353	3.182	5.841	12.941
4	0.941	1.533	2.132	2.776	4.604	8.610
5	0.920	1.476	2.015	2.571	4.032	6.859
6	0.906	1.440	1.943	2.447	3.707	5.959
7	0.895	1.415	1.895	2.365	3.499	5.405
8	0.889	1.397	1.860	2.306	3.355	5.041
9	0.883	1.383	1.833	2.262	3.250	4.781
10	0.879	1.372	1.812	2.228	3.169	4.587
11	0.876	1.363	1.796	2.201	3.106	4.437
12	0.873	1.356	1.782	2.179	3.055	4.318
13	0.870	1.350	1.771	2.160	3.012	4.221
14	0.868	1.345	1.761	2.145	2.977	4.140
15	0.866	1.341	1.753	2.131	2.947	4.073
16	0.865	1.337	1.746	2.120	2.921	4.015
17	0.863	1.333	1.740	2.110	2.898	3.965
18	0.862	1.330	1.734	2.101	2.878	3.922
19	0.861	1.328	1.729	2.093	2.861	3.883
20	0.860	1.325	1.725	2.086	2.845	3.850
21	0.859	1.323	1.721	2.080	2.831	3.819
22	0.858	1.321	1.717	2.074	2.819	3.792
23	0.858	1.319	1.714	2.069	2.807	3.767
24	0.857	1.318	1.711	2.064	2.797	3.745
25	0.856	1.316	1.708	2.060	2.787	3.725
26	0.856	1.315	1.706	2.056	2.779	3.707
27	0.855	1.314	1.703	2.052	2.771	3.690
28	0.855	1.313	1.701	2.048	2.763	3.674
29	0.854	1.311	1.699	2.045	2.756	3.659
30	0.854	1.310	1.697	2.042	2.750	3.646

TABLE A.2.
Continued.

Degrees of Freedom	Probability of Obtaining a Value as Large or Larger					
	0.400	0.200	0.100	0.050	0.010	0.001
35	0.852	1.306	1.690	2.030	2.724	3.591
40	0.851	1.303	1.684	2.021	2.704	3.551
45	0.850	1.301	1.680	2.014	2.690	3.520
50	0.849	1.299	1.676	2.008	2.678	3.496
55	0.849	1.297	1.673	2.004	2.669	3.476
60	0.848	1.296	1.671	2.000	2.660	3.460
70	0.847	1.294	1.667	1.994	2.648	3.435
80	0.847	1.293	1.665	1.989	2.638	3.416
90	0.846	1.291	1.662	1.986	2.631	3.402
100	0.846	1.290	1.661	1.982	2.625	3.390
120	0.845	1.289	1.658	1.980	2.617	3.373
∞	0.8416	1.2816	1.6448	1.9600	2.5758	3.2905

[a]Parts of this table are taken from Table III of Fisher and Yates: *Statistical Tables for Biological, Agricultural, and Medical Research*, published by Longman Group Ltd., London (previously published by Oliver & Boyd, Edinburgh), by permission of the authors and publishers. Other parts were calculated following Chen and Makowsky (see footnote to Table A.3).

TABLE A.3.
10%, 5% and 1% points for the F distribution.[a]

DF For Denom	P	Degrees of Freedom for Numerator (Greater Mean Square)																			
		1	2	3	4	5	6	7	8	9	10	11	12	13	14	15	16	17	18	19	20
1	.10	39.86	49.50	53.59	55.83	57.24	58.20	58.91	59.44	59.86	60.19	60.47	60.71	60.90	61.07	61.22	61.35	61.46	61.57	61.66	61.74
	.05	161	200	216	225	230	234	237	239	241	242	243	244	245	245	246	246	247	247	248	248
	.01	4,052	4,999	5,403	5,625	5,764	5,859	5,928	5,981	6,022	6,056	6,083	6,106	6,126	6,143	6,157	6,170	6,181	6,191	6,201	6,209
2	.10	8.53	9.00	9.16	9.24	9.29	9.33	9.35	9.37	9.38	9.39	9.40	9.41	9.41	9.42	9.42	9.43	9.43	9.44	9.44	9.44
	.05	18.51	19.00	19.16	19.25	19.30	19.33	19.35	19.37	19.38	19.40	19.40	19.41	19.42	19.42	19.43	19.43	19.44	19.44	19.44	19.45
	.01	98.50	99.00	99.17	99.25	99.30	99.33	99.36	99.37	99.39	99.40	99.41	99.42	99.42	99.43	99.43	99.44	99.44	99.44	99.45	99.45
3	.10	5.54	5.46	5.39	5.34	5.31	5.28	5.27	5.25	5.24	5.23	5.22	5.22	5.21	5.20	5.20	5.20	5.19	5.19	5.19	5.18
	.05	10.13	9.55	9.28	9.12	9.01	8.94	8.89	8.85	8.81	8.79	8.76	8.74	8.73	8.71	8.70	8.69	8.68	8.67	8.67	8.66
	.01	34.12	30.82	29.46	28.71	28.24	27.91	27.67	27.49	27.35	27.23	27.13	27.05	26.98	26.92	26.87	26.83	26.79	26.75	26.72	26.69
4	.10	4.54	4.32	4.19	4.11	4.05	4.01	3.98	3.95	3.94	3.92	3.91	3.90	3.89	3.88	3.87	3.86	3.86	3.85	3.85	3.84
	.05	7.71	6.94	6.59	6.39	6.26	6.16	6.09	6.04	6.00	5.96	5.94	5.91	5.89	5.87	5.86	5.84	5.83	5.82	5.81	5.80
	.01	21.20	18.00	16.69	15.98	15.52	15.21	14.98	14.80	14.66	14.55	14.45	14.37	14.31	14.25	14.20	14.15	14.11	14.08	14.05	14.02
5	.10	4.06	3.78	3.62	3.52	3.45	3.40	3.37	3.34	3.32	3.30	3.28	3.27	3.26	3.25	3.24	3.23	3.22	3.22	3.21	3.21
	.05	6.61	5.79	5.41	5.19	5.05	4.95	4.88	4.82	4.77	4.74	4.70	4.68	4.66	4.64	4.62	4.60	4.59	4.58	4.57	4.56
	.01	16.26	13.27	12.06	11.39	10.97	10.67	10.46	10.29	10.16	10.05	9.96	9.89	9.82	9.77	9.72	9.68	9.64	9.61	9.58	9.55
6	.10	3.78	3.46	3.29	3.18	3.11	3.05	3.01	2.98	2.96	2.94	2.92	2.90	2.89	2.88	2.87	2.86	2.85	2.85	2.84	2.84
	.05	5.99	5.14	4.76	4.53	4.39	4.28	4.21	4.15	4.10	4.06	4.03	4.00	3.98	3.96	3.94	3.92	3.91	3.90	3.88	3.87
	.01	13.75	10.92	9.78	9.15	8.75	8.47	8.26	8.10	7.98	7.87	7.79	7.72	7.66	7.60	7.56	7.52	7.48	7.45	7.42	7.40

[a]The points of this table were calculated from Hubert J. Chen and A. B. Makowsky, "On Approximations to the F-Distribution and Its Inverse," Report 76-3, Memphis State University, Department of Mathematical Sciences (1976).

TABLE A.3.
Continued.

DF For Denom	P	1	2	3	4	5	6	7	8	9	10	11	12	13	14	15	16	17	18	19	20
7	.10	3.59	3.26	3.07	2.96	2.88	2.83	2.78	2.75	2.72	2.70	2.68	2.67	2.65	2.64	2.63	2.62	2.61	2.61	2.60	2.59
	.05	5.59	4.74	4.35	4.12	3.97	3.87	3.79	3.73	3.68	3.64	3.60	3.57	3.55	3.53	3.51	3.49	3.48	3.47	3.46	3.44
	.01	12.25	9.55	8.45	7.85	7.46	7.19	6.99	6.84	6.72	6.62	6.54	6.47	6.41	6.36	6.31	6.28	6.24	9.21	6.18	6.16
8	.10	3.46	3.11	2.92	2.81	2.73	2.67	2.62	2.59	2.56	2.54	2.52	2.50	2.49	2.48	2.46	2.45	2.45	2.44	2.43	2.42
	.05	5.32	4.46	4.07	3.84	3.69	3.58	3.50	3.44	3.39	3.35	3.31	3.28	3.26	3.24	3.22	3.20	3.19	3.17	3.16	3.15
	.01	11.26	8.65	7.59	7.01	6.63	6.37	6.18	6.03	5.91	5.81	5.73	5.67	5.61	5.56	5.52	5.48	5.44	5.41	5.38	5.36
9	.10	3.36	3.01	2.81	2.69	2.61	2.55	2.51	2.47	2.44	2.42	2.40	2.38	2.36	2.35	2.34	2.33	2.32	2.31	2.30	2.30
	.05	5.12	4.26	3.86	3.63	3.48	3.37	3.29	3.23	3.18	3.14	3.10	3.07	3.05	3.03	3.01	2.99	2.97	2.96	2.95	2.94
	.01	10.56	8.02	6.99	6.42	6.06	5.80	5.61	5.47	5.35	5.26	5.18	5.11	5.05	5.01	4.96	4.92	4.89	4.86	4.83	4.81
10	.10	3.29	2.92	2.73	2.61	2.52	2.46	2.41	2.38	2.35	2.32	2.30	2.28	2.27	2.26	2.24	2.23	2.22	2.22	2.21	2.20
	.05	4.96	4.10	3.71	3.48	3.33	3.22	3.14	3.07	3.02	2.98	2.94	2.91	2.89	2.86	2.85	2.83	2.81	2.80	2.79	2.77
	.01	10.04	7.56	6.55	5.99	5.64	5.39	5.20	5.06	4.94	4.85	4.77	4.71	4.65	4.60	4.56	4.52	4.49	4.46	4.43	4.41
11	.10	3.23	2.86	2.66	2.54	2.45	2.39	2.34	2.30	2.27	2.25	2.23	2.21	2.19	2.18	2.17	2.16	2.15	2.14	2.13	2.12
	.05	4.84	3.98	3.59	3.36	3.20	3.09	3.01	2.95	2.90	2.85	2.82	2.79	2.76	2.74	2.72	2.70	2.69	2.67	2.66	2.65
	.01	9.65	7.21	6.22	5.67	5.32	5.07	4.89	4.74	4.63	4.54	4.46	4.40	4.34	4.29	4.25	4.21	4.18	4.15	4.12	4.10
12	.10	3.18	2.81	2.61	2.48	2.39	2.33	2.28	2.24	2.21	2.19	2.17	2.15	2.13	2.12	2.10	2.09	2.08	2.08	2.07	2.06
	.05	4.75	3.89	3.49	3.26	3.11	3.00	2.91	2.85	2.80	2.75	2.72	2.69	2.66	2.64	2.62	2.60	2.58	2.57	2.56	2.54
	.01	9.33	6.93	5.95	5.41	5.06	4.82	4.64	4.50	4.39	4.30	4.22	4.16	4.10	4.05	4.01	3.97	3.94	3.91	3.88	3.86

(Column header group: Degrees of Freedom for Numerator (Greater Mean Square))

13	.10	3.14	2.76	2.56	2.43	2.35	2.28	2.23	2.20	2.16	2.14	2.12	2.10	2.08	2.07	2.05	2.04	2.03	2.02	2.01	2.01
	.05	4.67	3.81	3.41	3.18	3.03	2.92	2.83	2.77	2.71	2.67	2.63	2.60	2.58	2.55	2.53	2.51	2.50	2.48	2.47	2.46
	.01	9.07	6.70	5.74	5.21	4.86	4.62	4.44	4.30	4.19	4.10	4.02	3.96	3.91	3.86	3.82	3.78	3.75	3.72	3.69	3.66
14	.10	3.10	2.73	2.52	2.39	2.31	2.24	2.19	2.15	2.12	2.10	2.07	2.05	2.04	2.02	2.01	2.00	1.99	1.98	1.97	1.96
	.05	4.60	3.74	3.34	3.11	2.96	2.85	2.76	2.70	2.65	2.60	2.57	2.53	2.51	2.48	2.46	2.44	2.43	2.41	2.40	2.39
	.01	8.86	6.51	5.56	5.04	4.69	4.46	4.28	4.14	4.03	3.94	3.86	3.80	3.75	3.70	3.66	3.62	3.59	3.56	3.53	3.51
15	.10	3.07	2.70	2.49	2.36	2.27	2.21	2.16	2.12	2.09	2.06	2.04	2.02	2.00	1.99	1.97	1.96	1.95	1.94	1.93	1.92
	.05	4.54	3.68	3.29	3.06	2.90	2.79	2.71	2.64	2.59	2.54	2.51	2.48	2.45	2.42	2.40	2.38	2.37	2.35	2.34	2.33
	.01	8.68	6.36	5.42	4.89	4.56	4.32	4.14	4.00	3.89	3.80	3.73	3.67	3.61	3.56	3.57	3.49	3.45	3.42	3.40	3.37
16	.10	3.05	2.67	2.46	2.33	2.24	2.18	2.13	2.09	2.06	2.03	2.01	1.99	1.97	1.95	1.94	1.93	1.92	1.91	1.90	1.89
	.05	4.49	3.63	3.24	3.01	2.85	2.74	2.66	2.59	2.54	2.49	2.46	2.42	2.40	2.37	2.35	2.33	2.32	2.30	2.29	2.28
	.01	8.53	6.23	5.29	4.77	4.44	4.20	4.03	3.89	3.78	3.69	3.62	3.55	3.50	3.45	3.41	3.37	3.34	3.31	3.28	3.26
17	.10	3.03	2.64	2.44	2.31	2.22	2.15	2.10	2.06	2.03	2.00	1.98	1.96	1.94	1.93	1.91	1.90	1.89	1.88	1.87	1.86
	.05	4.45	3.59	3.20	2.96	2.81	2.70	2.61	2.55	2.49	2.45	2.41	2.38	2.35	2.33	2.31	2.29	2.27	2.26	2.24	2.23
	.01	8.40	6.11	5.18	4.67	4.34	4.10	3.93	3.79	3.68	3.59	3.52	3.46	3.40	3.35	3.31	3.27	3.24	3.21	3.19	3.16
18	.10	3.01	2.62	2.42	2.29	2.20	2.13	2.08	2.04	2.00	1.98	1.95	1.93	1.92	1.90	1.89	1.87	1.86	1.85	1.84	1.84
	.05	4.41	3.55	3.16	2.93	2.77	2.66	2.58	2.51	2.46	2.41	2.37	2.34	2.31	2.29	2.27	2.25	2.23	2.22	2.20	2.19
	.01	8.29	6.01	5.09	4.58	4.25	4.01	3.84	3.71	3.60	3.51	3.43	3.37	3.32	3.27	3.23	3.19	3.16	3.13	3.10	3.08
19	.10	2.99	2.61	2.40	2.27	2.18	2.11	2.06	2.02	1.98	1.96	1.93	1.91	1.89	1.88	1.86	1.85	1.84	1.83	1.82	1.81
	.05	4.38	3.52	3.13	2.90	2.74	2.63	2.54	2.48	2.42	2.38	2.34	2.31	2.28	2.26	2.23	2.21	2.20	2.18	2.17	2.16
	.01	8.18	5.93	5.01	4.50	4.17	3.94	3.77	3.63	3.52	3.43	3.36	3.30	3.24	3.19	3.15	3.12	3.08	3.05	3.03	3.00
20	.10	2.97	2.59	2.38	2.25	2.16	2.09	2.04	2.00	1.96	1.94	1.91	1.89	1.87	1.86	1.84	1.83	1.82	1.81	1.80	1.79
	.05	4.35	3.49	3.10	2.87	2.71	2.60	2.51	2.45	2.39	2.35	2.31	2.28	2.25	2.22	2.20	2.18	2.17	2.15	2.14	2.12
	.10	8.10	5.85	4.94	4.43	4.10	3.87	3.70	3.56	3.46	3.37	3.29	3.23	3.18	3.13	3.09	3.05	3.02	2.99	2.96	2.94

TABLE A.3.
Continued.

DF For Denom	P	Degrees of Freedom for Numerator (Greater Mean Square)																			
		1	2	3	4	5	6	7	8	9	10	11	12	13	14	15	16	17	18	19	20
21	.10	2.96	2.57	2.36	2.23	2.14	2.08	2.02	1.98	1.95	1.92	1.90	1.87	1.86	1.84	1.83	1.81	1.80	1.79	1.78	1.78
	.05	4.32	3.47	3.07	2.84	2.68	2.57	2.49	2.42	2.37	2.32	2.28	2.25	2.22	2.20	2.18	2.16	2.14	2.12	2.11	2.10
	.01	8.02	5.78	4.87	4.37	4.04	3.81	3.64	3.51	3.40	3.31	3.24	3.17	3.12	3.07	3.03	2.99	2.96	2.93	2.90	2.88
22	.10	2.95	2.56	2.35	2.22	2.13	2.06	2.01	1.97	1.93	1.90	1.88	1.86	1.84	1.83	1.81	1.80	1.79	1.78	1.77	1.76
	.05	4.30	3.44	3.05	2.82	2.66	2.55	2.46	2.40	2.34	2.30	2.26	2.23	2.20	2.17	2.15	2.13	2.11	2.10	2.08	2.07
	.01	7.95	5.72	4.82	4.31	3.99	3.76	3.59	3.45	3.35	3.26	3.18	3.12	3.07	3.02	2.98	2.94	2.91	2.88	2.85	2.83
23	.10	2.94	2.55	2.34	2.21	2.11	2.05	1.99	1.95	1.92	1.89	1.87	1.84	1.83	1.81	1.80	1.78	1.77	1.76	1.75	1.74
	.05	4.28	3.42	3.03	2.80	2.64	2.53	2.44	2.37	2.32	2.27	2.24	2.20	2.18	2.15	2.13	2.11	2.09	2.08	2.06	2.05
	.01	7.88	5.66	4.76	4.26	3.94	3.71	3.54	3.41	3.30	3.21	3.14	3.07	3.02	2.97	2.93	2.89	2.86	2.83	2.80	2.78
24	.10	2.93	2.54	2.33	2.19	2.10	2.04	1.98	1.94	1.91	1.88	1.85	1.83	1.81	1.80	1.78	1.77	1.76	1.75	1.74	1.73
	.05	4.26	3.40	3.01	2.78	2.62	2.51	2.42	2.36	2.30	2.25	2.22	2.18	2.15	2.13	2.11	2.09	2.07	2.05	2.04	2.03
	.01	7.82	5.61	4.72	4.22	3.90	3.67	3.50	3.36	3.26	3.17	3.09	3.03	2.98	2.93	2.89	2.85	2.82	2.79	2.76	2.74
25	.10	2.92	2.53	2.32	2.18	2.09	2.02	1.97	1.93	1.89	1.87	1.84	1.82	1.80	1.79	1.77	1.76	1.75	1.74	1.73	1.72
	.05	4.24	3.39	2.99	2.76	2.60	2.49	2.40	2.34	2.28	2.24	2.20	2.16	2.14	2.11	2.09	2.07	2.05	2.04	2.02	2.01
	.01	7.77	5.57	4.68	4.18	3.85	3.63	3.46	3.32	3.22	3.13	3.06	2.99	2.94	2.89	2.85	2.81	2.78	2.75	2.72	2.70
26	.10	2.91	2.52	2.31	2.17	2.08	2.01	1.96	1.92	1.88	1.86	1.83	1.81	1.79	1.77	1.76	1.75	1.73	1.72	1.71	1.71
	.05	4.23	3.37	2.98	2.74	2.59	2.47	2.39	2.32	2.27	2.22	2.18	2.15	2.12	2.09	2.07	2.05	2.03	2.02	2.00	1.99
	.01	7.72	5.53	4.64	4.14	3.82	3.59	3.42	3.29	3.18	3.09	3.02	2.96	2.90	2.86	2.81	2.78	2.75	2.72	2.69	2.66

27	.10	2.90	2.51	2.30	2.17	2.07	2.00	1.95	1.91	1.87	1.85	1.82	1.80	1.78	1.76	1.75	1.74	1.72	1.71	1.70	1.70
	.05	4.21	3.35	2.96	2.73	2.57	2.46	2.37	2.31	2.25	2.20	2.17	2.13	2.10	2.08	2.06	2.04	2.02	2.00	1.99	1.97
	.01	7.68	5.49	4.60	4.11	3.78	3.56	3.39	3.26	3.15	3.06	2.99	2.93	2.87	2.82	2.78	2.75	2.71	2.68	2.66	2.63
28	.10	2.89	2.50	2.29	2.16	2.06	2.00	1.94	1.90	1.87	1.84	1.81	1.79	1.77	1.75	1.74	1.73	1.71	1.70	1.69	1.69
	.05	4.20	3.34	2.95	2.71	2.56	2.45	2.36	2.29	2.24	2.19	2.15	2.12	2.09	2.06	2.04	2.02	2.00	1.99	1.97	1.96
	.01	7.64	5.45	4.57	4.07	3.75	3.53	3.36	3.23	3.12	3.03	2.96	2.90	2.84	2.79	2.75	2.72	2.68	2.65	2.63	2.60
29	.10	2.89	2.50	2.28	2.15	2.06	1.99	1.93	1.89	1.86	1.83	1.80	1.78	1.76	1.75	1.73	1.72	1.71	1.69	1.68	1.68
	.05	4.18	3.33	2.93	2.70	2.55	2.43	2.35	2.28	2.22	2.18	2.14	2.10	2.08	2.05	2.03	2.01	1.99	1.97	1.96	1.94
	.01	7.60	5.42	4.54	4.04	3.73	3.50	3.33	3.20	3.09	3.00	2.93	2.87	2.81	2.77	2.73	2.69	2.66	2.63	2.60	2.57
30	.10	2.88	2.49	2.28	2.14	2.05	1.98	1.93	1.88	1.85	1.82	1.79	1.77	1.75	1.74	1.72	1.71	1.70	1.69	1.68	1.67
	.05	4.17	3.32	2.92	2.69	2.53	2.42	2.33	2.27	2.21	2.16	2.13	2.09	2.06	2.04	2.01	1.99	1.98	1.96	1.95	1.93
	.01	7.56	5.39	4.51	4.02	3.70	3.47	3.30	3.17	3.07	2.98	2.91	2.84	2.79	2.74	2.70	2.66	2.63	2.60	2.57	2.55
32	.10	2.87	2.48	2.26	2.13	2.04	1.97	1.91	1.87	1.83	1.81	1.78	1.76	1.74	1.72	1.71	1.69	1.68	1.67	1.66	1.65
	.05	4.15	3.29	2.90	2.67	2.51	2.40	2.31	2.24	2.19	2.14	2.10	2.07	2.04	2.01	1.99	1.97	1.95	1.94	1.92	1.91
	.01	7.50	5.34	4.46	3.97	3.65	3.43	3.26	3.13	3.02	2.93	2.86	2.80	2.74	2.70	2.65	2.62	2.58	2.55	2.53	2.50
34	.10	2.86	2.47	2.25	2.12	2.02	1.96	1.90	1.86	1.82	1.79	1.77	1.75	1.73	1.71	1.69	1.68	1.67	1.66	1.65	1.64
	.05	4.13	3.28	2.88	2.65	2.49	2.38	2.29	2.23	2.17	2.12	2.08	2.05	2.02	1.99	1.97	1.95	1.93	1.92	1.90	1.89
	.01	7.44	5.29	4.42	3.93	3.61	3.39	3.22	3.09	2.98	2.89	2.82	2.76	2.70	2.66	2.61	2.58	2.54	2.51	2.49	2.46
36	.10	2.85	2.46	2.24	2.11	2.01	1.94	1.89	1.85	1.81	1.78	1.76	1.73	1.71	1.70	1.68	1.67	1.66	1.65	1.64	1.63
	.05	4.11	3.26	2.87	2.63	2.48	2.36	2.28	2.21	2.15	2.11	2.07	2.03	2.00	1.98	1.95	1.93	1.92	1.90	1.88	1.87
	.01	7.40	5.25	4.38	3.89	3.57	3.35	3.18	3.05	2.95	2.86	2.79	2.72	2.67	2.62	2.58	2.54	2.51	2.48	2.45	2.43
38	.10	2.84	2.45	2.23	2.10	2.01	1.94	1.88	1.84	1.80	1.77	1.75	1.72	1.70	1.69	1.67	1.66	1.65	1.63	1.62	1.61
	.05	4.10	3.24	2.85	2.62	2.46	2.35	2.26	2.19	2.14	2.09	2.05	2.02	1.99	1.96	1.94	1.92	1.90	1.88	1.87	1.85
	.01	7.35	5.21	4.34	3.86	3.54	3.32	3.15	3.02	2.92	2.83	2.75	2.69	2.64	2.59	2.55	2.51	2.48	2.45	2.42	2.40

TABLE A.3.
Continued.

DF For Denom	P	Degrees of Freedom for Numerator (Greater Mean Square) 1	2	3	4	5	6	7	8	9	10	11	12	13	14	15	16	17	18	19	20
40	.10	2.84	2.44	2.23	2.09	2.00	1.93	1.87	1.83	1.79	1.76	1.74	1.71	1.70	1.68	1.66	1.65	1.64	1.62	1.61	1.61
	.05	4.08	3.23	2.84	2.61	2.45	2.34	2.25	2.18	2.12	2.08	2.04	2.00	1.97	1.95	1.92	1.90	1.89	1.87	1.85	1.84
	.01	7.31	5.18	4.31	3.83	3.51	3.29	3.12	2.99	2.89	2.80	2.73	2.66	2.61	2.56	2.52	2.48	2.45	2.42	2.39	2.37
42	.10	2.83	2.43	2.22	2.08	1.99	1.92	1.86	1.82	1.78	1.75	1.73	1.71	1.69	1.67	1.65	1.64	1.63	1.62	1.61	1.60
	.05	4.07	3.22	2.83	2.59	2.44	2.32	2.24	2.17	2.11	2.06	2.03	1.99	1.96	1.94	1.91	1.89	1.87	1.86	1.84	1.83
	.01	7.28	5.15	4.29	3.80	3.49	3.27	3.10	2.97	2.86	2.78	2.70	2.64	2.59	2.54	2.50	2.46	2.43	2.40	2.37	2.34
44	.10	2.82	2.43	2.21	2.08	1.98	1.91	1.86	1.81	1.78	1.75	1.72	1.70	1.68	1.66	1.65	1.63	1.62	1.61	1.60	1.59
	.05	4.06	3.21	2.82	2.58	2.43	2.31	2.23	2.16	2.10	2.05	2.01	1.98	1.95	1.92	1.90	1.88	1.86	1.84	1.83	1.81
	.01	7.25	5.12	4.26	3.78	3.47	3.24	3.08	2.95	2.84	2.75	2.68	2.62	2.56	2.52	2.47	2.44	2.40	2.37	2.35	2.32
46	.10	2.82	2.42	2.21	2.07	1.98	1.91	1.85	1.81	1.77	1.74	1.71	1.69	1.67	1.65	1.64	1.63	1.61	1.60	1.59	1.58
	.05	4.05	3.20	2.81	2.57	2.42	2.30	2.22	2.15	2.09	2.04	2.00	1.97	1.94	1.91	1.89	1.87	1.85	1.83	1.82	1.80
	.01	7.22	5.10	4.24	3.76	3.44	3.22	3.06	2.93	2.82	2.73	2.66	2.60	2.54	2.50	2.45	2.42	2.38	2.35	2.33	2.30
48	.10	2.81	2.42	2.20	2.07	1.97	1.90	1.85	1.80	1.77	1.73	1.71	1.69	1.67	1.65	1.63	1.62	1.61	1.59	1.58	1.57
	.05	4.04	3.19	2.80	2.57	2.41	2.29	2.21	2.14	2.08	2.03	1.99	1.96	1.93	1.90	1.88	1.86	1.84	1.82	1.81	1.79
	.01	7.19	5.08	4.22	3.74	3.43	3.20	3.04	2.91	2.80	2.71	2.64	2.58	2.53	2.48	2.44	2.40	2.37	2.33	2.31	2.28
50	.10	2.81	2.41	2.20	2.06	1.97	1.90	1.84	1.80	1.76	1.73	1.70	1.68	1.66	1.64	1.63	1.61	1.60	1.59	1.58	1.57
	.05	4.03	3.18	2.79	2.56	2.40	2.29	2.20	2.13	2.07	2.03	1.99	1.95	1.92	1.89	1.87	1.85	1.83	1.81	1.80	1.78
	.01	7.17	5.06	4.20	3.72	3.41	3.19	3.02	2.89	2.78	2.70	2.63	2.56	2.51	2.46	2.42	2.38	2.35	2.32	2.29	2.27

55	.10	2.80	2.40	2.19	2.05	1.95	1.88	1.83	1.78	1.75	1.72	1.69	1.67	1.65	1.63	1.61	1.60	1.59	1.58	1.56	1.55
	.05	4.02	3.16	2.77	2.54	2.38	2.27	2.18	2.11	2.06	2.01	1.97	1.93	1.90	1.88	1.85	1.83	1.81	1.79	1.78	1.76
	.01	7.12	5.01	4.16	3.68	3.37	3.15	2.98	2.85	2.75	2.66	2.59	2.53	2.47	2.42	2.38	2.34	2.31	2.28	2.25	2.23
60	.10	2.79	2.39	2.18	2.04	1.95	1.87	1.82	1.77	1.74	1.71	1.68	1.66	1.64	1.62	1.60	1.59	1.58	1.56	1.55	1.54
	.05	4.00	3.15	2.76	2.53	2.37	2.25	2.17	2.10	2.04	1.99	1.95	1.92	1.89	1.86	1.84	1.82	1.80	1.78	1.76	1.75
	.01	7.08	4.98	4.13	3.65	3.34	3.12	2.95	2.82	2.72	2.63	2.56	2.50	2.44	2.39	2.35	2.31	2.28	2.25	2.22	2.20
65	.10	2.78	2.39	2.17	2.03	1.94	1.87	1.81	1.77	1.73	1.70	1.67	1.65	1.63	1.61	1.59	1.58	1.57	1.55	1.54	1.53
	.05	3.99	3.14	2.75	2.51	2.36	2.24	2.15	2.08	2.03	1.98	1.94	1.90	1.87	1.85	1.82	1.80	1.78	1.76	1.75	1.73
	.01	7.04	4.95	4.10	3.62	3.31	3.09	2.93	2.80	2.69	2.61	2.53	2.47	2.42	2.37	2.33	2.29	2.26	2.23	2.20	2.17
70	.10	2.78	2.38	2.16	2.03	1.93	1.86	1.80	1.76	1.72	1.69	1.66	1.64	1.62	1.60	1.59	1.57	1.56	1.55	1.54	1.53
	.05	3.98	3.13	2.74	2.50	2.35	2.23	2.14	2.07	2.02	1.97	1.93	1.89	1.86	1.84	1.81	1.79	1.77	1.75	1.74	1.72
	.01	7.01	4.92	4.07	3.60	3.29	3.07	2.91	2.78	2.67	2.59	2.51	2.45	2.40	2.35	2.31	2.27	2.23	2.20	2.18	2.15
80	.10	2.77	2.37	2.15	2.02	1.92	1.85	1.79	1.75	1.71	1.68	1.65	1.63	1.61	1.59	1.57	1.56	1.55	1.53	1.52	1.51
	.05	3.96	3.11	2.72	2.49	2.33	2.21	2.13	2.06	2.00	1.95	1.91	1.88	1.84	1.82	1.79	1.77	1.75	1.73	1.72	1.70
	.01	6.96	4.88	4.04	3.56	3.26	3.04	2.87	2.74	2.64	2.55	2.48	2.42	2.36	2.31	2.27	2.23	2.20	2.17	2.14	2.12
100	.10	2.76	2.36	2.14	2.00	1.91	1.83	1.78	1.73	1.69	1.66	1.64	1.61	1.59	1.57	1.56	1.54	1.53	1.52	1.50	1.49
	.05	3.94	3.09	2.70	2.46	2.31	2.19	2.10	2.03	1.97	1.93	1.89	1.85	1.82	1.79	1.77	1.75	1.73	1.71	7.69	1.68
	.01	6.90	4.82	3.98	3.51	3.21	2.99	2.82	2.69	2.59	2.50	2.43	2.37	2.31	2.27	2.22	2.19	2.15	2.12	2.09	2.07
120	.10	2.75	2.35	2.13	1.99	1.90	1.82	1.77	1.72	1.68	1.65	1.63	1.60	1.58	1.56	1.55	1.53	1.52	1.50	1.49	1.48
	.05	3.92	3.07	2.68	2.45	2.29	2.18	2.09	2.02	1.96	1.91	1.87	1.83	1.80	1.78	1.75	1.73	1.71	1.69	1.67	1.66
	.01	6.85	4.79	3.95	3.48	3.17	2.96	2.79	2.66	2.56	2.47	2.40	2.34	2.28	2.23	2.19	2.15	2.12	2.09	2.06	2.03
150	.10	2.74	2.34	2.12	1.98	1.89	1.81	1.76	1.71	1.67	1.64	1.61	1.59	1.57	1.55	1.53	1.52	1.50	1.49	1.48	1.47
	.05	3.90	3.06	2.66	2.43	2.27	2.16	2.07	2.00	1.94	1.89	1.85	1.82	1.79	1.76	1.73	1.71	1.69	1.67	1.66	1.64
	.01	6.81	4.75	3.91	3.45	3.14	2.92	2.76	2.63	2.53	2.44	2.37	2.31	2.25	2.20	2.16	2.12	2.08	2.06	2.03	2.00

TABLE A.3.
Continued.

DF For Denom	P	Degrees of Freedom for Numerator (Greater Mean Square) 1	2	3	4	5	6	7	8	9	10	11	12	13	14	15	16	17	18	19	20
200	.10	2.73	2.33	2.11	1.97	1.88	1.80	1.75	1.70	1.66	1.63	1.60	1.58	1.56	1.54	1.52	1.51	1.49	1.48	1.47	1.46
	.05	3.89	3.01	2.65	2.42	2.26	2.14	2.06	1.98	1.93	1.88	1.84	1.80	1.77	1.74	1.72	1.69	1.67	1.66	1.64	1.62
	.01	6.76	4.71	3.88	3.41	3.11	2.89	2.73	2.60	2.50	2.41	2.34	2.27	2.22	2.17	2.13	2.09	2.06	2.03	2.00	1.97
400	.10	2.72	2.32	2.10	1.96	1.86	1.79	1.73	1.69	1.65	1.61	1.59	1.56	1.54	1.52	1.50	1.49	1.47	1.46	1.45	1.44
	.05	3.86	3.02	2.63	2.39	2.24	2.12	2.03	1.96	1.90	1.85	1.81	1.78	1.74	1.72	1.69	1.67	1.65	1.63	1.61	1.60
	.01	6.70	4.66	3.83	3.37	3.06	2.85	2.68	2.56	2.45	2.37	2.29	2.23	2.17	2.13	2.08	2.05	2.01	1.98	1.95	1.92
1000	.10	2.71	2.31	2.09	1.95	1.85	1.78	1.72	1.68	1.64	1.61	1.58	1.55	1.53	1.51	1.49	1.48	1.46	1.45	1.44	1.43
	.05	3.85	3.00	2.61	2.38	2.22	2.11	2.02	1.95	1.89	1.84	1.80	1.76	1.73	1.70	1.68	1.65	1.63	1.61	1.60	1.58
	.01	6.66	4.63	3.80	3.34	3.04	2.82	2.66	2.53	2.43	2.34	2.27	2.20	2.15	2.10	2.06	2.02	1.98	1.95	1.92	1.90
∞	.10	2.71	2.30	2.08	1.94	1.85	1.77	1.72	1.67	1.63	1.60	1.57	1.55	1.52	1.50	1.49	1.47	1.45	1.44	1.43	1.42
	.05	3.84	3.00	2.60	2.37	2.21	2.10	2.01	1.94	1.88	1.83	1.79	1.75	1.72	1.69	1.67	1.64	1.62	1.60	1.58	1.57
	.01	6.63	4.61	3.78	3.32	3.02	2.80	2.64	2.51	2.41	2.32	2.25	2.18	2.13	2.08	2.04	1.99	1.96	1.93	1.90	1.88

TABLE A.4.
Significant studentized factors (R) to multiply by LSD for testing means at various ranges (p), 5% level; n = degrees of freedom for "error."

n	p:															
	2	3	4	5	6	7	8	9	10	12	14	16	18	20	50	100
4	1.00	1.02	1.02	1.02	1.02	1.02	1.02	1.02	1.02	1.02	1.02	1.02	1.02	1.02	1.02	1.02
5	1.00	1.03	1.04	1.05	1.05	1.05	1.05	1.05	1.05	1.05	1.05	1.05	1.05	1.05	1.05	1.05
6	1.00	1.03	1.05	1.06	1.06	1.06	1.06	1.06	1.06	1.06	1.06	1.06	1.06	1.06	1.06	1.06
7	1.00	1.04	1.06	1.07	1.07	1.08	1.08	1.08	1.08	1.08	1.08	1.08	1.08	1.08	1.08	1.08
8	1.00	1.04	1.06	1.08	1.09	1.09	1.09	1.09	1.09	1.09	1.09	1.09	1.09	1.09	1.09	1.09
9	1.00	1.04	1.07	1.08	1.09	1.10	1.10	1.10	1.10	1.10	1.10	1.10	1.10	1.10	1.10	1.10
10	1.00	1.05	1.07	1.09	1.10	1.10	1.10	1.10	1.10	1.10	1.10	1.10	1.10	1.10	1.10	1.10
11	1.00	1.05	1.08	1.09	1.10	1.11	1.11	1.11	1.11	1.11	1.11	1.11	1.12	1.12	1.12	1.12
12	1.00	1.05	1.08	1.09	1.10	1.11	1.12	1.12	1.12	1.12	1.12	1.12	1.13	1.13	1.13	1.13
13	1.00	1.05	1.08	1.09	1.10	1.11	1.12	1.12	1.13	1.13	1.13	1.13	1.13	1.13	1.13	1.13
14	1.00	1.05	1.08	1.10	1.11	1.12	1.13	1.13	1.14	1.14	1.14	1.14	1.15	1.15	1.15	1.15
15	1.00	1.05	1.08	1.10	1.12	1.12	1.13	1.14	1.14	1.14	1.15	1.15	1.15	1.15	1.15	1.15
16	1.00	1.05	1.08	1.10	1.12	1.12	1.13	1.14	1.14	1.15	1.15	1.16	1.16	1.16	1.16	1.16
17	1.00	1.05	1.08	1.10	1.12	1.13	1.13	1.14	1.15	1.15	1.16	1.16	1.16	1.16	1.16	1.16
18	1.00	1.05	1.08	1.10	1.12	1.13	1.13	1.14	1.15	1.15	1.16	1.16	1.17	1.17	1.17	1.17
19	1.00	1.05	1.08	1.10	1.12	1.13	1.14	1.15	1.15	1.16	1.16	1.17	1.17	1.17	1.17	1.17
20	1.00	1.05	1.08	1.10	1.12	1.13	1.14	1.15	1.15	1.16	1.17	1.17	1.17	1.18	1.18	1.18
22	1.00	1.05	1.08	1.10	1.12	1.13	1.14	1.15	1.16	1.17	1.18	1.18	1.18	1.18	1.18	1.18
24	1.00	1.05	1.08	1.10	1.12	1.13	1.14	1.15	1.16	1.17	1.18	1.18	1.18	1.19	1.19	1.19
26	1.00	1.05	1.08	1.10	1.12	1.13	1.15	1.15	1.16	1.17	1.18	1.19	1.19	1.19	1.19	1.19
28	1.00	1.05	1.08	1.10	1.12	1.14	1.15	1.16	1.16	1.17	1.18	1.19	1.19	1.20	1.20	1.20
30	1.00	1.05	1.08	1.11	1.12	1.14	1.15	1.16	1.17	1.18	1.19	1.19	1.20	1.20	1.20	1.20
40	1.00	1.05	1.08	1.11	1.13	1.14	1.15	1.16	1.17	1.19	1.20	1.20	1.21	1.21	1.21	1.21
60	1.00	1.05	1.09	1.11	1.13	1.14	1.16	1.17	1.18	1.19	1.20	1.21	1.22	1.23	1.23	1.23
100	1.00	1.05	1.09	1.11	1.14	1.15	1.16	1.18	1.19	1.20	1.21	1.22	1.23	1.24	1.26	1.26
∞	1.00	1.05	1.09	1.12	1.14	1.15	1.17	1.18	1.19	1.21	1.22	1.23	1.24	1.25	1.30	1.32

TABLE A.5.

Significant studentized factors (E) to multiply by LSD for testing means at various ranges (p), 1% level; n = degrees of freedom for "error."

n	p: 2	3	4	5	6	7	8	9	10	12	14	16	18	20	50	100
3	1.00	1.03	1.04	1.05	1.07	1.08	1.08	1.09	1.09	1.09	1.10	1.11	1.13	1.13	1.13	1.13
4	1.00	1.04	1.06	1.08	1.09	1.09	1.11	1.11	1.12	1.12	1.14	1.14	1.15	1.15	1.15	1.15
5	1.00	1.05	1.07	1.08	1.10	1.11	1.12	1.13	1.14	1.16	1.16	1.18	1.18	1.19	1.19	1.19
6	1.00	1.05	1.08	1.09	1.11	1.12	1.14	1.15	1.15	1.16	1.18	1.18	1.20	1.20	1.20	1.20
7	1.00	1.05	1.08	1.10	1.12	1.13	1.15	1.16	1.17	1.17	1.19	1.19	1.21	1.21	1.21	1.21
8	1.00	1.05	1.08	1.10	1.12	1.14	1.15	1.16	1.17	1.18	1.20	1.20	1.22	1.22	1.22	1.23
9	1.00	1.06	1.08	1.10	1.12	1.14	1.16	1.17	1.17	1.20	1.20	1.22	1.24	1.24	1.24	1.24
10	1.00	1.06	1.09	1.11	1.13	1.15	1.16	1.17	1.18	1.20	1.21	1.22	1.24	1.24	1.24	1.24
11	1.00	1.05	1.09	1.11	1.13	1.14	1.15	1.17	1.17	1.19	1.20	1.22	1.23	1.23	1.23	1.23
12	1.00	1.05	1.08	1.10	1.12	1.14	1.15	1.16	1.17	1.19	1.20	1.21	1.21	1.22	1.22	1.22
13	1.00	1.05	1.08	1.10	1.12	1.14	1.15	1.16	1.17	1.18	1.19	1.20	1.21	1.21	1.21	1.21
14	1.00	1.05	1.08	1.10	1.12	1.14	1.15	1.16	1.17	1.18	1.19	1.20	1.20	1.20	1.20	1.20
15	1.00	1.05	1.08	1.10	1.11	1.13	1.14	1.15	1.16	1.18	1.18	1.19	1.20	1.20	1.20	1.20
16	1.00	1.05	1.08	1.10	1.11	1.13	1.14	1.15	1.16	1.17	1.18	1.19	1.19	1.20	1.20	1.20
17	1.00	1.05	1.08	1.10	1.11	1.13	1.14	1.15	1.16	1.17	1.18	1.19	1.19	1.19	1.19	1.19
18	1.00	1.05	1.08	1.10	1.11	1.13	1.14	1.15	1.16	1.17	1.18	1.18	1.19	1.19	1.19	1.19
19	1.00	1.05	1.07	1.09	1.11	1.13	1.14	1.15	1.16	1.17	1.18	1.18	1.19	1.19	1.19	1.19
20	1.00	1.05	1.07	1.09	1.11	1.13	1.14	1.15	1.15	1.17	1.18	1.18	1.19	1.19	1.19	1.19
22	1.00	1.05	1.07	1.09	1.11	1.12	1.14	1.15	1.15	1.17	1.17	1.18	1.19	1.19	1.19	1.19
24	1.00	1.05	1.07	1.09	1.11	1.12	1.13	1.15	1.15	1.17	1.17	1.18	1.19	1.19	1.20	1.20
26	1.00	1.05	1.07	1.09	1.11	1.12	1.13	1.15	1.15	1.17	1.18	1.18	1.19	1.19	1.20	1.20
28	1.00	1.04	1.07	1.09	1.11	1.12	1.13	1.14	1.15	1.17	1.18	1.18	1.19	1.19	1.21	1.21
30	1.00	1.04	1.07	1.09	1.11	1.12	1.13	1.14	1.15	1.17	1.18	1.19	1.19	1.20	1.21	1.21
40	1.00	1.04	1.07	1.09	1.11	1.12	1.14	1.14	1.15	1.17	1.18	1.19	1.20	1.20	1.23	1.23
60	1.00	1.04	1.07	1.09	1.11	1.12	1.14	1.15	1.15	1.17	1.18	1.19	1.20	1.20	1.24	1.24
100	1.00	1.04	1.07	1.09	1.11	1.12	1.14	1.15	1.15	1.17	1.18	1.19	1.20	1.21	1.25	1.25
∞	1.00	1.04	1.07	1.09	1.11	1.12	1.14	1.15	1.15	1.17	1.18	1.19	1.20	1.21	1.26	1.29

TABLE A.6.
Distribution of χ^2 (Chi-Square)[a]

Degrees of Freedom	Probability of Obtaining a Value as Large or Larger							
	.99	.95	.90	.50	.10	.05	.01	.001
1	.0002	.00393	.0158	.455	2.706	3.841	6.635	10.827
2	.0201	.103	.211	1.386	4.605	5.991	9.210	13.815
3	.115	.352	.584	2.366	6.251	7.815	11.345	16.268
4	.297	.711	1.064	3.357	7.779	9.488	13.277	18.465
5	.554	1.145	1.610	4.351	9.236	11.070	15.086	20.517
6	.872	1.635	2.204	5.348	10.645	12.592	16.812	22.457
7	1.239	2.167	2.833	6.346	12.017	14.067	18.475	24.322
8	1.646	2.733	3.490	7.344	13.362	15.507	20.090	26.125
9	2.088	3.325	4.168	8.343	14.684	16.919	21.666	27.877
10	2.558	3.940	4.865	9.342	15.987	18.307	23.209	29.588
11	3.053	4.575	5.578	10.341	17.275	19.675	24.725	31.264
12	3.571	5.226	6.304	11.340	18.549	21.026	26.217	32.909
13	4.107	5.892	7.042	12.340	19.812	22.362	27.688	34.528
14	4.660	6.571	7.790	13.339	21.064	23.685	29.141	36.123
15	5.229	7.261	8.547	14.339	22.307	24.996	30.578	37.697
16	5.812	7.962	9.312	15.338	23.542	26.296	32.000	29.252
17	6.408	8.672	10.085	16.338	24.769	27.587	33.409	40.790
18	7.015	9.390	10.865	17.338	25.989	28.869	34.805	42.312
19	7.633	10.117	11.651	18.338	27.204	30.144	36.191	43.820
20	8.260	10.851	12.443	19.337	28.412	31.410	37.566	45.315
21	8.897	11.591	13.240	20.337	29.615	32.671	38.932	46.797
22	9.542	12.338	14.041	21.337	30.813	33.924	40.289	48.268
23	10.196	13.091	14.848	22.337	32.007	35.172	41.638	49.728
24	10.856	13.848	15.659	23.337	33.196	36.415	42.980	51.179
25	11.524	14.611	16.473	24.337	34.382	37.652	44.314	52.620
26	12.198	15.379	17.292	25.336	35.563	38.885	45.642	54.052
27	12.879	16.151	18.114	26.336	36.741	40.113	46.963	55.476
28	13.565	16.928	18.939	27.336	37.916	41.337	48.278	56.893
29	14.256	17.708	19.768	28.336	39.087	42.557	49.588	58.302
30	14.953	18.493	20.599	29.336	40.256	43.773	50.892	59.703

[a]Table A.6 is abridged from Table IV of Fisher and Yates: *Statistical Tables for Biological, Agricultural and Medical Research*, published by Longman Group Ltd., London (previously published by Oliver and Boyd, Edinburgh), by permission of the authors and publishers.

TABLE A.7.
Values of the correlation coefficient, r, for certain levels of significance.[a]

Degrees of Freedom	Probability of Obtaining a Value as Large or Larger .1	.05	.01	.001
1	.9879	.9969	.9999	1.0000
2	.9000	.9500	.9900	.9990
3	.8054	.8783	.9587	.9912
4	.7293	.8114	.9172	.9741
5	.6694	.7545	.8745	.9507
6	.6215	.7067	.8343	.9249
7	.5822	.6664	.7977	.8982
8	.5494	.6319	.7646	.8721
9	.5214	.6021	.7348	.8471
10	.4973	.5760	.7079	.8233
11	.4762	.5529	.6835	.8010
12	.4575	.5324	.6614	.7800
13	.4409	.5139	.6411	.7603
14	.4259	.4973	.6226	.7420
15	.4124	.4821	.6055	.7246
16	.4000	.4683	.5897	.7084
17	.3887	.4555	.5751	.6932
18	.3783	.4438	.5614	.6787
19	.3687	.4329	.5487	.6652
20	.3598	.4227	.5368	.6524
25	.3233	.3809	.4869	.5974
30	.2960	.3494	.4487	.5541
35	.2746	.3246	.4182	.5189
40	.2573	.3044	.3032	.4806
45	.2428	.2875	.3721	.4648
50	.2306	.2732	.3541	.4433
60	.2108	.2500	.3248	.4078
70	.1954	.2319	.3017	.3799
80	.1829	.2172	.2830	.3568
90	.1726	.2050	.2673	.3375
100	.1638	.1946	.2540	.3211

[a]Table A.7 is abridged from Table VI of Fisher and Yates: *Statistical Tables for Biological, Agricultural, and Medical Research*, published by Longman Group Ltd., London (previously published by Oliver and Boyd, Edinburgh), by permission of the authors and publishers.

TABLE A.8.
The angular transformation of percentages to degrees[a]

%	0	1	2	3	4	5	6	7	8	9
0	0	5.7	8.1	10.0	11.5	12.9	14.2	15.3	16.4	17.5
10	18.4	19.4	20.3	21.1	22.0	22.8	23.6	24.4	25.1	25.8
20	26.6	27.3	28.0	28.7	29.3	30.0	30.7	31.3	31.9	32.6
30	33.2	33.8	34.4	35.1	35.7	36.3	36.9	37.5	38.1	38.6
40	39.2	39.8	40.4	41.0	41.6	42.1	42.7	43.3	43.9	44.4
50	45.0	45.6	46.1	46.7	47.3	47.9	48.4	49.0	49.6	50.2
60	50.8	51.4	51.9	52.5	53.1	53.7	54.3	54.9	55.6	56.2
70	56.8	57.4	58.1	58.7	59.3	60.0	60.7	61.3	62.0	62.7
80	63.4	64.2	64.9	65.6	66.4	67.2	68.0	68.9	69.7	70.6
90	71.6	72.5	73.6	74.7	75.8	77.1	78.5	80.0	81.9	84.3
100	90.0	—	—	—	—	—	—	—	—	—

[a]Table A.8 is abridged from Table X of Fisher and Yates: *Statistical Tables for Biological, Agricultural, and Medical Research*, published by Longman Group Ltd., London (previously published by Oliver and Boyd, Edinburgh), by permission of the authors and publishers.

TABLE A.9.
Logarithms.

Natural Numbers	0	1	2	3	4	5	6	7	8	9	Proportional Parts 1	2	3	4	5	6	7	8	9
10	0000	0043	0086	0128	0170	0212	0253	0294	0334	0374	4	8	12	17	21	25	29	33	37
11	0414	0453	0492	0531	0569	0607	0645	0682	0719	0755	4	8	11	15	19	23	26	30	34
12	0792	0828	0864	0899	0934	0969	1004	1038	1072	1106	3	7	10	14	17	21	24	28	31
13	1139	1173	1206	1239	1271	1303	1335	1367	1399	1430	3	6	10	13	16	19	23	26	29
14	1461	1492	1523	1553	1584	1614	1644	1673	1703	1732	3	6	9	12	15	18	21	24	27
15	1761	1790	1818	1847	1875	1903	1931	1959	1987	2014	3	6	8	11	14	17	20	22	25
16	2041	2068	2095	2122	2148	2175	2201	2227	2253	2279	3	5	8	11	13	16	18	21	24
17	2304	2330	2355	2380	2405	2430	2455	2480	2504	2529	2	5	7	10	12	15	17	20	22
18	2553	2577	2601	2625	2648	2672	2695	2718	2742	2765	2	5	7	9	12	14	16	19	21
19	2788	2810	2833	2856	2878	2900	2923	2945	2967	2989	2	4	7	9	11	13	16	18	20
20	3010	3032	3054	3075	3096	3118	3139	3160	3181	3201	2	4	6	8	11	13	15	17	19
21	3222	3243	3263	3284	3304	3324	3345	3365	3385	3404	2	4	6	8	10	12	14	16	18
22	3424	3444	3464	3483	3502	3522	3541	3560	3579	3598	2	4	6	8	10	12	14	15	17
23	3617	3636	3655	3674	3692	3711	3729	3747	3766	3784	2	4	6	7	9	11	13	15	17
24	3802	3820	3838	3856	3874	3892	3909	3927	3945	3962	2	4	5	7	9	11	12	14	16
25	3979	3997	4014	4031	4048	4065	4082	4099	4116	4133	2	3	5	7	9	10	12	14	15
26	4150	4166	4183	4200	4216	4232	4249	4265	4281	4298	2	3	5	7	8	10	11	13	15
27	4314	4330	4346	4362	4378	4393	4409	4425	4440	4456	2	3	5	6	8	9	11	13	14
28	4472	4487	4502	4518	4533	4548	4564	4579	4594	4609	2	3	5	6	8	9	11	12	14
29	4624	4639	4654	4669	4683	4698	4713	4728	4742	4757	1	3	4	6	7	9	10	12	13

30	4771	4786	4800	4814	4829	4843	4857	4871	4886	4900	1	3	4	6	7	9	10	11	13
31	4914	4928	4924	4955	4969	4983	4997	5011	5024	5038	1	3	4	6	7	8	10	11	12
32	5051	5065	5079	5092	5105	5119	5132	5145	5159	5172	1	3	4	5	7	8	9	11	12
33	5185	5198	5211	5224	5237	5250	5263	5276	5289	5302	1	3	4	5	6	8	9	10	12
34	5315	5328	5340	5353	5366	5378	5391	5403	5416	5428	1	3	4	5	6	8	9	10	11
35	5441	5453	5465	5478	5490	5502	5514	5527	5539	5551	1	2	4	5	6	7	9	10	11
36	5563	5575	5587	5599	5611	5623	5635	5647	5658	5670	1	2	4	5	6	7	8	10	11
37	5682	5694	5705	5717	5729	5740	5752	5763	5775	5786	1	2	3	5	6	7	8	9	10
38	5798	5809	5821	5832	5843	5855	5866	5877	5888	5899	1	2	3	5	6	7	8	9	10
39	5911	5922	5933	5944	5955	5966	5977	5988	5999	6010	1	2	3	4	5	7	8	9	10
40	6021	6031	6042	6053	6064	6075	6085	6096	6107	6117	1	2	3	4	5	6	8	9	10
41	6128	6138	6149	6160	6170	6180	6191	6201	6212	6222	1	2	3	4	5	6	7	8	9
42	6232	6243	6253	6263	6274	6284	6294	6304	6314	6325	1	2	3	4	5	6	7	8	9
43	6335	6345	6355	6365	6375	6385	6395	6405	6415	6425	1	2	3	4	5	6	7	8	9
44	6435	6444	6454	6464	6474	6484	6493	6503	6513	6522	1	2	3	4	5	6	7	8	9
45	6532	6542	6551	6561	6571	6580	6590	6599	6609	6618	1	2	3	4	5	6	7	8	9
46	6628	6637	6646	6656	6665	6675	6684	6693	6702	6712	1	2	3	4	8	6	7	7	8
47	6721	6730	6739	6749	6758	6767	6776	6785	6794	6803	1	2	3	4	5	5	6	7	8
48	6812	6821	6830	6839	6848	6857	6866	6875	6884	6893	1	2	3	4	4	5	6	7	8
49	6902	6911	6920	6928	6937	6946	6955	6964	6972	6981	1	2	3	4	4	5	6	7	8
50	6990	6998	7007	7016	7024	7033	7042	7050	7059	7067	1	2	3	3	4	5	6	7	8
51	7076	7084	7093	7101	7110	7118	7126	7135	7143	7152	1	2	3	3	4	5	6	7	8
52	7160	7168	7177	7185	7193	7202	7210	7218	7226	7235	1	2	2	3	4	5	6	7	7
53	7243	7251	7259	7267	7275	7284	7292	7300	7308	7316	1	2	2	3	4	5	6	6	7
54	7324	7332	7340	7348	7356	7364	7372	7380	7388	7396	1	2	2	3	4	5	6	6	7

TABLE A.9.
Continued.

Natural Numbers	0	1	2	3	4	5	6	7	8	9	Proportional Parts								
											1	2	2	4	5	6	7	8	9
55	7404	7412	7419	7427	7435	7443	7451	7459	7466	7474	1	2	2	3	4	5	5	6	7
56	7482	7490	7497	7505	7513	7520	7528	7536	7543	7551	1	2	2	3	4	5	5	6	7
57	7559	7566	7574	7582	7589	7597	7604	7612	7619	7627	1	2	2	3	4	5	5	6	7
58	7634	7642	7649	7657	7664	7672	7979	7686	7694	7701	1	1	2	3	4	4	5	6	7
59	7709	7716	7723	7731	7738	7745	7752	7760	7767	7774	1	1	2	3	4	4	5	6	7
60	7782	7789	7796	7803	7810	7818	7825	7832	7839	7846	1	1	2	3	4	4	5	6	6
61	7853	7860	7868	7875	7882	7889	7896	7903	7910	7917	1	1	2	3	4	4	5	6	6
63	7924	7931	7938	7945	7952	7959	7966	7973	7980	7987	1	1	2	3	3	4	5	6	6
63	7993	8000	8007	8014	8021	8028	8035	8041	8048	8055	1	1	2	3	3	4	5	5	6
64	8062	8069	8075	8082	8089	8096	8102	8109	8116	8122	1	1	2	3	3	4	5	5	6
65	8129	8136	8142	8149	8156	8162	8169	8176	8182	8189	1	1	2	3	3	4	5	5	6
66	8195	8202	8209	8215	8222	8228	8235	8241	8248	8254	1	1	2	3	3	4	5	5	6
67	8261	8267	8274	8280	8287	8293	8299	8306	8312	8319	1	1	2	3	3	4	5	5	6
68	8325	8331	8338	8344	8351	8357	8363	8370	8376	8382	1	1	2	3	3	4	4	5	6
69	8388	8395	8401	8407	8414	8420	8426	8432	8439	8445	1	1	2	2	3	4	4	5	6
70	8451	8457	8463	8470	8476	8482	8488	8494	8500	8506	1	1	2	2	3	4	4	5	6
71	8513	8519	8525	8531	8537	8543	8549	8555	8561	8567	1	1	2	2	3	4	4	5	5
72	8573	8579	8585	8591	8597	8603	8609	8615	8621	8627	1	1	2	2	3	4	4	5	5
73	8633	8639	8645	8651	8657	8663	8669	8675	8681	8686	1	1	2	2	3	4	4	5	5
74	8692	8698	8704	8710	8716	8722	8727	8733	8739	8745	1	1	2	2	3	4	4	5	5

75	8751	8756	8762	8768	8774	8779	8785	8791	8797	8802	1	1	2	2	3	3	4	5	5
76	8808	8814	8820	8825	8831	8837	8842	8848	8854	8859	1	1	2	2	3	3	4	5	5
77	8865	8871	8876	8882	8887	8893	8899	8904	8910	8915	1	1	2	2	3	3	4	4	5
78	8921	8927	8932	8938	8943	8949	8954	8960	8965	8971	1	1	2	2	3	3	4	4	5
79	8976	8982	8987	8993	8998	9004	9009	9015	9020	9025	1	1	2	2	3	3	4	4	5
80	9031	9036	9042	9047	9053	9058	9063	9069	9074	9079	1	1	2	2	3	3	4	4	5
81	9085	9090	9096	9101	9106	9112	9117	9122	9128	9133	1	1	2	2	3	3	4	4	5
82	9138	9143	9149	9154	9159	9165	9170	9175	9180	9186	1	1	2	2	3	3	4	4	5
83	9191	9196	9201	9206	9212	9217	9222	9227	9232	9238	1	1	2	2	3	3	4	4	5
84	9243	9248	9253	9258	9263	9269	9274	9279	9284	9289	1	1	2	2	3	3	4	4	5
85	9294	9299	9304	9309	9315	9320	9325	9330	9335	9340	1	1	2	2	3	3	4	4	5
86	9345	9350	9355	9360	9365	9370	9375	9380	9385	9390	1	1	2	2	3	3	4	4	5
87	9395	9400	9405	9410	9415	9420	9425	9430	9435	9440	0	1	1	2	2	3	3	4	4
88	9445	9450	9455	9460	9465	9469	9474	9479	9484	9489	0	1	1	2	2	3	3	4	4
89	9494	9499	9504	9509	9513	9518	9523	9528	9533	9538	0	1	1	2	2	3	3	4	4
90	9542	9547	9552	9557	9562	9566	9571	9576	9581	9586	0	1	1	2	2	3	3	4	4
91	9590	9595	9600	9605	9609	9614	9619	9624	9628	9633	0	1	1	2	2	3	3	4	4
92	9638	9643	9647	9652	9657	9661	9666	9671	9675	9680	0	1	1	2	2	3	3	4	4
93	9685	9689	9694	9699	9703	9708	9713	9717	9722	9727	0	1	1	2	2	3	3	4	4
94	9731	9736	9741	9745	9750	9754	9759	9763	9768	9773	0	1	1	2	2	3	3	4	4
95	9777	9782	9786	9791	9795	9800	9805	9809	9814	9818	0	1	1	2	2	3	3	4	4
96	9823	9827	9832	9836	9841	9845	9850	9854	9859	9863	0	1	1	2	2	3	3	4	4
97	9868	9872	9877	9881	9886	9890	9894	9899	9903	9908	0	1	1	2	2	3	3	4	4
98	9912	9917	9921	9926	9930	9934	9939	9943	9948	9952	0	1	1	2	2	3	3	4	4
99	9956	9961	9965	9969	9974	9978	9983	9987	9991	9996	0	1	1	2	2	3	3	3	4

TABLE A.10.
Squares and square roots.

N	N^2	$\sqrt{N}$	$\sqrt{10N}$	N	N^2	$\sqrt{N}$	$\sqrt{10N}$
1.00	1.0000	1.00000	3.16228	1.30	1.6900	1.14018	3.60555
1.01	1.0201	1.00499	3.17805	1.31	1.7161	1.14455	3.61939
1.02	1.0404	1.00995	3.19374	1.32	1.7424	1.14891	3.63318
1.03	1.0609	1.01489	3.20936	1.33	1.7689	1.15326	3.64692
1.04	1.0816	1.01980	3.22490	1.34	1.7956	1.15758	3.66060
1.05	1.1025	1.02470	3.24037	1.35	1.8225	1.16190	3.67423
1.06	1.1236	1.02956	3.25576	1.36	1.8496	1.16619	3.68782
1.07	1.1449	1.03441	3.27109	1.37	1.8769	1.17047	3.70135
1.08	1.1664	1.03923	3.28634	1.38	1.9044	1.17473	3.71484
1.09	1.1881	1.04403	3.30151	1.39	1.9321	1.17898	3.72827
1.10	1.2100	1.04881	3.31662	1.40	1.9600	1.18322	3.74166
1.11	1.2321	1.05357	3.33167	1.41	1.9881	1.18743	3.75500
1.12	1.2544	1.05830	3.34664	1.42	2.0164	1.19164	3.76829
1.13	1.2769	1.06301	3.36155	1.43	2.0449	1.19583	3.78153
1.14	1.2996	1.06771	3.37639	1.44	2.0736	1.20000	3.79473
1.15	1.3225	1.07238	3.39116	1.45	2.1025	1.20416	3.80789
1.16	1.3456	1.07703	3.40588	1.46	2.1316	1.20830	3.82099
1.17	1.3689	1.08167	3.42053	1.47	2.1609	1.21244	3.83406
1.18	1.3924	1.08628	3.43511	1.48	2.1904	1.21655	3.84708
1.19	1.4161	1.09087	3.44964	1.49	2.2201	1.22066	3.86005
1.20	1.4400	1.09545	3.46410	1.50	2.2500	1.22474	3.87298
1.21	1.4641	1.10000	3.47851	1.51	2.2801	1.22882	3.88587
1.22	1.4884	1.10454	3.49285	1.52	2.3104	1.23288	3.89872
1.23	1.5129	1.10905	3.50714	1.53	2.3409	1.23693	3.91152
1.24	1.5376	1.11355	3.52136	1.54	2.3716	1.24097	3.92428
1.25	1.5625	1.11803	3.53553	1.55	2.4025	1.24499	3.93700
1.26	1.5876	1.12250	3.54965	1.56	2.4336	1.24900	3.94968
1.27	1.6129	1.12694	3.56371	1.57	2.4649	1.25300	3.96232
1.28	1.6384	1.13137	3.57771	1.58	2.4964	1.25698	3.97492
1.29	1.6641	1.13578	3.59166	1.59	2.5281	1.26095	3.98748

TABLE A.10.
Continued.

N	N^2	$\sqrt{N}$	$\sqrt{10N}$	N	N^2	$\sqrt{N}$	$\sqrt{10N}$
1.60	2.5600	1.26491	4.00000	1.90	3.6100	1.37840	4.35890
1.61	2.5921	1.26886	4.01248	1.91	3.6481	1.38203	4.37035
1.62	2.6244	1.27279	4.02492	1.92	3.6864	1.38564	4.38178
1.63	2.6569	1.27671	4.03733	1.93	3.7249	1.38924	4.39318
1.64	2.6896	1.28062	4.04969	1.94	3.7636	1.39284	4.40454
1.65	2.7225	1.28452	4.06202	1.95	3.8025	1.39642	4.41588
1.66	2.7556	1.28841	4.07431	1.96	3.8416	1.40000	4.42719
1.67	2.7889	1.29228	4.08656	1.97	3.8809	1.40357	4.43847
1.68	2.8224	1.29615	4.09878	1.98	3.9204	1.40712	4.44972
1.69	2.8561	1.30000	4.11096	1.99	3.9601	1.41067	4.46094
1.70	2.8900	1.30384	4.12311	2.00	4.0000	1.41421	4.47214
1.71	2.9241	1.30767	4.13521	2.01	4.0401	1.41774	4.48330
1.72	2.9584	1.31149	4.14729	2.02	4.0804	1.42127	4.49444
1.73	2.9929	1.31529	4.15933	2.03	4.1209	1.42478	4.50555
1.74	3.0276	1.31909	4.17133	2.04	4.1616	1.42829	4.51664
1.75	3.0625	1.32288	4.18330	2.05	4.2025	1.43178	4.52769
1.76	3.0976	1.32665	4.19524	2.06	4.2436	1.43527	4.53872
1.77	3.1329	1.33041	4.20714	2.07	4.2849	1.43875	4.54973
1.78	3.1684	1.33417	4.21900	2.08	4.3264	1.44222	4.56070
1.79	3.2041	1.33791	4.23084	2.09	4.3681	1.44568	4.57165
1.80	3.2400	1.34164	4.24264	2.10	4.4100	1.44914	4.58258
1.81	3.2761	1.34536	4.25441	2.11	4.4521	1.45258	4.59347
1.82	3.3124	1.34907	4.26615	2.12	4.4944	1.45602	4.60435
1.83	3.3489	1.35277	4.27785	2.13	4.5369	1.45945	4.61519
1.84	3.3856	1.35647	4.28952	2.14	4.5796	1.46287	4.62601
1.85	3.4225	1.36015	4.30116	2.15	4.6225	1.46629	4.63681
1.86	3.4596	1.36382	4.31277	2.16	4.6656	1.46969	4.64758
1.87	3.4969	1.36748	4.32435	2.17	4.7089	1.47309	4.65833
1.88	3.5344	1.37113	4.33590	2.18	4.7524	1.47648	4.66905
1.89	3.5721	1.37477	4.34741	2.19	4.7961	1.47986	4.67974

TABLE A.10.
Continued.

N	N^2	$\sqrt{N}$	$\sqrt{10N}$	N	N^2	$\sqrt{N}$	$\sqrt{10N}$
2.20	4.8400	1.48324	4.69042	2.50	6.2500	1.58114	5.00000
2.21	4.8841	1.48661	4.70106	2.51	6.3001	1.58430	5.00999
2.22	4.9284	1.48997	4.71169	2.52	6.3504	1.58745	5.01996
2.23	4.9729	1.49332	4.72229	2.53	6.4009	1.59060	5.02991
2.24	5.0176	1.49666	4.73286	2.54	6.4516	1.59374	5.03984
2.25	5.0625	1.50000	4.74342	2.55	6.5025	1.59687	5.04975
2.26	5.1076	1.50333	4.75395	2.56	6.5536	1.60000	5.05964
2.27	5.1529	1.50665	4.76445	2.57	6.6049	1.60312	5.06952
2.28	5.1984	1.50997	4.77493	2.58	6.6564	1.60624	5.07937
2.29	5.2441	1.51327	4.78539	2.59	6.7081	1.60935	5.08920
2.30	5.2900	1.51658	4.79583	2.60	6.7600	1.61245	5.09902
2.31	5.3361	1.51987	4.80625	2.61	6.8121	1.61555	5.10882
2.32	5.3824	1.52315	4.81664	2.62	6.8644	1.61864	5.11859
2.33	5.4289	1.52643	4.82701	2.63	6.9169	1.62173	5.12835
2.34	5.4756	1.52971	4.83735	2.64	6.9696	1.62481	5.13809
2.35	5.5225	1.53297	4.84768	2.65	7.0225	1.62788	5.14782
2.36	5.5696	1.53623	4.85798	2.66	7.0756	1.63095	5.15752
2.37	5.6169	1.53948	4.86826	2.67	7.1289	1.63401	5.16720
2.38	5.6644	1.54272	4.87852	2.68	7.1824	1.63707	5.17687
2.39	5.7121	1.54596	4.88876	2.69	7.2361	1.64012	5.18652
2.40	5.7600	1.54919	4.89898	2.70	7.2900	1.64317	5.19615
2.41	5.8081	1.55242	4.90918	2.71	7.3441	1.64621	5.20577
2.42	5.8564	1.55563	4.91935	2.72	7.3984	1.64924	5.21536
2.43	5.9049	1.55885	4.92950	2.73	7.4529	1.65227	5.22494
2.44	5.9536	1.56205	4.93964	2.74	7.5076	1.65529	5.23450
2.45	6.0025	1.56525	4.94975	2.75	7.5625	1.65831	5.24404
2.46	6.0516	1.56844	4.95984	2.76	7.6176	1.66132	5.25357
2.47	6.1009	1.57162	4.96991	2.77	7.6729	1.66433	5.26308
2.48	6.1504	1.57480	4.97996	2.78	7.7284	1.66733	5.27257
2.49	6.2001	1.57797	4.98999	2.79	7.7841	1.67033	5.28205

TABLE A.10.
Continued.

N	N^2	$\sqrt{N}$	$\sqrt{10N}$	N	N^2	$\sqrt{N}$	$\sqrt{10N}$
2.80	7.8400	1.67332	5.29150	3.10	9.6100	1.76068	5.56776
2.81	7.8961	1.67631	5.30094	3.11	9.6721	1.76352	5.57674
2.82	7.9524	1.67929	5.31037	3.12	9.7344	1.76635	5.58570
2.83	8.0089	1.68226	5.31977	3.13	9.7969	1.76918	5.59464
2.84	8.0656	1.68523	5.32917	3.14	9.8596	1.77200	5.60357
2.85	8.1225	1.68819	5.33854	3.15	9.9225	1.77482	5.61249
2.86	8.1796	1.69115	5.34790	3.16	9.9856	1.77764	5.62139
2.87	8.2369	1.69411	5.35724	3.17	10.0489	1.78045	5.63028
2.88	8.2944	1.69706	5.36656	3.18	10.1124	1.78326	5.63915
2.89	8.3521	1.70000	5.37587	3.19	10.1761	1.78606	5.64801
2.90	8.4100	1.70294	5.38516	3.20	10.2400	1.78885	5.65685
2.91	8.4681	1.70587	5.39444	3.21	10.3041	1.79165	5.66569
2.92	8.5264	1.70880	5.40370	3.22	10.3684	1.79444	5.67450
2.93	8.5849	1.71172	5.41295	3.23	10.4329	1.79722	5.68331
2.94	8.6436	1.71464	5.42218	3.24	10.4976	1.80000	5.69210
2.95	8.7025	1.71756	5.43139	3.25	10.5625	1.80278	5.70088
2.96	8.7616	1.72047	5.44059	3.26	10.6276	1.80555	5.70964
2.97	8.8209	1.72337	5.44977	3.27	10.6929	1.80831	5.71839
2.98	8.8804	1.72627	5.45894	3.28	10.7584	1.81108	5.72713
2.99	8.9401	1.72916	5.46809	3.29	10.8241	1.81384	5.73585
3.00	9.0000	1.73205	5.47723	3.30	10.8900	1.81659	5.74456
3.01	9.0601	1.73494	5.48635	3.31	10.9561	1.81934	5.75326
3.02	9.1204	1.73781	5.49545	3.32	11.0224	1.82209	5.76194
3.03	9.1809	1.74069	5.50454	3.33	11.0889	1.82483	5.77062
3.04	9.2416	1.74356	5.51362	3.34	11.1556	1.82757	5.77927
3.05	9.3025	1.74642	5.52268	3.35	11.2225	1.83030	5.78792
3.06	9.3636	1.74929	5.53173	3.36	11.2896	1.83303	5.79655
3.07	9.4249	1.75214	5.54076	3.37	11.3569	1.83576	5.80517
3.08	9.4864	1.75499	5.54977	3.38	11.4244	1.83848	5.81378
3.09	9.5481	1.75784	5.55878	3.39	11.4921	1.84120	5.82237

TABLE A.10.
Continued.

N	N^2	$\sqrt{N}$	$\sqrt{10N}$	N	N^2	$\sqrt{N}$	$\sqrt{10N}$
3.40	11.5600	1.84391	5.83095	3.70	13.6900	1.92354	6.08276
3.41	11.6281	1.84662	5.83952	3.71	13.7641	1.92614	6.09098
3.42	11.6964	1.84932	5.84808	3.72	13.8384	1.92873	6.09918
3.43	11.7649	1.85203	5.85662	3.73	13.9129	1.93132	6.10737
3.44	11.8336	1.85472	5.86515	3.74	13.9876	1.93391	6.11555
3.45	11.9025	1.85742	5.87367	3.75	14.0625	1.93649	6.12372
3.46	11.9716	1.86011	5.88218	3.76	14.1376	1.93907	6.13188
3.47	12.0409	1.86279	5.89067	3.77	14.2129	1.94165	6.14003
3.48	12.1104	1.86548	5.89915	3.78	14.2884	1.94422	6.14817
3.49	12.1801	1.86815	5.90762	3.79	14.3641	1.94679	6.15630
3.50	12.2500	1.87083	5.91608	3.80	14.4400	1.94936	6.16441
3.51	12.3201	1.87350	5.92453	3.81	14.5161	1.95192	6.17252
3.52	12.3904	1.87617	5.93296	3.82	14.5924	1.95448	6.18061
3.53	12.4609	1.87883	5.94138	3.83	14.6689	1.95704	6.18870
3.54	12.5316	1.88149	5.94979	3.84	14.7456	1.95959	6.19677
3.55	12.6025	1.88414	5.95819	3.85	14.8225	1.96214	6.20484
3.56	12.6736	1.88680	5.96657	3.86	14.8996	1.96469	6.21289
3.57	12.7449	1.88944	5.97495	3.87	14.9769	1.96723	6.22093
3.58	12.8164	1.89209	5.98331	3.88	15.0544	1.96977	6.22896
3.59	12.8881	1.89473	5.99166	3.89	15.1321	1.97231	6.23699
3.60	12.9600	1.89737	6.00000	3.90	15.2100	1.97484	6.24500
3.61	13.0321	1.90000	6.00833	3.91	15.2881	1.97737	6.25300
3.62	13.1044	1.90263	6.01664	3.92	15.3664	1.97990	6.26099
3.63	13.1769	1.90526	6.02495	3.93	15.4449	1.98242	6.26897
3.64	13.2496	1.90788	6.03324	3.94	15.5236	1.98494	6.27694
3.65	13.3225	1.91050	6.04152	3.95	15.6025	1.98746	6.28490
3.66	13.3956	1.91311	6.04979	3.96	15.6816	1.98997	6.29285
3.67	13.4689	1.91572	6.05805	3.97	15.7609	1.99249	6.30079
3.68	13.5424	1.91833	6.06630	3.98	15.8404	1.99499	6.30872
3.69	13.6161	1.92094	6.07454	3.99	15.9201	1.99750	6.31664

TABLE A.10.
Continued.

N	N^2	$\sqrt{N}$	$\sqrt{10N}$	N	N^2	$\sqrt{N}$	$\sqrt{10N}$
4.00	16.0000	2.00000	6.32456	4.30	18.4900	2.07364	6.55744
4.01	16.0801	2.00250	6.33246	4.31	18.5761	2.07605	6.56506
4.02	16.1604	2.00499	6.34035	4.32	18.6624	2.07846	6.57267
4.03	16.2409	2.00749	6.34823	4.33	18.7489	2.08087	6.58027
4.04	16.3216	2.00998	6.35610	4.34	18.8356	2.08327	6.58787
4.05	16.4025	2.01246	6.36396	4.35	18.9225	2.08567	6.59545
4.06	16.4836	2.01494	6.37181	4.36	19.0096	2.08806	6.60303
4.07	16.5649	2.01742	6.37966	4.37	19.0969	2.09045	6.61060
4.08	16.6464	2.01990	6.38749	4.38	19.1844	2.09284	6.61816
4.09	16.7281	2.02237	6.39531	4.39	19.2721	2.09523	6.62571
4.10	16.8100	2.02485	6.40312	4.40	19.3600	2.09762	6.63325
4.11	16.8921	2.02731	6.41093	4.41	19.4481	2.10000	6.64078
4.12	16.9744	2.02978	6.41872	4.42	19.5364	2.10238	6.64831
4.13	17.0569	2.03224	6.42651	4.43	19.6249	2.10476	6.65582
4.14	17.1396	2.03470	6.43428	4.44	19.7136	2.10713	6.66333
4.15	17.2225	2.03715	6.44205	4.45	19.8025	2.10950	6.67083
4.16	17.3056	2.03961	6.44981	4.46	19.8916	2.11187	6.67832
4.17	17.3889	2.04206	6.45755	4.47	19.9809	2.11424	6.68581
4.18	17.4724	2.04450	6.46529	4.48	20.0704	2.11660	6.69328
4.19	17.5561	2.04695	6.47302	4.49	20.1601	2.11896	6.70075
4.20	17.6400	2.04939	6.48074	4.50	20.2500	2.12132	6.70820
4.21	17.7241	2.05183	6.48845	4.51	20.3401	2.12368	6.71565
4.22	17.8084	2.05426	6.49615	4.52	20.4304	2.12603	6.72309
4.23	17.8929	2.05670	6.50385	4.53	20.5209	2.12838	6.73053
4.24	17.9776	2.05913	6.51153	4.54	20.6116	2.13073	6.73795
4.25	18.0625	2.06155	6.51920	4.55	20.7025	2.13307	6.74537
4.26	18.1476	2.06398	6.52687	4.56	20.7936	2.13542	6.75278
4.27	18.2329	2.06640	6.53452	4.57	20.8849	2.13776	6.76018
4.28	18.3184	2.06882	6.54217	4.58	20.9764	2.14009	6.76757
4.29	18.4041	2.07123	6.54981	4.59	21.0681	2.14243	6.77495

TABLE A.10.
Continued.

N	N^2	$\sqrt{N}$	$\sqrt{10N}$	N	N^2	$\sqrt{N}$	$\sqrt{10N}$
4.60	21.1600	2.14476	6.78233	4.90	24.0100	2.21359	7.00000
4.61	21.2521	2.14709	6.78970	4.91	24.1081	2.21585	7.00714
4.62	21.3444	2.14942	6.79706	4.92	24.2064	2.21811	7.01427
4.63	21.4369	2.15174	6.80441	4.93	24.3049	2.22036	7.02140
4.64	21.5296	2.15407	6.81175	4.94	24.4036	2.22261	7.02851
4.65	21.6225	2.15639	6.81909	4.95	24.5025	2.22486	7.03562
4.66	21.7156	2.15870	6.82642	4.96	24.6016	2.22711	7.04273
4.67	21.8089	2.16102	6.83374	4.97	24.7009	2.22935	7.04982
4.68	21.9024	2.16333	6.84105	4.98	24.8004	2.23159	7.05691
4.69	21.9961	2.16564	6.84836	4.99	24.9001	2.23383	7.06399
4.70	22.0900	2.16795	6.85565	5.00	25.0000	2.23607	7.07107
4.71	22.1841	2.17025	6.86294	5.01	25.1001	2.23830	7.07814
4.72	22.2784	2.17256	6.87023	5.02	25.2004	2.24054	7.08520
4.73	22.3729	2.17486	6.87750	5.03	25.3009	2.24277	7.09225
4.74	22.4676	2.17715	6.88477	5.04	25.4016	2.24499	7.09930
4.75	22.5625	2.17945	6.89202	5.05	25.5025	2.24722	7.10634
4.76	22.6576	2.18174	6.89928	5.06	25.6036	2.24944	7.11337
4.77	22.7529	2.18403	6.90652	5.07	25.7049	2.25167	7.12039
4.78	22.8484	2.18632	6.91375	5.08	25.8064	2.25389	7.12741
4.79	22.9441	2.18861	6.92098	5.09	25.9081	2.25610	7.13442
4.80	23.0400	2.19089	6.92820	5.10	26.0100	2.25832	7.14143
4.81	23.1361	2.19317	6.93542	5.11	26.1121	2.26053	7.14843
4.82	23.2324	2.19545	6.94262	5.12	26.2144	2.26274	7.15542
4.83	23.3289	2.19773	6.94982	5.13	26.3169	2.26495	7.16240
4.84	23.4256	2.20000	6.95701	5.14	26.4196	2.26716	7.16938
4.85	23.5225	2.20227	6.96419	5.15	26.5225	2.26936	7.17635
4.86	23.6196	2.20454	6.97137	5.16	26.6256	2.27156	7.18331
4.87	23.7169	2.20681	6.97854	5.17	26.7289	2.27376	7.19027
4.88	23.8144	2.20907	6.98570	5.18	26.8324	2.27596	7.19722
4.89	23.9121	2.21133	6.99285	5.19	26.9361	2.27816	7.20417

TABLE A.10.
Continued.

N	N^2	$\sqrt{N}$	$\sqrt{10N}$	N	N^2	$\sqrt{N}$	$\sqrt{10N}$
5.20	27.0400	2.28035	7.21110	5.50	30.2500	2.34521	7.41620
5.21	27.1441	2.28254	7.21803	5.51	30.3601	2.34734	7.42294
5.22	27.2484	2.28473	7.22496	5.52	30.4704	2.34947	7.42967
5.23	27.3529	2.28692	7.23187	5.53	30.5809	2.35160	7.43640
5.24	27.4576	2.28910	7.23878	5.54	30.6916	2.35372	7.44312
5.25	27.5625	2.29129	7.24569	5.55	30.8025	2.35584	7.44983
5.26	27.6676	2.29347	7.25259	5.56	30.9136	2.35797	7.45654
5.27	27.7729	2.29565	7.25948	5.57	31.0249	2.36008	7.46324
5.28	27.8784	2.29783	7.26636	5.58	31.1364	2.36220	7.46994
5.29	27.9841	2.30000	7.27324	5.59	31.2481	2.36432	7.47663
5.30	28.0900	2.30217	7.28011	5.60	31.3600	2.36643	7.48331
5.31	28.1961	2.30434	7.28697	5.61	31.4721	2.36854	7.48999
5.32	28.3024	2.30651	7.29383	5.62	31.5844	2.37065	7.49667
5.33	28.4089	2.30868	7.30068	5.63	31.6969	2.37276	7.50333
5.34	28.5156	2.31084	7.30753	5.64	31.8096	2.37487	7.50999
5.35	28.6225	2.31301	7.31437	5.65	31.9225	2.37697	7.51665
5.36	28.7296	2.31517	7.32120	5.66	32.0356	2.37908	7.52330
5.37	28.8369	2.31733	7.32803	5.67	32.1489	2.38118	7.52994
5.38	28.9444	2.31948	7.33485	5.68	32.2624	2.38328	7.53658
5.39	29.0521	2.32164	7.34166	5.69	32.3761	2.38537	7.54321
5.40	29.1600	2.32379	7.34847	5.70	32.4900	2.38747	7.54983
5.41	29.2681	2.32594	7.35527	5.71	32.6041	2.38956	7.55645
5.42	29.3764	2.32809	7.36206	5.72	32.7184	2.39165	7.56307
5.43	29.4849	2.33024	7.36885	5.73	32.8329	2.39374	7.56968
5.44	29.5936	2.33238	7.37564	5.74	32.9476	2.39583	7.57628
5.45	29.7025	2.33452	7.38241	5.75	33.0625	2.39792	7.58288
5.46	29.8116	2.33666	7.38918	5.76	33.1776	2.40000	7.58947
5.47	29.9209	2.33880	7.39594	5.77	33.2929	2.40208	7.59605
5.48	30.0304	2.34094	7.40270	5.78	33.4084	2.40416	7.60263
5.49	30.1401	2.34307	7.40945	5.79	33.5241	2.40624	7.60920

TABLE A.10.
Continued.

N	N^2	$\sqrt{N}$	$\sqrt{10N}$	N	N^2	$\sqrt{N}$	$\sqrt{10N}$
5.80	33.6400	2.40832	7.61577	6.10	37.2100	2.46982	7.81025
5.81	33.7561	2.41039	7.62234	6.11	37.3321	2.47184	7.81665
5.82	33.8724	2.41247	7.62889	6.12	37.4544	2.47386	7.82304
5.83	33.9889	2.41454	7.63544	6.13	37.5769	2.47588	7.82943
5.84	34.1056	2.41661	7.64199	6.14	37.6996	2.47790	7.83582
5.85	34.2225	2.41868	7.64853	6.15	37.8225	2.47992	7.84219
5.86	34.3396	2.42074	7.65506	6.16	37.9456	2.48193	7.84857
5.87	34.4569	2.42281	7.66159	6.17	38.0689	2.48395	7.85493
5.88	34.5744	2.42487	7.66812	6.18	38.1924	2.48596	7.86130
5.89	34.6921	2.42693	7.67463	6.19	38.3161	2.48797	7.86766
5.90	34.8100	2.42899	7.68115	6.20	38.4400	2.48998	7.87401
5.91	34.9281	2.43105	7.68765	6.21	38.5641	2.49199	7.88036
5.92	35.0464	2.43311	7.69415	6.22	38.6884	2.49399	7.88670
5.93	35.1649	2.43516	7.70065	6.23	38.8129	2.49600	7.89303
5.94	35.2836	2.43721	7.70714	6.24	38.9376	2.49800	7.89937
5.95	35.4025	2.43926	7.71362	6.25	39.0625	2.50000	7.90569
5.96	35.5216	2.44131	7.72010	6.26	39.1876	2.50200	7.91202
5.97	35.6409	2.44336	7.72658	6.27	39.3129	2.50400	7.91833
5.98	35.7604	2.44540	7.73305	6.28	39.4384	2.50599	7.92465
5.99	35.8801	2.44745	7.73951	6.29	39.5641	2.50799	7.93095
6.00	36.0000	2.44949	7.74597	6.30	39.6900	2.50998	7.93725
6.01	36.1201	2.45153	7.75242	6.31	39.8161	2.51197	7.94355
6.02	36.2404	2.45357	7.75887	6.32	39.9424	2.51396	7.94984
6.03	36.3609	2.45561	7.76531	6.33	40.0689	2.51595	7.95613
6.04	36.4816	2.45764	7.77174	6.34	40.1956	2.51794	7.96241
6.05	36.6025	2.45967	7.77817	6.35	40.3225	2.51992	7.96869
6.06	36.7236	2.46171	7.78460	6.36	40.4496	2.52190	7.97496
6.07	36.8449	2.46374	7.79102	6.37	40.5769	2.52389	7.98123
6.08	36.9664	2.46577	7.79744	6.38	40.7044	2.52587	7.98749
6.09	37.0881	2.46779	7.80385	6.39	40.8321	2.52784	7.99375

TABLE A.10.
Continued.

N	N^2	$\sqrt{N}$	$\sqrt{10N}$	N	N^2	$\sqrt{N}$	$\sqrt{10N}$
6.40	40.9600	2.52982	8.00000	6.70	44.8900	2.58844	8.18535
6.41	41.0881	2.53180	8.00625	6.71	45.0241	2.59037	8.19146
6.42	41.2164	2.53377	8.01249	6.72	45.1584	2.59230	8.19756
6.43	41.3449	2.53574	8.01873	6.73	45.2929	2.59422	8.20366
6.44	41.4736	2.53772	8.02496	6.74	45.4276	2.59615	8.20975
6.45	41.6025	2.53969	8.03119	6.75	45.5625	2.59808	8.21584
6.46	41.7316	2.54165	8.03741	6.76	45.6976	2.60000	8.22192
6.47	41.8609	2.54362	8.04363	6.77	45.8329	2.60192	8.22800
6.48	41.9904	2.54558	8.04984	6.78	45.9684	2.60384	8.23408
6.49	42.1201	2.54755	8.05605	6.79	46.1041	2.60576	8.24015
6.50	42.2500	2.54951	8.06226	6.80	46.2400	2.60768	8.24621
6.51	42.3801	2.55147	8.06846	6.81	46.3761	2.60960	8.25227
6.52	42.5104	2.55343	8.07465	6.82	46.5124	2.61151	8.25833
6.53	42.6409	2.55539	8.08084	6.83	46.6489	2.61343	8.26438
6.54	42.7716	2.55734	8.08703	6.84	46.7856	2.61534	8.27043
6.55	42.9025	2.55930	8.09321	6.85	46.9225	2.61725	8.27647
6.56	43.0336	2.56125	8.09938	6.86	47.0596	2.61916	8.28251
6.57	43.1649	2.56320	8.10555	6.87	47.1969	2.62107	8.28855
6.58	43.2964	2.56515	8.11172	6.88	47.3344	2.62298	8.29458
6.59	43.4281	2.56710	8.11788	6.89	47.4721	2.62488	8.30060
6.60	43.5600	2.56905	8.12404	6.90	47.6100	2.62679	8.30662
6.61	43.6921	2.57099	8.13019	6.91	47.7481	2.62869	8.31264
6.62	43.8244	2.57294	8.13634	6.92	47.8864	2.63059	8.31865
6.63	43.9569	2.57488	8.14248	6.93	48.0249	2.63249	8.32466
6.64	44.0896	2.57682	8.14862	6.94	48.1636	2.63439	8.33067
6.65	44.2225	2.57876	8.15475	6.95	48.3025	2.63629	8.33667
6.66	44.3556	2.58070	8.16088	6.96	48.4416	2.63818	8.34266
6.67	44.4889	2.58263	8.16701	6.97	48.5809	2.64008	8.34865
6.68	44.6224	2.58457	8.17313	6.98	48.7204	2.64197	8.35464
6.69	44.7561	2.58650	8.17924	6.99	48.8601	2.64386	8.36062

TABLE A.10.
Continued.

N	N^2	$\sqrt{N}$	$\sqrt{10N}$	N	N^2	$\sqrt{N}$	$\sqrt{10N}$
7.00	49.0000	2.64575	8.36660	7.30	53.2900	2.70185	8.54400
7.01	49.1401	2.64764	8.37257	7.31	53.4361	2.70370	8.54985
7.02	49.2804	2.64953	8.37854	7.32	53.5824	2.70555	8.55570
7.03	49.4209	2.65141	8.38451	7.33	53.7289	2.70740	8.56154
7.04	49.5616	2.65330	8.39047	7.34	53.8756	2.70924	8.56738
7.05	49.7025	2.65518	8.39643	7.35	54.0225	2.71109	8.57321
7.06	49.8436	2.65707	8.40238	7.36	54.1696	2.71293	8.57904
7.07	49.9849	2.65895	8.40833	7.37	54.3169	2.71477	8.58487
7.08	50.1264	2.66083	8.41427	7.38	54.4644	2.71662	8.59069
7.09	50.2681	2.66271	8.42021	7.39	54.6121	2.71846	8.59651
7.10	50.4100	2.66458	8.42615	7.40	54.7600	2.72029	8.60233
7.11	50.5521	2.66646	8.43208	7.41	54.9081	2.72213	8.60814
7.12	50.6944	2.66833	8.43801	7.42	55.0564	2.72397	8.61394
7.13	50.8369	2.67021	8.44393	7.43	55.2049	2.72580	8.61974
7.14	50.9796	2.67208	8.44985	7.44	55.3536	2.72764	8.62554
7.15	51.1225	2.67395	8.45577	7.45	55.5025	2.72947	8.63134
7.16	51.2656	2.67582	8.46168	7.46	55.6516	2.73130	8.63713
7.17	51.4089	2.67769	8.46759	7.47	55.8009	2.73313	8.64292
7.18	51.5524	2.67955	8.47349	7.48	55.9504	2.73496	8.64870
7.19	51.6961	2.68142	8.47939	7.49	56.1001	2.73679	8.65448
7.20	51.8400	2.68328	8.48528	7.50	56.2500	2.73861	8.66025
7.21	51.9841	2.68514	8.49117	7.51	56.4001	2.74044	8.66603
7.22	52.1284	2.68701	8.49706	7.52	56.5504	2.74226	8.67179
7.23	52.2729	2.68887	8.50294	7.53	56.7000	2.74408	8.67756
7.24	52.4176	2.69072	8.50882	7.54	56.8516	2.74591	8.68332
7.25	52.5625	2.69258	8.51469	7.55	57.0025	2.74773	8.68907
7.26	52.7076	2.69444	8.52056	7.56	57.1536	2.74955	8.69483
7.27	52.8529	2.69629	8.52643	7.57	57.3049	2.75136	8.70057
7.28	52.9984	2.69815	8.53229	7.58	57.4564	2.75318	8.70632
7.29	53.1441	2.70000	8.53815	7.59	57.6081	2.75500	8.71206

TABLE A.10.
Continued.

N	N^2	$\sqrt{N}$	$\sqrt{10N}$	N	N^2	$\sqrt{N}$	$\sqrt{10N}$
7.60	57.7600	2.75681	8.71780	7.90	62.4100	2.81069	8.88819
7.61	57.9121	2.75862	8.72353	7.91	62.5681	2.81247	8.89382
7.62	58.0644	2.76043	8.72926	7.92	62.7264	2.81425	8.89944
7.63	58.2169	2.76225	8.73499	7.93	62.8849	2.81603	8.90505
7.64	58.3696	2.76405	8.74071	7.94	63.0436	2.81780	8.91067
7.65	58.5225	2.76586	8.74643	7.95	63.2025	2.81957	8.91628
7.66	58.6756	2.76767	8.75214	7.96	63.3616	2.82135	8.92188
7.67	58.8289	2.76948	8.75785	7.97	63.5209	2.82312	8.92749
7.68	58.9824	2.77128	8.76356	7.98	63.6804	2.82489	8.93308
7.69	59.1361	2.77308	8.76926	7.99	63.8401	2.82666	8.93868
7.70	59.2900	2.77489	8.77496	8.00	64.0000	2.82843	8.94427
7.71	59.4441	2.77669	8.78066	8.01	64.1601	2.83019	8.94986
7.72	59.5984	2.77849	8.78635	8.02	64.3204	2.83196	8.95545
7.73	59.7529	2.78029	8.79204	8.03	64.4809	2.83373	8.96103
7.74	59.9076	2.78209	8.79773	8.04	64.6416	2.83549	8.96660
7.75	60.0625	2.78388	8.80341	8.05	64.8025	2.83725	8.97218
7.76	60.2176	2.78568	8.80909	8.06	64.9636	2.83901	8.97775
7.77	60.3729	2.78747	8.81476	8.07	65.1249	2.84077	8.98332
7.78	60.5284	2.78927	8.82043	8.08	65.2864	2.84253	8.98888
7.79	60.6841	2.79106	8.82610	8.09	65.4481	2.84429	8.99444
7.80	60.8400	2.79285	8.83176	8.10	65.6100	2.84605	9.00000
7.81	60.9961	2.79464	8.83742	8.11	65.7721	2.84781	9.00555
7.82	61.1524	2.79643	8.84308	8.12	65.9344	2.84956	9.01110
7.83	61.3089	2.79821	8.84873	8.13	66.0969	2.85132	9.01665
7.84	61.4656	2.80000	8.85438	8.14	66.2596	2.85307	9.02219
7.85	61.6225	2.80179	8.86002	8.15	66.4225	2.85482	9.02774
7.86	61.7796	2.80357	8.86566	8.16	66.5856	2.85657	9.03327
7.87	61.9369	2.80535	8.87130	8.17	66.7489	2.85832	9.03881
7.88	62.0944	2.80713	8.87694	8.18	66.9124	2.86007	9.04434
7.89	62.2521	2.80891	8.88257	8.19	67.0761	2.86182	9.04986

TABLE A.10.
Continued.

N	N^2	$\sqrt{N}$	$\sqrt{10N}$	N	N^2	$\sqrt{N}$	$\sqrt{10N}$
8.20	67.2400	2.86356	9.05539	8.50	72.2500	2.91548	9.21954
8.21	67.4041	2.86531	9.06091	8.51	72.4201	2.91719	9.22497
8.22	67.5684	2.86705	9.06642	8.52	72.5904	2.91890	9.23038
8.23	67.7329	2.86880	9.07193	8.53	72.7609	2.92062	9.23580
8.24	67.8976	2.87054	9.07744	8.54	72.9316	2.92233	9.24121
8.25	68.0625	2.87228	9.08295	8.55	73.1025	2.92404	9.24662
8.26	68.2276	2.87402	9.08845	8.56	73.2736	2.92575	9.25203
8.27	68.3929	2.87576	9.09395	8.57	73.4449	2.92746	9.25743
8.28	68.5584	2.87750	9.09945	8.58	73.6164	2.92916	9.26283
8.29	68.7241	2.87924	9.10494	8.59	73.7881	2.93087	9.26823
8.30	68.8900	2.88097	9.11043	8.60	73.9600	2.93258	9.27362
8.31	69.0561	2.88271	9.11592	8.61	74.1321	2.93428	9.27901
8.32	69.2224	2.88444	9.12140	8.62	74.3044	2.93598	9.28440
8.33	69.3889	2.88617	9.12688	8.63	74.4769	2.93769	9.28978
8.34	69.5556	2.88791	9.13236	8.64	74.6496	2.93939	9.29516
8.35	69.7225	2.88964	9.13783	8.65	74.8225	2.94109	9.30054
8.36	69.8896	2.89137	9.14330	8.66	74.9956	2.94279	9.30591
8.37	70.0569	2.89310	9.14877	8.67	75.1689	2.94449	9.31128
8.38	70.2244	2.89482	9.15423	8.68	75.3424	2.94618	9.31665
8.39	70.3921	2.89655	9.15969	8.69	75.5161	2.94788	9.32202
8.40	70.5600	2.89828	9.16515	8.70	75.6900	2.94958	9.32738
8.41	70.7281	2.90000	9.17061	8.71	75.8641	2.95127	9.33274
8.42	70.8964	2.90172	9.17606	8.72	76.0384	2.95296	9.33809
8.43	71.0649	2.90345	9.18150	8.73	76.2129	2.95466	9.34345
8.44	71.2336	2.90517	9.18695	8.74	76.3876	2.95635	9.34880
8.45	71.4025	2.90689	9.19239	8.75	76.5625	2.95804	9.35414
8.46	71.5716	2.90861	9.19783	8.76	76.7376	2.95973	9.35949
8.47	71.7409	2.91033	9.20326	8.77	76.9129	2.96142	9.36483
8.48	71.9104	2.91204	9.20869	8.78	77.0884	2.96311	9.37017
8.49	72.0801	2.91376	9.21412	8.79	77.2641	2.96479	9.37550

TABLE A.10.
Continued.

N	N^2	$\sqrt{N}$	$\sqrt{10N}$	N	N^2	$\sqrt{N}$	$\sqrt{10N}$
8.80	77.4400	2.96648	9.38083	9.10	82.8100	3.01662	9.53939
8.81	77.6161	2.96816	9.38616	9.11	82.9921	3.01828	9.54463
8.82	77.7924	2.96985	9.39149	9.12	83.1744	3.01993	9.54987
8.83	77.9689	2.97153	9.39681	9.13	83.3569	3.02159	9.55510
8.84	78.1456	2.97321	9.40213	9.14	83.5396	3.02324	9.56033
8.85	78.3225	2.97489	9.40744	9.15	83.7225	3.02490	9.56556
8.86	78.4996	2.97658	9.41276	9.16	83.9056	3.02655	9.57079
8.87	78.6769	2.97825	9.41807	9.17	84.0889	3.02820	9.57601
8.88	78.8544	2.97993	9.42338	9.18	84.2724	3.02985	9.58123
8.89	79.0321	2.98161	9.42868	9.19	84.4561	3.03150	9.58645
8.90	79.2100	2.98329	9.43398	9.20	84.6400	3.03315	9.59166
8.91	79.3881	2.98496	9.43928	9.21	84.8241	3.03480	9.59687
8.92	79.5664	2.98664	9.44458	9.22	85.0084	3.03645	9.60208
8.93	79.7449	2.98831	9.44987	9.23	85.1929	3.03809	9.60729
8.94	79.9236	2.98998	9.45516	9.24	85.3776	3.03974	9.61249
8.95	80.1025	2.99166	9.46044	9.25	85.5625	3.04138	9.61769
8.96	80.2816	2.99333	9.46573	9.26	85.7476	3.04302	9.62289
8.97	80.4609	2.99500	9.47101	9.27	85.9329	3.04467	9.62808
8.98	80.6404	2.99666	9.47629	9.28	86.1184	3.04631	9.63328
8.99	80.8201	2.99833	9.48156	9.29	86.3041	3.04795	9.63846
9.00	81.0000	3.00000	9.48683	9.30	86.4900	3.04959	9.64365
9.01	81.1801	3.00167	9.49210	9.31	86.6761	3.05123	9.64883
9.02	81.3604	3.00333	9.49737	9.32	86.8624	3.05287	9.65401
9.03	81.5409	3.00500	9.50263	9.33	87.0489	3.05450	9.65919
9.04	81.7216	3.00666	9.50789	9.34	87.2356	3.05614	9.66437
9.05	81.9025	3.00832	9.51315	9.35	87.4225	3.05778	9.66954
9.06	82.0836	3.00998	9.51840	9.36	87.6096	3.05941	9.67471
9.07	82.2649	3.01164	9.52365	9.37	87.7969	3.06105	9.67988
9.08	82.4464	3.01330	9.52890	9.38	87.9844	3.06268	9.68504
9.09	82.6281	3.01496	9.53415	9.39	88.1721	3.06431	9.69020

TABLE A.10.
Continued.

N	N^2	$\sqrt{N}$	$\sqrt{10N}$	N	N^2	$\sqrt{N}$	$\sqrt{10N}$
9.40	88.3600	3.06594	9.69536	9.70	94.0900	3.11448	9.84886
9.41	88.5481	3.06757	9.70052	9.71	94.2841	3.11609	9.85393
9.42	88.7364	3.06920	9.70567	9.72	94.4784	3.11769	9.85901
9.43	88.9249	3.07083	9.71082	9.73	94.6729	3.11929	9.86408
9.44	89.1136	3.07246	9.71597	9.74	94.8676	3.12090	9.86914
9.45	89.3025	3.07409	9.72111	9.75	95.0625	3.12250	9.87421
9.46	89.4916	3.07571	9.72625	9.76	95.2576	3.12410	9.87927
9.47	89.6809	3.07734	9.73139	9.77	95.4529	3.12570	9.88433
9.48	89.8704	3.07896	9.73653	9.78	95.6484	3.12730	9.88939
9.49	90.0601	3.08058	9.74166	9.79	95.8441	3.12890	9.89444
9.50	90.2500	3.08221	9.74679	9.80	96.0400	3.13050	9.89949
9.51	90.4401	3.08383	9.75192	9.81	96.2361	3.13209	9.90454
9.52	90.6304	3.08545	9.75705	9.82	96.4324	3.13369	9.90959
9.53	90.8209	3.08707	9.76217	9.83	96.6289	3.13528	9.91464
9.54	91.0116	3.08869	9.76729	9.84	96.8256	3.13688	9.91968
9.55	91.2025	3.09031	9.77241	9.85	97.0225	3.13847	9.92472
9.56	91.3936	3.09192	9.77753	9.86	97.2196	3.14006	9.92975
9.57	91.5849	3.09354	9.78264	9.87	97.4169	3.14166	9.93479
9.58	91.7764	3.09516	9.78775	9.88	97.6144	3.14325	9.93982
9.59	91.9681	3.09677	9.79285	9.89	97.8121	3.14484	9.94485
9.60	92.1600	3.09839	9.79796	9.90	98.0100	3.14643	9.94987
9.61	92.3521	3.10000	9.80306	9.91	98.2081	3.14802	9.95490
9.62	92.5444	3.10161	9.80816	9.92	98.4064	3.14960	9.95992
9.63	92.7369	3.10322	9.81326	9.93	98.6049	3.15119	9.96494
9.64	92.9296	3.10483	9.81835	9.94	98.8036	3.15278	9.96995
9.65	93.1225	3.10644	9.82344	9.95	99.0025	3.15436	9.97497
9.66	93.3156	3.10805	9.82853	9.96	99.2016	3.15595	9.97998
9.67	93.5089	3.10966	9.83362	9.97	99.4009	3.15753	9.98499
9.68	93.7024	3.11127	9.83870	9.98	99.6004	3.15911	9.98999
9.69	93.8961	3.11288	9.84378	9.99	99.8001	3.16070	9.99500

TABLE A.11.
Coefficients, divisors, and K values for fitting up to quartic curves to equally spaced data, and partitioning the sum of squares.

n:	3		4			5				6			
	c_1	c_2	c_1	c_2	c_3	c_1	c_2	c_3	c_4	c_1	c_2	c_3	c_4
	−1	1	−3	1	−1	−2	2	−1	1	−5	5	−5	1
	0	−2	−1	−1	3	−1	−1	2	−4	−3	−1	7	−3
	1	1	1	−1	−3	0	−2	0	6	−1	−4	4	2
			3	1	1	1	−1	−2	−4	1	−4	−4	2
						2	2	1	1	3	−1	−7	−3
										5	5	5	1
Divisors	2	6	20	4	20	10	14	10	70	70	84	180	28
K_1		1/3			5/16				1/7				5/96
K_2		1/2			1/20				1/10				1/70
K_3					41/240				17/60				101/4320
K_4		1/2			1/16				1/14				1/224
K_5					1/48				1/12				1/864
K_6									1/24				1/768
K_7									31/168				95/2688
K_8									3/35				27/256

TABLE A.11.
Continued.

n:	7				8				9			
	c_1	c_2	c_3	c_4	c_1	c_2	c_3	c_4	c_1	c_2	c_3	c_4
	−3	5	−1	3	−7	7	−7	7	−4	28	−14	14
	−2	0	1	−7	−5	1	5	−13	−3	7	7	−21
	−1	−3	1	1	−3	−3	7	−3	−2	−8	13	−11
	0	−4	0	6	−1	−5	3	9	−1	−17	9	9
	1	−3	−1	1	1	−5	−3	9	0	−20	0	18
	2	0	−1	−7	3	−3	−7	−3	1	−17	−9	9
	3	5	1	3	5	1	−5	−13	2	−8	−13	−11
					7	7	7	7	3	7	−7	−21
									4	28	14	14
Divisors												
	28	84	6	154	168	168	264	616	60	2772	990	2002
K_1				1/21				1/32				5/693
K_2				1/28				1/68				1.60
K_3				7/36				37/3168				59/5940
K_4				1/84				1/672				1/924
K_5				1/36				1/3168				1/1188
K_6				1/264				1/16896				1/3432
K_7				67/1848				179/59136				115/24024
K_8				3/77				9/512				9/1001

TABLE A.11.
Continued.

	n: 10				11		
c_1	c_2	c_3	c_4	c_1	c_2	c_3	c_4
−9	6	−42	18	−5	15	−30	6
−7	2	14	−22	−4	6	6	−6
−5	−1	35	−17	−3	−1	22	−6
−3	−3	31	3	−2	−6	23	−1
−1	−4	12	18	−1	−9	14	4
1	−4	−12	18	0	−10	0	6
3	−3	−31	3	1	−9	−14	4
5	−1	−35	−17	2	−6	−23	−1
7	2	−14	−22	3	−1	−22	−6
9	6	42	18	4	6	−6	−6
				5	15	30	6
Divisors							
330	132	8580	2860	110	858	4290	286
K_1			1/32			5/429	
K_2			1/330			1/110	
K_3			293/205920			89/25740	
K_4			1/1056			1/858	
K_5			1/41184			1/5148	
K_6			1/109824			1/3432	
K_7			41/54912			25/3432	
K_8			9/1280			3/143	

TABLE A.11.
Continued.

n: 12				13			
c_1	c_2	c_3	c_4	c_1	c_2	c_3	c_4
−11	55	−33	33	−6	22	−11	99
−9	25	3	−27	−5	11	0	−66
−7	1	21	−33	−4	2	6	−96
−5	−17	25	−13	−3	−5	8	−54
−3	−29	19	12	−2	−10	7	11
−1	−35	7	28	−1	−13	4	64
1	−35	−7	28	0	−14	0	84
3	−29	−19	12	1	−13	−4	64
5	−17	−25	−13	2	−10	−7	11
7	1	−21	−33	3	−5	−8	−54
9	25	−3	−27	4	2	−6	−96
11	55	33	33	5	11	0	−66
				6	22	11	99
Divisors							
572	12012	5148	8008	182	2002	572	68068
K_1			1/336				1/143
K_2			1/572				1/182
K_3			85/61776				25/3432
K_4			1/16016				1/2002
K_5			1/61776				1/3432
K_6			1/439296				1/116688
K_7			419/1537536				19/62832
K_8			27/7168				3/2431

TABLE A.11.
Continued.

n:	14				15			
	c_1	c_2	c_3	c_4	c_1	c_2	c_3	c_4
	−13	13	−143	143	−7	91	−91	1001
	−11	7	−11	−77	−6	52	−13	−429
	−9	2	66	−132	−5	19	35	−869
	−7	−2	98	−92	−4	−8	58	−704
	−5	−5	95	−13	−3	−29	61	−249
	−3	−7	63	63	−2	−44	49	251
	−1	−8	24	108	−1	−53	27	621
	1	−8	−24	108	0	−56	0	756
	3	−7	−67	63	1	−53	−27	621
	5	−5	−95	−13	2	−44	−49	251
	7	−2	−98	−92	3	−29	−61	−249
	9	2	−66	−132	4	−8	−58	−704
	11	7	11	−77	5	19	−35	−869
	13	13	143	143	6	52	13	−429
					7	91	91	1001
Divisors								
	910	728	97240	136136	280	37128	39780	6466460
K_1				5/448				1/663
K_2				1/910				1/280
K_3				581/2333760				167/238680
K_4				1/5824				1/12376
K_5				1/466752				1/47736
K_6				1/3734016				1/2217072
K_7				575/13069056				331/15519504
K_8				3/3584				27/230945

TABLE A.11.
Continued.

n: 16				17			
c_1	c_2	c_3	c_4	c_1	c_2	c_3	c_4
−15	35	−455	273	−8	40	−28	52
−13	21	−91	−91	−7	25	−7	−13
−11	9	143	−221	−6	12	7	−39
−9	−1	267	−201	−5	1	15	−39
−7	−9	301	−101	−4	−8	18	−24
−5	−15	265	23	−3	−15	17	−3
−3	−19	179	129	−2	−20	13	17
−1	−21	63	189	−1	−23	7	31
1	−21	−63	189	0	−24	0	36
3	−19	−179	129	1	−23	−7	31
5	−15	−265	23	2	−20	−13	17
7	−9	−301	−101	3	−15	−17	−3
9	−1	−267	−201	4	−8	−18	−24
11	9	−143	−221	5	1	−15	−39
13	21	91	−91	6	12	−7	−39
15	35	455	273	7	25	7	−13
				8	40	28	52
Divisors							
1360	5712	1007760	470288	408	7752	3876	16796
K_1			5/1344				1/323
K_2			1/1360				1/408
K_3			761/12093120				43/23256
K_4			1/22848				1/7752
K_5			1/2418624				1/23256
K_6			1/12899328				1/201552
K_7			755/45147648				61/201552
K_8			3/7168				9/4199

TABLE A.11.
Continued.

n: 18				19			
c_1	c_2	c_3	c_4	c_1	c_2	c_3	c_4
−17	68	−68	68	−9	51	−204	612
−15	44	−20	−12	−8	34	−68	−68
−13	23	13	−47	−7	19	28	−388
−11	5	33	−51	−6	6	89	−453
−9	−10	42	−36	−5	−5	120	−354
−7	−22	42	−12	−4	−14	126	−168
−5	−31	35	13	−3	−21	112	42
−3	−37	23	33	−2	−26	83	227
−1	−40	8	44	−1	−29	44	352
1	−40	−8	44	0	−30	0	396
3	−37	−23	33	1	−29	−44	352
5	−31	−35	13	2	−26	−83	227
7	−22	−42	−12	3	−21	−112	42
9	−10	−42	−36	4	−14	−126	−168
11	5	−33	−51	5	−5	−120	−354
13	23	−13	−47	6	6	−89	−453
15	44	20	−12	7	19	−28	−388
17	68	68	68	8	34	68	−68
				9	51	204	612
Divisors							
1938	23256	23256	28424	570	13466	213180	2288132
K_1			1/576				5/2261
K_2			1/1938				1/570
K_3			193/558144				269/1279080
K_4			1/62016				1/13566
K_5			1/558144				1/255816
K_6			1/5457408				1/3922512
K_7			137/2728704				535/27457584
K_8			9/5632				9/52003

TABLE A.11.
Continued.

n: 20				21			
c_1	c_2	c_3	c_4	c_1	c_2	c_3	c_4
−19	57	−969	1938	−10	190	−285	969
−17	39	−357	−102	−9	133	−114	0
−15	23	85	−1122	−8	82	12	−510
−13	9	377	−1402	−7	37	98	−680
−11	−3	539	−1187	−6	−2	149	−615
−9	−13	591	−687	−5	−35	170	−406
−7	−21	553	−77	−4	−62	166	−130
−5	−27	445	503	−3	−83	142	150
−3	−31	287	948	−2	−98	103	385
−1	−33	99	1188	−1	−107	54	540
1	−33	−99	1188	0	−110	0	594
3	−31	−287	948	1	−107	−54	540
5	−27	−445	503	2	−98	−103	385
7	−21	−553	−77	3	−83	−142	150
9	−13	−591	−687	4	−62	−166	−130
11	−3	−539	−1187	5	−35	−170	−406
13	9	−377	−1402	6	−2	−149	−615
15	23	−85	−1122	7	37	−98	−680
17	39	357	−102	8	82	−12	−510
19	57	969	1938	9	133	114	0
				10	190	285	969
Divisors							
2660	17556	4903140	22881320	770	201894	432630	5720330
K_1			1/528				5/9177
K_2			1/2660				1/770
K_3			1193/58837680				329/2595780
K_4			1/70224				1/67298
K_5			1/11767536				1/519156
K_6			1/251040768				1/9806280
K_7			1187/878642688				131/13728792
K_8			3/56320				27/260015

TABLE A.11.
Continued.

n: 22				23			
c_1	c_2	c_3	c_4	c_1	c_2	c_3	c_4
−21	35	−133	1197	−11	77	−77	1463
−19	25	−57	57	−10	56	−35	133
−17	16	0	−570	−9	37	−3	−627
−15	8	40	−810	−8	20	20	−950
−13	1	65	−775	−7	5	35	−955
−11	−5	77	−563	−6	−8	43	−747
−9	−10	78	−258	−5	−19	45	−417
−7	−14	70	70	−4	−28	42	−42
−5	−17	−55	365	−3	−35	35	315
−3	−19	35	585	−2	−40	25	605
−1	−20	12	702	−1	−43	13	793
1	−20	−12	702	0	−44	0	858
3	−19	−35	585	1	−43	−13	793
5	−17	−55	365	2	−40	−25	605
7	−14	−70	70	3	−35	−35	315
9	−10	−78	−258	4	−28	−42	−42
11	−5	−77	−563	5	−19	−45	−417
13	1	−65	−775	6	−8	−43	−747
15	8	−40	−810	7	5	−35	−955
17	16	0	−570	8	20	−20	−950
19	25	57	57	9	37	3	−627
21	35	133	1197	10	56	35	133
				11	77	77	1463
Divisors							
3542	7084	96140	8748740	1012	35420	32890	131231100
K_1			1/352				1/805
K_2			1/3542				1/1012
K_3			289/2307360				79/197340
K_4			1/56672				1/35520
K_5			1/2307360				1/197340
K_6			1/239965440				1/22496760
K_7			1439/839879040				787/157477320
K_8			3/36608				1/15295

TABLE A.11.
Continued.

n: 24				25			
c_1	c_2	c_3	c_4	c_1	c_2	c_3	c_4
−23	252	−1771	253	−12	92	−506	1518
−21	187	−847	33	−11	69	−253	253
−19	127	−133	−97	−10	48	−55	−517
−17	73	391	−157	−9	29	93	−897
−15	25	745	−165	−8	12	196	−982
−13	−17	949	−137	−7	−3	259	−857
−11	−53	1023	−87	−6	−16	287	−597
−9	−83	987	−27	−5	−27	285	−267
−7	−107	861	33	−4	−36	258	78
−5	−125	665	85	−3	−43	211	393
−3	−137	419	123	−2	−48	149	643
−1	−143	143	143	−1	−51	77	803
1	−143	−143	143	0	−52	0	858
3	−137	−419	123	1	−51	−77	803
5	−125	−665	85	2	−48	−149	643
7	−107	−861	33	3	−43	−211	393
9	−83	−987	−27	4	−36	−258	78
11	−53	−1023	−87	5	−27	−285	−267
13	−17	−949	−137	6	−16	−287	−597
15	25	−745	−165	7	−3	−259	−857
17	73	−391	−157	8	12	−196	−982
19	127	133	−97	9	29	−93	−897
21	187	847	33	10	48	55	−517
23	253	1771	253	11	69	253	253
				12	92	506	1518
Divisors							
4600	394680	17760600	394680	1300	53820	1480050	14307150
K_1			5/13728				1/1035
K_2			1/4600				1/1300
K_3			1721/213127200				467/8880300
K_4			1/526240				1/53820
K_5			1/42625440				1/1776060
K_6			1/75778560				1/34337160
K_7			49/7577856				133/34337160
K_8			27/73216				1/16675

TABLE A.11A.
Coefficients and divisors for some selected sets of unequally spaced treatments

X	c_1	c_2
1	−13	5
5	−1	−9
10	14	4
	366	122

X	c_1	c_2
0	−7	3
2	−1	−5
5	8	2
	114	38

X	c_1	c_2
1	−5	3
2	−2	−4
5	7	1
	78	26

X	c_1	c_2
2	−11	5
5	−2	−8
10	13	3
	294	98

X	c_1	c_2
0	−2	4
1	−1	−5
5	3	1
	14	42

X	c_1	c_2
1	−4	2
2	−1	−3
4	5	1
	42	14

X	c_1	c_2	c_3
1	−11	20	−8
2	−7	−4	14
4	1	−29	−7
8	17	13	1
	460	1426	310

X	c_1	c_2	c_3
0	−7	7	−3
1	−3	−4	8
2	1	−8	−6
4	9	5	1
	140	154	110

X	c_1	c_2	c_3
1	−2	1	−1
2	−1	−1	2
4	1	−1	−2
5	2	1	1
	10	4	10

X	c_1	c_2	c_3
0	−2	43	−6
1	−1	−17	15
2	0	−49	−10
5	3	23	1
	14	5068	362

X	c_1	c_2	c_3
0	−5	9	−5
1	−3	−3	9
3	1	−13	−5
6	7	7	1
	84	308	132

X	c_1	c_2	c_3
1	−3	3	−9
2	−2	−1	14
5	1	−5	−7
8	4	3	2
	30	44	330

X	c_1	c_2	c_3
0	−4	107	−18
1	−3	−5	25
5	1	−205	−9
10	6	103	2
	62	64108	1034

X	c_1	c_2	c_3
1	−7	63	−10
2	−5	−4	15
5	1	−107	−6
10	11	48	1
	196	17738	362

X	c_1	c_2	c_3	c_4
0	−3	31	−837	21
1	−2	2	752	−64
2	−1	−19	916	56
4	1	−37	−1016	−14
8	5	23	185	1
	40	3224	3171610	4870

X	c_1	c_2	c_3	c_4
1	−26	30	−176	−31454
2	−21	11	76	63213
4	−11	−19	252	−37667
8	9	−47	−181	4854
16	49	25	29	1054
	3720	4216	133858	6428823890

X	c_1	c_2	c_3
0	−31	93	−35433
1	−25	40	4600
2	−19	−6	28078
4	−7	−77	35289
8	17	−135	−39829
16	65	85	7295
	6510	41664	4949911360

TABLE A.12.
Coefficients for fitting periodic curves and partitioning sums of squares for data taken at equal time intervals throughout a complete cycle.

X values for n[a]												
4	6	8	12	24	U_1	V_1	U_2	V_2	U_3	V_3	U_4	V_4
0	0	0	0	0	1.000	.000	1.000	.000	1.000	.000	1.000	.000
				1	.966	.259	.866	.500	.707	.707	.500	.866
			1	2	.866	.500	.500	.866	.000	1.000	−.500	.866
		1		3	.707	.707	.000	1.000	−.707	.707	−1.000	.000
	1		2	4	.500	.866	−.500	.866	−1.000	.000	−.500	−.866
				5	.259	.966	−.866	.500	−.707	−.707	.500	−.866
1		2	3	6	.000	1.000	−1.000	.000	.000	−1.000	1.000	.000
				7	−.259	.966	−.866	−.500	.707	−.707	.500	.866
	2		4	8	−.500	.866	−.500	−.866	1.000	.000	−.500	.866
		3		9	−.707	.707	.000	−1.000	.707	.707	−1.000	.000
			5	10	−.866	.500	.500	−.866	.000	1.000	−.500	−.866
				11	−.966	.259	.866	−.500	−.707	.707	.500	−.866
2	3	4	6	12	−1.000	.000	1.000	.000	−1.000	.000	1.000	.000
				13	−.966	−.259	.866	.500	−.707	−.707	.500	.866
			7	14	−.866	−.500	.500	.866	.000	−1.000	−.500	.866
		5		15	−.707	−.707	.000	1.000	.707	−.707	−1.000	.000
	4		8	16	−.500	−.866	−.500	.866	1.000	.000	−.500	−.866
				17	−.259	−.966	−.866	.500	.707	.707	.500	−.866
3		6	9	18	.000	−1.000	−1.000	.000	.000	1.000	1.000	.000
				19	.259	−.966	−.866	−.500	−.707	.707	.500	.866
	5		10	20	.500	−.866	−.500	−.866	−1.000	.000	−.500	.866
		7		21	.707	−.707	.000	−1.000	−.707	−.707	−1.000	.000
			11	22	.866	−.500	.500	−.866	.000	−1.000	−.500	−.866
				23	.966	−.259	.866	−.500	.707	−.707	.500	−.866

[a]For a given value of n, use only the lines of the table for which X values are given. When n=4, use only columns to U_2. When n=6, use only columns to U_3. When n=8, use only columns to U_4.

X values for n=7

	U_1	V_1	U_2	V_2	U_3	V_3
0	1.000	.000	1.000	.000	1.000	.000
1	.623	.782	−.223	.975	−.901	.434
2	−.223	.975	−.901	−.434	.623	−.782
3	−.901	.434	.623	−.782	−.223	.975
4	−.901	−.434	.623	.782	−.223	−.975
5	−.223	−.975	−.901	.434	.623	.782
6	.623	−.782	−.223	−.975	−.901	−.434

TABLE A.12.
Continued.

X values for n = 52

	U_1	V_1	U_2	V_2	U_3	V_3	U_4	V_4
0	1.000	.000	1.000	.000	1.000	.000	1.000	.000
1	.993	.121	.971	.239	.935	.355	.885	.465
2	.971	.239	.885	.465	.749	.663	.568	.823
3	.935	.355	.749	.663	.465	.885	.121	.993
4	.885	.465	.568	.823	.121	.993	−.355	.935
5	.823	.568	.355	.935	−.239	.971	−.749	.663
6	.749	.663	.121	.993	−.598	.823	−.971	.239
7	.663	.749	−.121	.993	−.823	.568	−.971	−.239
8	.568	.823	−.355	.935	−.971	.239	−.749	−.663
9	.465	.885	−.568	.823	−.993	−.121	−.355	−.935
10	.335	.935	−.749	.885	−.885	−.465	.121	−.993
11	.239	.971	−.885	.465	−.663	−.749	.568	−.823
12	.121	.993	−.971	.239	−.355	−.935	.885	−.465
13	.000	1.000	−1.000	.000	.000	−1.000	1.000	.000
14	−.121	.993	−.971	−.239	.355	−.935	.885	.465
15	−.239	.971	−.885	−.465	.663	−.749	.568	.823
16	−.355	.935	−.749	−.663	.885	−.465	.121	.993
17	−.465	.885	−.568	−.823	.993	−.121	−.355	.935
18	−.568	.823	−.355	−.935	.971	.239	−.749	.663
19	−.663	.749	−.121	−.993	.823	.568	−.971	.239
20	−.749	.663	.121	−.993	.568	.823	−.971	−.239
21	−.823	.568	.355	−.935	.239	.971	−.749	−.663
22	−.885	.465	.568	−.823	−.121	.993	−.355	−.935
23	−.935	.355	.749	−.663	−.465	.885	.121	−.993
24	−.971	.239	.885	−.465	−.749	.663	.568	−.823
25	−.993	.121	.971	−.239	−.935	.355	.885	−.465
26	−1.000	.000	1.000	.000	−1.000	.000	1.000	.000
27	−.993	−.121	.971	.239	−.935	−.355	.885	.465
28	−.971	−.239	.885	.465	−.749	−.663	.568	.823
29	−.935	−.355	.749	.663	−.465	−.885	.121	.993
30	−.885	−.465	.568	.823	−.121	−.121	−.355	.935
31	−.823	−.568	.355	.935	.239	−.971	−.749	.663
32	−.749	−.663	.121	.993	.568	−.823	−.971	.239
33	−.663	−.749	−.121	.993	.823	−.568	−.971	−.239
34	−.568	−.823	−.355	.935	.971	−.239	−.749	−.663
35	−.465	−.885	−.568	.823	.993	.121	−.355	−.935
36	−.355	−.935	−.749	.663	.885	.465	.121	−.993
37	−.239	−.971	−.885	.465	.663	.749	.568	−.823

TABLE A.12.
Continued.

X values for n=52

	U_1	V_1	U_2	V_2	U_3	V_3	U_4	V_4
38	−.121	−.993	−.971	.239	.355	.935	.885	−.465
39	.000	−1.000	−1.000	.000	.000	1.000	1.000	.000
40	.121	−.993	−.971	−.239	−.355	.935	.885	.465
41	.239	−.971	−.885	−.465	−.663	.749	.568	.823
42	.355	−.935	−.749	−.663	−.885	.465	.121	.993
43	.465	−.885	−.568	−.823	−.993	.121	−.355	.935
44	.568	−.823	−.355	−.935	−.971	−.239	−.749	.663
45	.663	−.749	−.121	−.993	−.823	−.568	−.971	.239
46	.749	−.663	.121	−.993	−.568	−.823	−.971	−.239
47	.823	−.568	.355	−.935	−.239	−.971	−.749	−.663
48	.885	−.465	.568	−.823	.121	−.993	−.355	−.935
49	.935	−.355	.749	−.663	.465	−.885	.121	−.993
50	.971	−.239	.885	−.465	.749	−.663	.568	−.823
51	.993	−.121	.971	−.239	.935	−.355	.885	−.465

INDEX

Abscissa, 168
Accuracy and precision, 41
Additivity, 143, 148–149, 153–154, 158
Adjusted treatment means, 290
Amplitude, 220
Analysis of Covariance, *see* Covariance analysis
Analysis of variance, 18, 31–45
  assumptions, 139
  basic principles, 31–33
  completely randomized design, 48–52
  curvilinear regression, 212, 215, 218
  and experimental designs, 44
  latin squares, 80–85
  linear regression, 181–187
  multiple regression, 254, 256
  periodic regression, 223
  randomized complete blocks, 54–60
  repeated observations, 128–130, 133–135
  split-blocks, 118–121
  split-plots, 90–94
  split-split plots, 101–110
  transformed data, 151, 156, 160
ANOVA, 18, 22
Arithmetic mean, 15
Asymptote, 206
Asymptotic curve, 206–207

Bartlett's test for homogeneity of variance, 146–147, 152–153, 157, 159, 160
Binomial distribution, 159, 268
Bliss, C. I., 242
Block, 53–54, 87–89, 115
Block effects, 58–60, 148, 287–288

Calculating machines, 18, 36, 81, 93, 107
Cause and effect, 176, 188
Chance, 3
Chi-square, 146, 267, 268
  adjusted, 147
  definition, 268
  degrees of freedom, 268, 275
  genetic ratios, 269–273, 275–277
  heterogeneity, 279–281
  independence, 274–278
  table of, 309
  unadjusted, 146
Class comparisons, 68–70
Coded values of X, 230–231, 233
Coding, 146, 230, 259
Coefficient, of alienation, 182
  of correlation, 169
  of determination, 170
  of multiple correlation, 248, 257
  of multiple determination, 212, 248, 257
  of partial correlation, 248, 257
  of partial regression, 249
  regression, 169, 178, 180–182, 249
  of variation, 18
Coefficients, high order partial, 257
  orthogonal, 65–76, 83, 94, 109, 121–122, 236, 235
  periodic, 238, 342–344
  polynomial, 121, 229–237, 331–341
Combining curve types, 218
Combining data for two or more years, 132–137
Computers, electronic, 215, 225, 226, 258
Confidence belts, 183–184
Confidence limits, of estimated Y, 183
  for mean, 21, 23–24
  for mean difference, 39–40
  of regression coefficient, 183
Contingency table, 274, 276–277
  collapsing, 278
Continuity, correction for, 270

Cook book procedure, 33
Correction term, 17, 35, 49, 56
Correlation, 167–266
  definition, 168
  direct, 167
  inverse, 167
  linear, 167–194
  between means and variances, 157
  more than three variables, 256–264
  multiple, 247–266
  negative, 167
  partial, 247
  part-whole, 189–190
  pitfalls, 187–192
  positive, 167
  product-moment method, 174–176
  rank difference method, 173
  versus regression, 170–171
  scatter diagrams, 171–172
  shortcut method, 173–174
  simple, 247
  spurious, 190, 255—256
  standard method, 174–176
  total, 247
Correlation coefficient, 169
  calculation, 171–176
  range and interpretation, 174
  Spearman's, 173
  test of significance, 176
Counts, analysis of, 267–282
  transformations, 154–159
Covariance analysis, 285–293
  adjusted treatment sum of squares, 289, 292
  adjusting more than one source of variation, 290–292
  adjustment of treatment means, 290
  application to reduction of error, 285
  approximate method, 292
  calculation of sums of cross-products, 286–287
  example, 286
  interpretation, 293
  partitioning treatment effects, 290
  standard errors for comparing adjusted means, 292
  sum of squares of adjusted treatment means, 292
  tests of significance, 292
Cubic response, 214–216
Curves, 195
  asymptotic, 206–207
  combination, 218–219
  cubic, 207
  decay, 202
  exponential, 202–206
  Fourier, 220
  growth, 202
  normal frequency, 13
  periodic, 220–225, 238–243
  polynomial, 207–218, 229–238
  power, 196–202
  quadratic, 207–208, 211–213
Curvilinear relations, 195–227

Degrees of freedom, 17, 176, 268
  in chi-square analysis, 268, 275
  for coefficient of correlation, 176–177
  partitioning of, 65
  single, 70, 73, 83
Design, completely randomized, 47–52
  latin square, 77–85
  randomized complete block, 53–60
  split-block, 115–124
  split-plot, 87–100
  split-split plot, 101–113
  subplots as repeated observations, 125–137
Deviation mean square, 182, 236 237
Difference, highly significant, 24
  significant, 24
  standard error of, 37
Distribution, binomial, 159, 268
  bivariate normal, 168, 170
  continuous, 268
  discrete, 268
  F, 25
  normal, 13, 268
  Poisson, 154

of sample means, 21
t, 22–23
Z, 23
Doolittle method, 215
Duncan's multiple range test, 63–65, 157, 160

Enumeration data, 267–282
Equally spaced treatments, 229–245
Equation, cubic, 207
linear, 169, 178–179, 186, 191, 207
logarithmic, 196–207
normal, 207, 221, 249, 259
simultaneous, 234
periodic, 220–225, 238–243
polynomial, 207–218, 229–237
quadratic, 207
Error, experimental, 5, 31
Error terms, 59, 288
distribution of, 145, 152
Exact probability, 269
Experiment, 2, 3, 11
characteristics of, 6–7
steps, 7
Experimental designs, 8, 44, 293
Experimental error, 5, 31
Experimental material, selection of, 7, 285
Experimental unit, 8, 11
Extrapolation, 190–192

F, ratio, 25, 36, 50, 60, 82
table, 299–306
test, 50, 60, 61
planned, 61, 65–76
Factorial, 258, 269
Factorial experiments, 42
Fisher, R. A., 25
Frequency, histogram, 12
polygon, 12
table, 12

Genetic ratios, 269–273, 275–277
table of, 273
Graph, 168, 171, 196
Graph paper, log, 196
semilog, 202

Harmonics, 240
Heterogeneity, test for, 279–281
Homogeneity of variance, 140–142
Bartlett's test for, 146–147, 152–153, 160
Hypothesis testing, 6, 267

Improving precision, 283–294
Independence, chi-square tests, 274–278
of means and variances, 142–143
Individual degrees of freedom, 70, 73, 83, 272
Interaction, 42, 72
partitioning of, 72–74, 95, 108–110
Intercept, 169, 178

K values for fitting polynomials, 229–231, 331–340

Latin square designs, 77–85, 117
analysis, 80–85
example, 79
randomization, 78–79
Least significant difference, 40, 61–63, 98–100, 110–113, 123–124, 141
Least squares method, 178, 206–207
Line, best fitting, 177–178, 180–181
Linear regression, analysis, 167–194
coefficient, 169, 178, 180, 181, 182, 249
graphs, 168, 171–172
interpretation, 176, 179, 188–189
tests for deviation from, 187
Linear response, 72
Local control, 6

Main plots, 87, 102, 115
Mean, arithmetic, 15
detransformed, 156
weighted, 156–157, 162, 220, 224
of differences, 37
geometric, 170
population, 13, 15
sample, 15
weighted, 162, 220, 224

Mean deviation, 18
Mean separation, 61–76, 82–85, 94–100, 107–110, 121–123
Median, 16
Mode, 16
Model, randomized complete block, 58
  regression, 170
Multiple correlation, 247–266
  calculation, 248, 257
  interpretation, 254–256
Multiple range tests, 63
Multiple regression, 247–266
  equation, 249, 258
  more than three variables, 256–264
  response surfaces, 258–264
  visualization, 257–258

Non-additivity, 149, 154, 158
Non-linear response, 72
Normal distribution, 13
  frequency curve, 13
Normal equations, multiple correlation, 249
  periodic curve fitting, 221, 238
  polynomial curve fitting, 207
  response surface, 259
Normality, 139, 140, 154
Null hypothesis, 24, 176

Objectives of experiment, 7
Ordinate, 168
Original values, 233, 261–262
Orthogonal coefficients, 65–76, 83, 94, 109, 121, 122, 136, 229, 235, 259

P, values, 230, 235
Paired values, t test, 40
Parabola, 207
Parameters, 12, 15
Partial correlation, 247
Partial regression coefficient, 249
Partitioning of sums of squares, 235–237, 241–242
Periodic curve fitting, 220–225, 238–243
Phase angle, 220, 224, 242–243
Pitfalls, 187
Planned F tests, 65
Plot, definition, 11
Poisson distribution, 154
Polynomial, curve fitting, 207–219, 229–235
  equations, 207
  in replicated experiments, 216–218
Population, concept of, 11
  of individuals, 21
  of mean differences, 37–38
  of means, 21
Precision, 6, 41, 283–294
Problem definition, 7
PU and PV, values in periodic regression, 240–241

Quadratic equation, 207
Quadratic response, 212, 218

Randomization, 5, 9, 47
  in completely randomized design, 47–48
  in latin square, 78–79
  in randomized complete blocks, 54
  in split-blocks, 115
  in split-plots, 90
  in split-split plots, 101
Randomized complete block design, 53–60
  analysis, 54–60
  arrangement of blocks, 53–54
  example, 54
  randomization, 54
Random numbers, table, 296
Range, 18
Reasoning, deductive and inductive, 1
Recording data, 41
References, selected, 294
Refinement of technique, 285
Regression, 169
  coefficient, 169, 178, 249
    variance of, 182
  curvilinear, 188, 216–218

equations, 169
linear, 167–194
model I, 170
model II, 170
more than three variables, 256
multiple, 247–266
partial, 247
in replicated experiments, 185, 216–218
Relation between r and F tests, 182
Replication, 5, 9, 11
required number, 283–284
Research, 2, 6
Response surfaces, 258
Rounding and reporting numbers, 41
Rule of signs, 220, 224

Sample, 11
random, 11
size, 272–273
Scales, pretransformed, 162–164
Scatter diagrams, 171–172
Scientific method, 6
Selected references, 294
Selection of, experimental material, 7, 285
experimental unit, 285
treatments, 284
Semi-amplitude, 220, 224, 242
Shortcut methods, 173, 229–245
Significance, statistical, 24, 44, 176
Significant digits, 41
Simultaneous equations, 234
Single degrees of freedom, 70, 73, 83, 272
Slope, 169
Snedecor, G. W., 25, 207
Split-block design, 115–124
analysis, 118–124
example, 116–117
randomization, 115
standard errors, 123
sums of squares, 119–120
Split-plot design, 87–100, 115, 116
analysis, 90–94
example, 89
randomization, 90
standard errors, 98
sums of squares, 92–93
Split-split plot design, analysis, 101–113
example, 103
randomization, 101
standard errors, 110–111
sums of squares, 106
Standard deviation, 16
population, 13, 16
sample, 16
Standard error, of difference, 37
of estimate, 180, 254
of mean, 21
of mean differences, 37
for repeated observations, 130–132, 137
for split-blocks, 123
for split-plots, 98
for split-split plots, 110–113
Statistic, definition, 12, 15
Statistical evaluation, 5
Statistics, basic concepts, 11–29
Subplots, 87, 101
as repeated observations, 125–137
standard errors, 130–132, 137
Subscript notation, 18–19
Sub-subplots, 101
Summation notation, 15, 20
Sum of products, in covariance analysis, 286–287
of orthogonal coefficients, 66, 83
in regression analysis, 175
Sum of squares, 17
Sum of squares and mean squares, in completely randomized experiments, 49–50
of individual degrees of freedom, 70, 73, 83
in latin squares, 80–81
in randomized complete blocks, 56–59
in split-blocks, 119–120
in split-plots, 92–93
in split-split plots, 106–107

t, distribution, 22–23
table of, 297–298
test, 31–45, 38–39
comparison with F test, 39, 96
for paired plots, 40
variances different, 39
Table, angular (arcsine) transformations, 311
chi-square, 309
coefficients, for equally spaced treatments, 331–340
for fitting periodic curves, 342–344
for unequally spaced treatments, 341
contingency, 274, 276–277
F, 299–306
genetic ratios, 273
logarithms, 312–315
pretransformed scales, 163
random numbers, 296
squares and square roots, 316–330
studentized factors, 307–308
t, 297–298
two way, 18–19
values, of coefficient of correlation, 310
Tests, of independence, 274–278
of comparisons, 66
degrees of freedom, 275
of significance, analysis of covariance, 292
correlation coefficient, 176, 182
distribution of t, 38
F ratio, 36
mean difference, 38–39
in regression, 176, 182, 187
Transformations, 139–165
angular or arcsine, 158–162
log, 150–154, 196
square root, 154–158
Treatment, 11
effects, 5, 148, 287–288
equally spaced, 229, 238
means, adjusted, 290–292
selection of, 7, 284
unequally spaced, 238
Trend comparisons, 70–74, 121, 229
Tukey's test for additivity, 148–149, 153–154, 158

Variable, 11
continuous, 11
dependent, 167
discrete, 11
fixed, 170
independent, 167
random, 170
Variance, analysis of, 18. *See also* Analysis of variance
definition, 16
of estimated Y, 183
of mean difference, 37
of means, 21
of regression coefficient, 182
Variate, definition, 11
Variation, assignable causes, 3, 44
coefficient of, 18
unassignable causes, 3, 5, 44

Weighted means, 156–157, 162, 220, 224

Yates correction for continuity, 270

Z values, 23